Fluid Power
Educational
Series

Industrial Hydraulics - Basic Level

(In the English Units)

Joji Parambath

Industrial Hydraulics –Basic Level
(In the English Units)

Copyright © 2025 Joji Parambath

All rights reserved

No part of this book may be reproduced or transmitted in any form or by any means, electronic or mechanical, including photocopying, recording, or by any information storage and retrieval system, without written permission from the publisher.

Paperback: ISBN: 9798653613388

https://jojibooks.com

First Edition – 2020
Revised Edition - 2021
Revised Edition - 2024
Revised Edition - 2025

Disclaimer of Liability

The contents of this book have been checked for accuracy. We cannot guarantee full agreement, as deviations cannot be entirely precluded. Only qualified personnel should be allowed to install and work on pneumatic and hydraulic equipment. Qualified persons are authorized to commission, ground, and tag circuits, equipment, and systems in accordance with established safety practices and standards.

Dedicated to

my mother-in-law, Kamalamma

Table of Contents

Chapter	Description	Page No
--	Preface	vii
1	Industrial power systems	1
2	Hydraulic fundamentals	7
3	Hydraulic fluids – Functions, characteristics, & categories	23
4	Hydraulic fluid contamination & its control	30
5	Hydraulic filters	35
6	Hydraulic reservoirs	50
7	Hydraulic pumps	63
8	Cavitation	88
9	Pressure intensifier (Pressure booster)	91
10	Pressure regulators	94
11	Hydraulic cylinders	103
12	Hydraulic motors	123
13	Directional control valves and control circuits	142
14	Non-return valves and control circuits	159
15	Flow control valves and circuits	166
16	Pressure control valves and control circuits	183
17	Hydraulic accumulators and circuits	198
18	Hydraulic seals	212
19	Fluid conductors	217
20	Hydraulic applications	237
21	Maintenance, troubleshooting, and safety of hydraulic systems	242
22	Summary of controls for hydraulic systems - Basic level	257
Appendix 1	Graphic Symbols for Hydraulic Components as per ISO 1219	259
Appendix 2	Unit Conversions	263
Appendix 3	Viscosity Grades and Viscosity Ranges as per ISO 3348 & Viscosity Comparison	267
Appendix 4	Important Standards for Hydraulic Systems	269
Appendix 5	Hydraulic Fluid Additives and Elements & Properties of Some Hydraulic Fluids	270
Appendix 6	Contamination Code Rating & Recommended Fluid Cleanliness Levels	271
Appendix 7	The SAE Aerospace Standard AS4059	273
Appendix 8	Mesh to Micron Conversion	275
Appendix 9	Theoretical Cylinder Forces & Hydraulic Cylinder Standards	276
Appendix 10	Seal Materials and their Temperature Ranges	277
Appendix 11	Pipe Specifications, Tube Specifications, Hose Specifications, & Standards Relevant to Hydraulic Fluid Conductors	278
22	References	282

Preface

Industrial hydraulic systems are expanding and are as fascinating as ever. They provide the muscle power to run machines with the smoothest control possibilities. Many professionals design, construct, and maintain hydraulic systems every day. Several budding engineers are introduced to fluid power technology from time to time. Therefore, disseminating information about fluid power technology is essential to advancing it. The textbook on 'Industrial Hydraulics—Basic Level (in the English Units)' is written to meet this objective.

The textbook deals with the components and circuits of hydraulic systems. It initially provides the fundamentals required to understand the core topics. The book then describes power packs, hydraulic actuators, and control valves. Further, it presents the maintenance, troubleshooting, and safety aspects of hydraulic systems.

The topics are presented in a logical sequence and simple-to-understand language. Whenever possible, critical positions in the circuits are highlighted to help the reader better understand the control circuits. Many exercises are provided at the end of each chapter to test the reader's understanding of the subject.

The same author covers many fluid power topics in other textbooks under the fluid power educational series. A list of all the books is given at the end of the book. Also, please see the details at https://jojibooks.com.

Enjoy reading the book.
Your feedback is most welcome.

JOJI Parambath

About the Author ....

Joji Parambath has been a resource person in Pneumatics, Hydraulics, and PLC for over 25 years. During his career, he has trained numerous professionals from various industries, as well as faculty members and students of engineering institutions.

He is the key trainer at Fluidsys Training Center, Bangalore, India (https://fluidsys.org), which provides training in Pneumatics and Hydraulics. He has already written two books on these subjects. The publication of the present series of 32 books is intended to restructure and update the existing books.

The author wishes to thank all trainees for their lively interaction and many useful suggestions during the training programs that prompted the author to write the present series of books. You may send your feedback to joji.p@hotmail.com

10th June 2020

Chapter 1 | Industrial Power Systems

Modern industrial production systems and mobile systems are designed to perform a wide range of work operations. A prime mover provides the muscle power to perform a work operation on a stationary production machine or mobile equipment. The prime mover is, in fact, an actuator that is part of a power transmission system consisting of a power source and a control system. The block diagram of an industrial power transmission system is given in Figure 1.1. Remember, the power must be conveyed to the load at the machine's point of work through the power transmission system in a controlled manner.

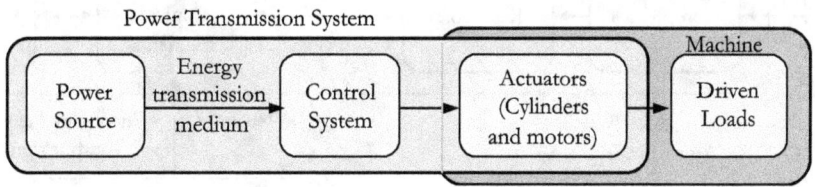

Figure 1.1 | Block diagram of an industrial power system

Apart from the cumbersome mechanical linkages, power can be transmitted easily through electrical, air, or oil media. Accordingly, there are three main types of power transmission systems. They are: (1) Electrical power transmission system, (2) Pneumatic power transmission system, and (3) Hydraulic power transmission system. Pneumatic and hydraulic power systems are commonly categorized under fluid power transmission systems.

Electrical Power Transmission System

The block diagram of an electrical power system is given in Figure 1.2. An electric generator produces electrical power. The power is then transmitted to electrical loads, such as an electric motor, lamp, or heater, through a final control element, such as a contactor. A relay controller or PLC controller controls the final control element. Electric power is cheaper and can be easily transmitted, controlled, and converted to other forms of energy. However, obtaining linear motions through electrical systems is cumbersome.

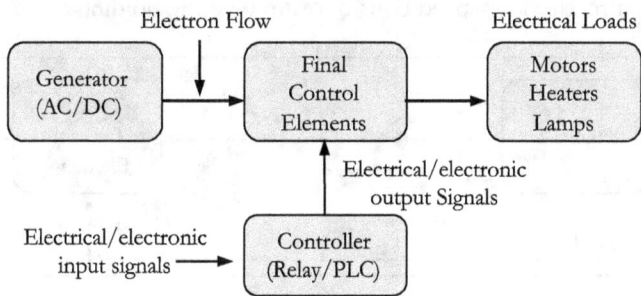

Figure 1.2 | Block diagram of an electrical power transmission system

Pneumatic Power Transmission System

Figure 1.3 shows the block diagram of a pneumatic power transmission system. This system involves generating compressed air with a compressor driven by a prime mover, such as an electric motor. The compressed air is stored in a reservoir. The stored compressed air, serving as the energy source, is then directed to actuators, including cylinders and air motors, through final control elements such as directional control valves. A pneumatic, relay, or PLC controller controls the final control elements. The controller receives the system's variables in the form of pneumatic, electrical, or electronic input signals and generates corresponding output signals to control the final control elements. Pneumatics is the engineering science of gaseous pressure and flow. Pneumatic systems are generally designed as low-pressure systems, suitable for systems involving low forces and/or high-speed linear motions.

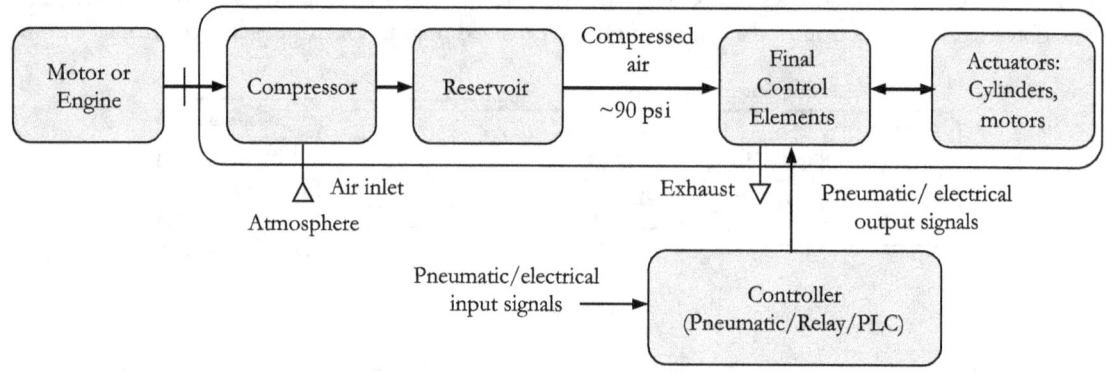

Figure 1.3 | Block diagram of a pneumatic power transmission system

Hydraulic Power Transmission System

Figure 1.4 shows the block diagram of a hydraulic power transmission system. This system stores incompressible hydraulic fluid in a reservoir (tank). A pump, driven by a prime mover like an electric motor, generates pressurized fluid. The pressure is limited by a pressure relief valve (PRV). This pressurized fluid transfers energy to the system actuators — such as cylinders and hydraulic motors — through final control elements, such as directional control valves. A relay, an electronic (non-programmable) controller, or a PLC controller controls the final control elements. The controller receives the system's variables in the form of electrical or electronic input signals and generates electrical or electronic output signals to control the final control elements. Hydraulics is the engineering science of liquid pressure and flow. Hydraulic systems are designed to operate at high pressure. They are suitable for systems that require precise slow-speed control or involve the holding of heavy loads.

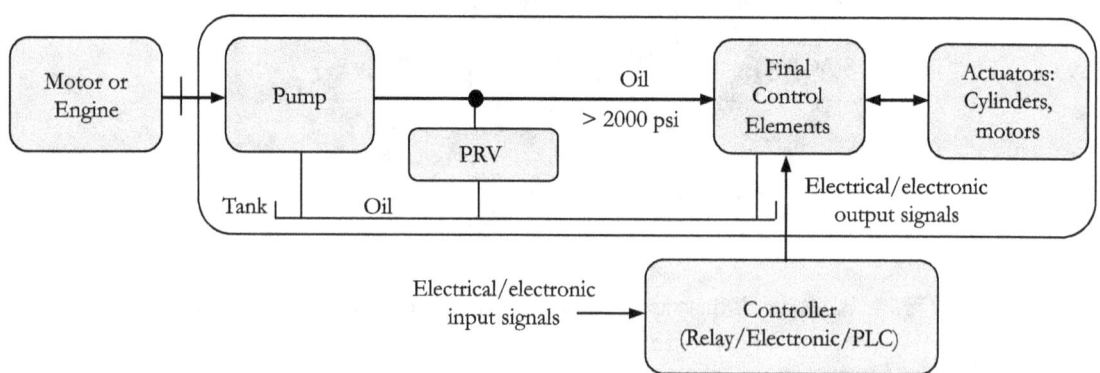

Figure 1.4 | Block diagram of a hydraulic power transmission system

Significant Criteria for the Selection of Power Transmission Systems

Electrical motors are the optimal devices for producing rotary motion, so they are used in systems that predominantly involve rotary motion. Pneumatic power suits systems with high-speed linear motions but low forces. Another major drawback of pneumatic systems is that they are unsuitable for obtaining uniform motions. Hydraulic power is suitable for systems that require very smooth position control or demand holding of heavy loads.

Evolution of Fluid Power Transmission Systems

Fluid power systems evolved from manual to mechanization to automation. In mechanization, a machine takes over mechanical work. In automation, a machine is controlled automatically, with limited or no human intervention.

Control Systems

In a power transmission system, the control part modulates the power part using final control elements. A final control element can be operated directly by manual force or via pneumatic or electrical signals from input elements, such as pushbuttons and sensors (Figure 1.5). It can also be operated indirectly through a pneumatic, relay, electronic, or PLC controller.

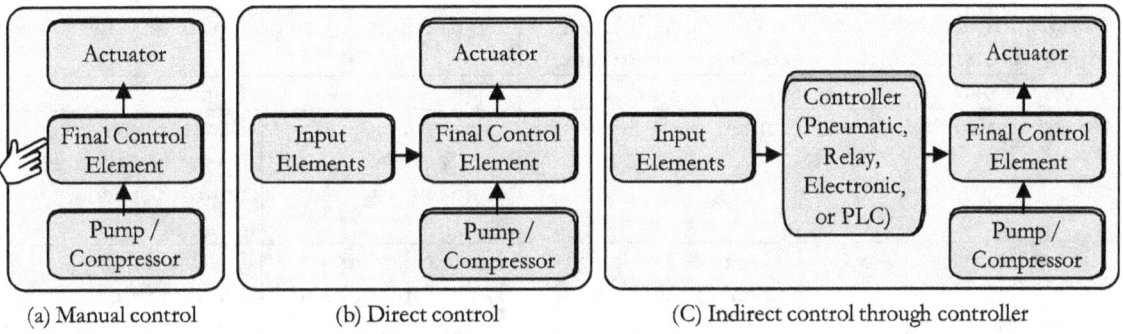

Figure 1.5 | Various configurations of control systems

Usually, the actuators' direction, position, speed, and/or force are to be controlled. The system may also be controlled based on pressure conditions in certain parts. In less sophisticated systems, control systems are typically designed as ON/OFF discrete control systems. In advanced systems, infinitely variable control systems can be employed in open-loop or closed-loop configurations. Closed-loop control systems offer enhanced precision and accuracy in governing position, speed, and force.

Concept of Automation

The work process for an industrial task typically involves many recurring steps. These steps can be carried out manually or automatically. In the manual system, an operator is always present to decide every process step. In an automatic system, the process controls itself, partially or entirely, based on feedback from its condition. Therefore, an automatic system can be semi-automatic or fully automatic.

In semi-automation, a machine automatically performs several recurring partial steps during the processing of a workpiece. Here, an operator is required to initiate every cycle of operations. In complete automation, a machine performs cyclic operations to process many jobs. Sensors or transducers are invariably used in automatic control systems.

Motion Control Systems

The motion control system is an extension of modern automation systems. It mainly controls the position, velocity, and/or force associated with work operations. A motion controller is the brain of the motion control system. Next, it computes and generates the output commands for the required motion path or trajectory. Further, motion control is an intricate part of robotics and modern CNC machines.

Comparison of Different Power Transmission Systems

Choosing the right and efficient form of energy for the industry's drive system is not easy. Its selection depends on various factors. A comparative study of power systems is now most appropriate. Table 1.1 compares different power transmission systems based on several essential criteria.

Table 1.1 | Comparison of different power transmission systems

Criteria / Power system	Electrical	Hydraulics	Pneumatics
Energy production	Hydro, thermal, atomic	Pump	Compressor
Availability of the energy transmission medium	Available everywhere	Obtaining and disposing of oil is expensive	Air is freely available
Maximum distance for energy transmission	Considerable distance, even beyond 600 miles	Up to 300 ft	Up to 3000 ft
Energy Cost	Smallest	High	Highest
Speed control	Limited	Good for slow speed, precise control	Best for high-speed operation, obtaining a uniform speed is difficult
Linear force	Using rotary to linear conversion devices	Using cylinders - Large forces due to high pressure	Using cylinders - Limited forces due to low pressure
Rotary force (Torque)	Using electric motors	Using hydraulic motors	Using air motors
Overloading	A severe problem	loadable until standstill with relief valves	Loadable until standstill
Sensitivity to temperature variations	Insensitive	Sensitive	Relatively insensitive
Leakage	Lethal accident risk at high voltages	Loss of energy and environmental fouling	Loss of energy

A Chapter Concluding Note

It is difficult to find a new product that has not been processed or handled by these power transmission systems at some stages of its production or distribution. The growing dependence of manufacturing systems on these power transmission systems is driven by the need for greater muscle power, faster production, higher quality, less waste, and lower costs. At the same time, these power transmission systems are found to be quite amenable to automation and lend themselves to design simplification. That means these power transmission systems will continue to grow in the times to come.

Objective-type Questions

1. The large magnitude of linear forces can be obtained easily in:
    a) Mechanical power transmission systems
    b) Electrical power transmission systems
    c) Pneumatic power transmission systems
    d) Hydraulic power transmission systems

2. Which of the following power transmission systems provides a fast-acting production system?
    a) Mechanical power transmission system
    b) Electrical power transmission system
    c) Pneumatic power transmission system
    d) Hydraulic power transmission system

3. Which of the following statements is incorrect?
    a) Pneumatic systems are overload-safe.
    b) Hydraulic systems are insensitive to variations in temperature.
    c) Pneumatic systems are capable of performing high-speed operations.
    d) Hydraulic energy can be transmitted economically, typically up to 100 m.

4. Which of the following statements is correct?
    a) The electrical power system provides linear motions in an optimum manner.
    b) The pneumatic power system provides uniform motion of its actuators.
    c) The hydraulic power system is not suitable for getting rotary motions.
    d) A motion control system calculates and generates output commands for the desired trajectory of motion.

5. The function of a controller in a power system is to:
    a) Transmit power through the system.
    b) Regulate the pressure in the system.
    c) Govern the primary power transmission system through commands.
    d) Sense the output parameter of the system.

6. Which power transmission system has a maximum distance of approximately 100 m for energy transmission?
    a) Mechanical power transmission systems
    b) Electrical power transmission systems
    c) Pneumatic power transmission systems
    d) Hydraulic power transmission systems

7. Which power transmission system is more sensitive to variations in temperature?
    a) Mechanical power transmission systems
    b) Electrical power transmission systems
    c) Pneumatic power transmission systems
    d) Hydraulic power transmission systems

Review Questions
1) What is an industrial prime mover?
2) What are the essential components of industrial power transmission systems?
3) Name three important power transmission systems.
4) Draw the block diagram of an industrial power transmission system.
5) Draw the block diagram of an electrical power transmission system.
6) Explain the function of an electrical power transmission system.
7) Draw the block diagram of a pneumatic power transmission system.
8) Explain the function of a pneumatic power transmission system.
9) Draw the block diagram of a hydraulic power transmission system.
10) Explain the function of a hydraulic power transmission system.
11) What is the final control element in a power transmission system?
12) What is the primary function of power transmission systems?
13) What is a fluid power system? Explain briefly.
14) What are the primary divisions of fluid power systems?
15) List some important essential functions performed by fluid power systems.
16) What is the primary advantage of fluid power systems?
17) What are the drawbacks of fluid power systems?
18) List any four applications of fluid power systems.
19) List two applications of oil hydraulics.
20) Describe some unique problems faced by fluid power systems.
21) Compare hydraulic and pneumatic power transmission systems.
22) Explain why you require 'control systems' in power transmission systems.
23) What are the various control options in fluid power systems?
24) Briefly describe the evolution of industrial work processes.
25) What are mechanization and automation?
26) What is the basic concept of automation?
27) Differentiate between semi-automation and complete automation.
28) Give one example of 'semi-automatic control' and 'fully automatic control.'
29) What is a motion control system? Explain briefly.
30) Mention three advantages of hydraulic systems compared to other power systems.
31) Draw the essential blocks of the pneumatic energy transmission system and explain them.
32) Depict the essential elements of the hydraulic energy transmission system with the help of a block diagram and describe the primary function of each element.
33) Compare electrical, pneumatic, and hydraulic power transmission systems in terms of the speed control function.
34) Compare electrical, pneumatic, and hydraulic power transmission systems in terms of linear force.
35) Compare electrical, pneumatic, and hydraulic power transmission systems based on the overloading function.
36) Why are fluid power systems important for modern industrial production systems?

Objective-type questions - answer key: *1-d, 2-c, 3-b, 4-d, 5-c, 6-d, 7-d*

Chapter 2 | Hydraulic Fundamentals

Hydraulics is the branch of engineering sciences concerned with the transmission of energy using incompressible fluids (oils). Hydraulic systems conventionally generate pressure and develop and control enormous forces through an enclosed, incompressible fluid medium.

How Do We Define Hydraulics?
Hydraulics may be defined as the science of transmitting force, motion, or both through an incompressible medium, such as pressurized oil, to power and control machines. The study of hydraulics is, therefore, all about knowing how to produce positive pressure through the oil medium using a force, and the reverse process of developing force and controlling a load using the developed pressure.

The Advent of Industrial Hydraulics
Initially, water was used as the energy-transfer medium in industrial hydraulic systems because it was abundant. It was found to have the main advantage of fire resistance. However, it was found to have many shortcomings, including low lubricity and a narrow working temperature range. Further, it was found to promote rusting and corrosion of the associated metal parts.

Petroleum-based oils, developed in the late nineteenth century, were proven to be highly incompressible. They were found to operate at high pressures. They were also determined to possess the appropriate viscosity range to meet the conflicting requirements of excellent lubricating and sealing properties. Given the overwhelming advantages of petroleum oils, system designers began using them as a medium for energy transfer in industrial hydraulic systems. That marked the beginning of 'industrial hydraulics' or 'oil hydraulics.'

Pascal's Law
Pascal's law states, 'Pressure at any one point in a static fluid is the same in every direction' and 'pressure exerted on a confined fluid is transmitted equally in all directions, acting with equal force on equal areas.' The law is central to the development of many hydraulic devices, such as brakes, presses, and jacks.

Pressure Development in a Hydraulic System
An application of force can squeeze an enclosed oil medium, as shown in Figure 2.1. The pressure results from the resistance offered to compression by the oil medium. Remember, 'pressure' is the force acting per unit area. That is, $P = F/A$.

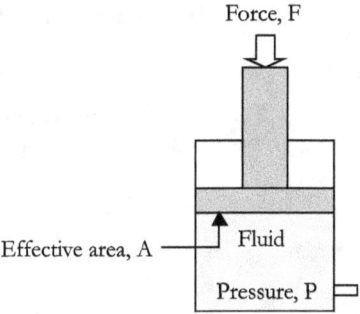

Figure 2.1 | The diagram of a fluid-filled cylinder illustrating pressure development

Units of Pressure

1 Pascal	= 1 Newton per square meter
1 bar	= 10^5 Pascal
1 Mega Pascal (MPa)	= 10^6 Pascal (10 bar)
1 bar	= 0.1 MPa
1 bar	= 14.5 Pounds per square inch (psi) [(lb/in²)]
1 kgf/cm²	= 0.981 bar ~ 1 bar
1 bar	= 1.02 kgf/cm²

Sections 2C and 2D of Appendix 2 give conversion factors for units of pressure.

Absolute and Gauge pressures

Every object on the earth's surface is subjected to significant pressure from the weight of the air above the object. This pressure is called 'atmosphere' (abbreviated as atm) and is approximately equal to 14.7 psi at sea level. The absolute pressure is a scale measured relative to a complete vacuum. Gauge pressure ignores atmospheric pressure. Gauge pressure is measured relative to atmospheric pressure. The relationship between the absolute pressure and the gauge pressure is as follows: $P_{absolute} = P_{atmospheric} + P_{gauge}$

Pressure Levels in Hydraulic Systems

A significant advantage of hydraulic systems is their ability to develop high pressures instantly. Various materials used to construct their constituent parts must withstand high system pressure. With advances in materials and component designs, there is a trend towards higher pressures in hydraulic systems. As we know, higher pressures mean a better power-to-weight ratio. Moreover, high-pressure fluid media can enable compact, efficient hydraulic systems.

Operating Pressures

Standard operating pressures in industrial hydraulic systems range up to 5000 psi. Next, high-pressure hydraulic systems operate up to 10000 psi. Experimental extra-high-pressure hydraulic systems operate up to 50,000 psi. Standard pressure is again subdivided into three ranges, as shown in Table 2.1.

Table 2.1 | Operating pressures in hydraulics

Class	Range (Up to)	Subdivision	Range
Standard Pressure	5000 psi	Low standard pressure	<1500 psi
		Medium standard pressure	1500 to 3000 psi
		High standard pressure	3000 to 5000 psi

Problem 2.1

A cutting machine requires a force of 22500 pounds. The piston area of the cylinder that provides the muscle power for the cutting operation is 28.27 in². What must be the minimum pressure in the cylinder?

Solution

Force	= 22500 lb
Area of the cylinder	= 28.27 in²
Pressure	= F/A
	= 22500 / 28.27 = 795.9 psi

Hydraulic Force

Let us now understand how to develop force in a hydraulic system. Figure 2.2 shows a schematic diagram of a hydraulic cylinder with a piston. When pressure (P) is applied to the piston's area (A), a force (F) develops. The force developed equals the applied pressure multiplied by the piston's area. That is, $F = P \times A$.

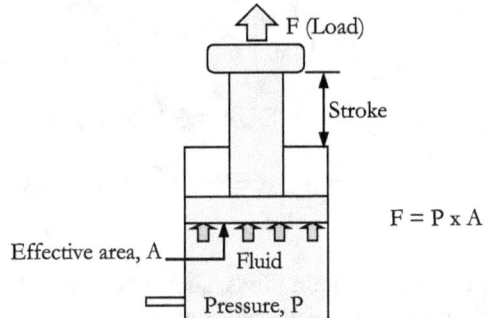

Figure 2.2 | The diagram of a fluid-filled cylinder illustrating the development of force

Force Multiplication

A hydraulic system can be designed to multiply force using a positive, rigid fluid medium. Figure 2.3 shows the arrangement of cylinders A and B with piston areas A1 (0.06 in²) and A2 (0.6 in²) (A2 > A1), respectively. A pipeline connects these two cylinders, as shown. The cylinders and pipeline are filled with fluid. When cylinder A is subjected to a force F1 (2 lb), a definite pressure (33.3 psi) is generated in the fluid medium. According to Pascal's law, the same pressure P acts on the piston of cylinder B. This pressure causes the development of force F2 (20 lb) by cylinder B. The governing equations for the forces developed in the cylinders are as follows:

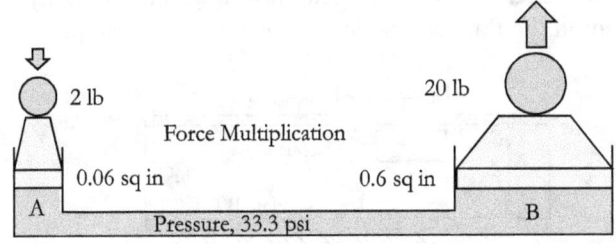

Figure 2.3 | Illustrating the force multiplication concept

$F1 = P \times A1$
$F2 = P \times A2$

Therefore,

$F2 = F1 \times (A2/A1)$

Assume the area ratio A2: A1 as 10: 1 (0.6/0.06). As illustrated in Figure 2.3, with a force of 2 lb exerted on the piston of cylinder A, cylinder B can lift a load of 20 lb. The hydraulic system's ability to multiply force can be considered a form of leverage. However, it may be noted that the force multiplication is achieved by sacrificing distance. If cylinder A moves by 4 in, then cylinder B moves by 0.4 in.

Problem 2.2

The piston of a hydraulic cylinder has a 5-inch bore diameter. What force does the piston exert when a pressure of 1450 psi acts on it?

Solution

Bore diameter	= 5 inches
Pressure (P)	= 1450 psi
Area of the piston (A)	= $\pi \times D^2/4$ = 3.14 x 5^2 / 4 = 19.63 in^2
Force (F)	= P x A = 1450 x 19.63 = 28464 lb

Viscosity

It measures the internal resistance to flow. Thin fluids, such as water and alcohol, flow quickly and have low viscosity. Thick fluids, such as molasses and cold honey, pour slowly and have high viscosity.

The viscosity of a fluid can be measured in terms of its resistance to movement when subjected to an external force or the gravitational force. Accordingly, there are two viewpoints on viscosity: absolute (dynamic) viscosity and kinematic viscosity.

The absolute viscosity is the property that represents the resistive movement of different layers of a fluid stream when subjected to an external force. The kinematic viscosity is the property that describes the difficulty with which the fluid moves under the force of gravity.

Absolute Viscosity (μ)

A thin plate A of surface area 'a' is located at a distance 'd' from a stationary reference plate B, as shown in Figure 2.4. Plate A is subjected to a force 'F_d' and moves with the velocity 'v.' For small values of v and d, the velocity gradient in the fluid layers tends to be linear, with slope v/d.

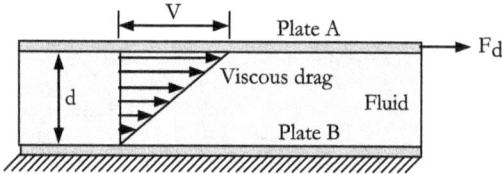

Figure 2.4 | Fluid velocity profile between two parallel plates due to viscosity

The force F is proportional to the area a and velocity v, and inversely proportional to the distance d. That is,

$$F_d \propto a \cdot \frac{v}{d}$$

$$F_d = \mu \cdot a \cdot \frac{v}{d}$$

$$\text{Absolute viscosity, } \mu = \frac{(F_d/a)}{(v/d)} = \frac{\text{Shear stress}}{\text{Shear strain}}$$

Units of Absolute Viscosity

1 Poise	= 1 dyne.s/cm² [In the CGS system]
1 Poise	= 0.0020885 lb.s/ft²
1 centipoise (cP)	= 0.01 Poise
1 Poise	= 0.1 Pa.s

Kinematic viscosity (ν)

Kinematic viscosity measures a fluid's resistance to flow under gravity. At a given temperature, it is given by the absolute viscosity (μ) divided by the fluid density (ϱ).

$$\text{Kinematic viscosity, } \nu = \frac{\mu}{\varrho}$$

Units of Kinematic Viscosity

1 Stoke	= 1 cm²/s
1 centi Stoke (cSt)	= 0.01 Stoke
1 cSt	= 1 mm²/s

In English unit: 1 ft²/s [1 Stoke = 0.00107639 ft²/s]

Note: In the CGS system, density equals specific gravity (SG). Hence, the kinematic viscosity in cSt can also be found from the following equation:

$$\text{Kinematic viscosity, } \nu \text{ (cSt)} = \frac{\mu \text{ (cP)}}{SG}$$

Problem 2.3

The absolute viscosity of a moving fluid is 0.01399 lb.s/ft², and the density of the fluid is 1.7 slug/ft³. Calculate its kinematic viscosity.

Solution

Absolute viscosity, μ	= 0.01399 ft.s/ft²
Density, ϱ	= 1.7 slug/ft³
Kinematic viscosity, ν	= μ / ϱ = 0.01399 /1.7 = 0.008229 ft²/s

Saybolt Universal Seconds (SUS)

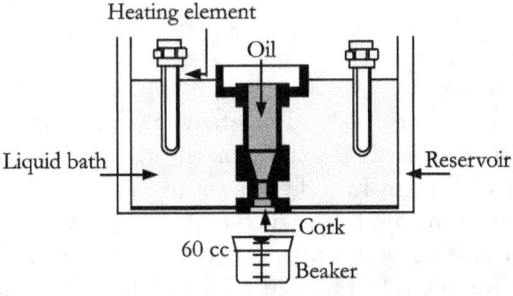

Figure 2.5 | Saybolt universal viscometer

In addition to the basic units for measuring kinematic viscosity, other units, such as Saybolt Universal Seconds, are used to specify it. Saybolt Universal Seconds (SUS) is the time measured in seconds required for 60 ml of oil to flow, under gravity, through the calibrated orifice of a Saybolt Universal viscometer at a specified temperature. Since thick oil flows slowly, its SUS value will be higher than that of thin oil. Figure 2.5 shows the Saybolt universal viscometer.

Section 2E of Appendix 2 gives viscosity unit conversions.

Viscometers
Apart from the Saybolt universal viscometers, there are primarily two types: glass capillary and rotational viscometers.

Glass Capillary Viscometer
Kinematic viscosity is usually measured using the glass capillary tube viscometer with a known diameter and length. Figure 2.6(a) shows the glass capillary viscometer. The viscometer measures the time required for a defined volume of fluid to flow under gravity through the glass capillary tube. The time is then multiplied by the viscometer's calibration constant to obtain the kinematic viscosity in centiStokes (cSt).

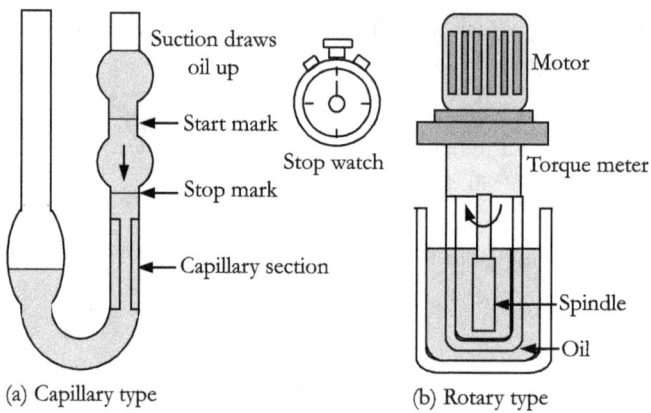

Figure 2.6 | Viscometers

Rotational viscometer
Absolute viscosity is typically measured using a rotary viscometer. This device uses the principle that the power required to turn a disk in a fluid can indicate the fluid's absolute viscosity. Figure 2.6(b) shows the rotational viscometer.

Viscosity Classification Systems
The Society of Automotive Engineers (SAE) developed the first viscosity classification standard in 1911. In 1975, the International Standards Organization (ISO), in unison with the American Society for Testing and Materials (ASTM) and many other standards organizations, adopted an approach to establish a viscosity measurement method. It is known as the ISO Viscosity Grade (VG) according to ISO 3448:1992. The ISO VG classification consists of 20 viscosity grades. Some of the Viscosity Grades are: 2, 3, 5, 7, 10, 15, 22, 32, 46, 68, 100, 150, 220, 320, 460, 680, 1000, and 1500. Table 2.2 provides details on some of the ISO viscosity grades. Also, see Table A3.1 of Appendix 3 for more details.

Table 2.2 | ISO viscosity grades

ISO VG	Viscosity in cSt at 40°C (100°F)		
	Mid-Point	Minimum	Maximum
2	2.2	1.98	2.42
3	3.2	2.88	3.52
5	4.6	4.14	5.06
7	6.8	6.12	7.48
10	10	9.0	11.0
15	15	13.5	16.5
22	22	19.8	24.2
32	32	28.8	35.2
46	46	41.4	50.6
68	68	61.2	74.8
100	100	90	110
150	150	135	165
220	220	198	242
320	320	288	352

Table A3.2 of Appendix 3 gives a viscosity comparison table.

The Effect of Variation in Pressure on Viscosity
An increase in the system pressure can increase the fluid's viscosity. Figure 2.7(a) shows the relationship between viscosity and pressure.

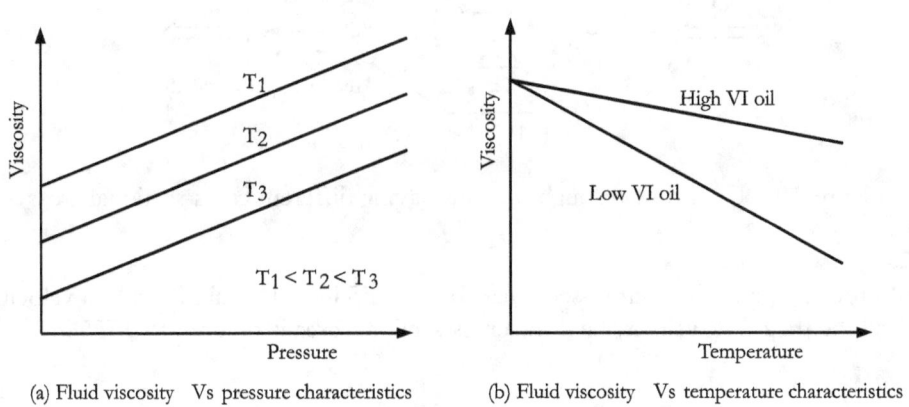

(a) Fluid viscosity Vs pressure characteristics   (b) Fluid viscosity Vs temperature characteristics

Figure 2.7 | Fluid viscosity characteristics

The Effect of Variation in Temperature on Viscosity and Viscosity Index
The viscosity of fluids can change appreciably with temperature changes. Fluids have higher viscosity when cold and lower viscosity when hot. The change in viscosity with temperature is measured with an arbitrary measure called the Viscosity Index (VI). A fluid with a low VI exhibits a significant change in viscosity with temperature. A High VI fluid has relatively stable viscosity, which does not change appreciably with temperature change. Self-explanatory: Figure 2.7(b) shows the relationship between temperature and viscosity.

Flow Rate
Volumetric flow rate measures the volume of oil passing a cross-section per unit of time, usually measured in gpm. This vital parameter decides the speed of hydraulic actuators.

Flow Velocity
The velocity of fluid flow is the average speed at which its particles move past a given cross-section. It is a design parameter that decides the internal diameter of the pipelines and the type of fluid flow. Certain flow velocities are found to be most favorable for hydraulic lines. They are:

For suction lines,	1.5 to 4 ft/s
For pressure lines,	13 to 25 ft/s
For return lines,	6.5 to 10 ft/s

Flow Rate vs. Velocity of Flow
Figure 2.8 illustrates the fluid flow through pipe sections having different cross-sectional areas. The flow rate (Q) equals the pipe area (A) multiplied by the velocity (v) of the fluid flow. That is,

$$Q = A \cdot v$$

Therefore, with a constant flow rate, the velocity of the fluid passing through the narrow section increases. The provided relation is valuable for determining the sizes of components and fluid conductors (a = Q/v).

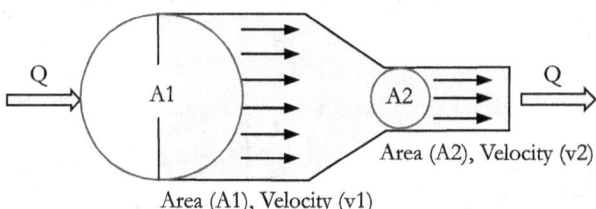

Figure 2.8 | Fluid flow through sections having different cross-sectional areas

Problem 2.4
Fluid flows through a pipe with a cross-sectional area of 1.55 in². The fluid flows at a velocity of 0.656 ft/s. What will the fluid flow velocity be if the cross-sectional area is reduced to 0.755 in²?

Solution

Cross-sectional area of the larger pipe section, A1	= 1.55 in²
Velocity of fluid flow at the larger cross-section, v1	= 0.656 ft/s
Cross-sectional area of the larger pipe section, A2	= 0.755 in²

$$A1 \times v1 = A2 \times v2$$

Velocity of fluid flow at the larger cross-section, v2	= A1 x v1 / A2
	= 1.55 x 0.656 / 0.755
	= 1.347 ft/s

It can be observed that the fluid velocity through the narrow section increases.

Laminar Flow

Figure 2.9 shows the fluid flow through a pipe having different cross-sectional areas. The flow through the pipe tends to become laminar when the viscosity forces dominate. Remember, viscosity forces dominate at high fluid viscosities or low fluid velocities. All the fluid particles then move parallel to each other, and a particle in a given layer tends to remain in that layer. Heat loss from the friction is minimal in laminar flow. If the fluid velocity exceeds a certain threshold, turbulent flow develops.

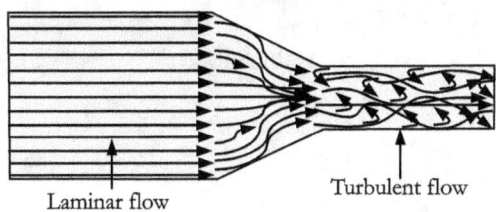

Figure 2.9 | Laminar and turbulent flows

Turbulent Flow

The flow tends to be turbulent when the inertia forces dominate. Inertia forces dominate at high fluid velocities or low fluid viscosity. When the fluid velocity exceeds a critical value, particles in the fluid collide, forming cross currents. The result of turbulent flow is a significant increase in friction, flow resistance, pressure drop, and energy loss compared to laminar flow. An essential objective of the hydraulic system is to avoid turbulent flow to keep the friction minimal. Figure 2.9 illustrates the laminar and turbulent flows.

Reynolds Number

Osborn Reynolds conducted a series of pioneering studies to determine the governing condition for the transition of fluid flow in a pipe from the laminar to the turbulent state. In 1883, he derived the Reynolds number Re.

$$\text{Reynolds number, Re} = v\,D\,\rho/\mu = v\,D/\nu$$

Where,
- V = Velocity of fluid flow, ft/s
- D = Inside diameter of the pipe, in
- ρ = Density of the fluid, slug/ft^3
- μ = Absolute viscosity of the fluid, lb.s/ft^2
- ν = Kinematic viscosity of the fluid, ft^2/s

The number 2000 is the decisive value for marking a borderline between the laminar and turbulent flows. Experiments have demonstrated that, for Reynolds numbers below 2000, the flow is found to be laminar, and above 4000, it is turbulent. The Reynolds number regime between 2000 and 4000 is the critical zone, and the associated flow can be considered turbulent for all practical purposes. Analyzing laminar and turbulent flow is crucial for properly designing hydraulic systems.

Section 2A of Appendix 2 gives Metric to English unit conversions, and Section 2B of Appendix 2 gives English to Metric unit conversions. Table A4.1 of Appendix 4 gives important standards for hydraulic systems.

Problem 2.5

An incompressible fluid with a density of 1.59 slug/ft³ and an absolute viscosity of 0.00668 lb.s/ft² flows through a pipe of diameter 0.082 inches at the rate of 8.2 ft/s. Is the flow laminar or turbulent?

Solution

Fluid velocity v	= 8.2 ft/s
Pipe diameter, D	= 0.082 in
Density, ϱ	= 1.59 slug/ft³
Absolute viscosity, μ	= 0.00668 lb.s/ft²
Reynolds Number	= v D ϱ / μ
	= 8.2 x 0.082 x 1.59 / 0.00668 = 160

The Reynolds number is below 2000. Therefore, the resulting flow will be laminar.

Conservation of energy

According to the principle of energy conservation, the total energy in a system can exist in different forms, but it must remain constant.

Forms of Energy in Hydraulic Power Transmission Systems

There are three primary energy states for the fluid contained in a hydraulic system. They are (1) potential energy (gravitational) due to elevation, (2) potential energy (hydrostatic energy) due to static pressure, and (3) kinetic energy (hydrodynamic energy).

Potential Energy due to Elevation: It is the energy by virtue of the position of the fluid having a definite mass (m, in Slug) and is dependent on the height (h, in ft) of the column of the fluid (elevation head) and acceleration due to gravity (g, in 32.2 ft/s²). The potential energy relation is:

  Potential energy = mgh ft.lb (In SI system, Joule)

Potential Energy due to Static Pressure: It is the energy due to the fluid's static pressure. This energy depends on the pressure (P, in lb/ft²) applied to the fluid (pressure head). The fluid molecules can absorb, store, and release this energy. The fluid pressure is equivalent to the fluid's energy per unit volume. Let V, in ft³, be the volume. The pressure energy relation is:

  Pressure energy = P x V ft.lb

Kinetic energy: It is the energy by virtue of the motion of the fluid through the pipe and is dependent on the velocity of the moving fluid (the velocity head). Let 'm' be the mass (m in Slug) and 'v' be the fluid velocity (v in ft/s). The kinetic energy relation is:

  Kinetic energy = ½ mv² ft.lb

Bernoulli's Equation

Bernoulli's equation, named after Daniel Bernoulli, is an enunciation of the law of energy conservation for an ideal fluid stream. It states that the sum of its kinetic and potential energies at various points in the fluid stream must remain constant if the fluid flow remains constant. The total energy can be expressed mathematically in the form of an equation as:

$$\text{Total Energy} = mgh + \tfrac{1}{2} mv^2 + PV = \text{Constant}$$

The Bernoulli equation assumes that (1) the flow is steady, (2) the fluid is incompressible, and (3) the fluid is inviscid (zero viscosity). Note that the potential energy due to elevation is usually negligible in hydraulic systems, since they are never constructed at great heights. Kinetic energy is also negligible in hydraulic systems since the volume of fluid moving through thin pipes is small, and the velocity is only a few meters per second. The energy of the hydraulic fluid is derived mainly from its pressure.

Bernoulli's equation is the mathematical representation of the fact that as the velocity of the fluid stream increases, its pressure decreases, and vice versa. For instance, its speed increases when the fluid flows through a restricted passage. Considering the corresponding Bernoulli equations as equal at two different points along the streamline, an unknown parameter at one point can be calculated from the remaining known parameters of the equations.

Compressibility and Bulk Modulus

Hydraulic fluids exhibit some degree of compressibility when pressure increases. Compressibility is the degree to which a fluid reduces in volume under increased pressure. Bulk modulus is another term representing the resistance to compressibility. That is, the bulk modulus is the reciprocal of compressibility. The bulk modulus of a newly purchased hydraulic fluid is typically about 240000 psi.

Advantages and Disadvantages of Hydraulic Systems

Hydraulic systems have many advantages and disadvantages. These factors dictate the way how hydraulics is used in applications.

Advantages of Hydraulic Systems

The advantages of hydraulic systems are listed below:

- **High power density:** Hydraulic systems transmit energy at high power-to-weight ratios. Hydraulic systems with comparatively compact components can develop high forces or torques.

- **Multi-function control:** A single hydraulic pump can provide power and control to many actuators in a system.

- **Simpler design:** A few simple pre-engineered components replace complicated mechanical linkages in many hydraulic machines.

- **High flexibility:** Hydraulic components can be positioned with considerable flexibility and convenience. Instead of mechanical linkages, pipes, and hoses virtually eliminate component location problems.

- **Oily fluid medium:** As a lubricant, the fluid reduces friction in the components of a hydraulic system, helping to prolong their life.

- **Heat dissipation:** The movement of the fluid through a hydraulic system helps to draw heat away from 'hot spots' in the system.

- **Accurate control:** Components such as discrete valves, proportional valves, and servo valves can quickly and accurately control the direction of motion, speed, and force or torque of hydraulic actuators. A large force can be controlled easily with a small force.

- **Electronic Interface:** Hydraulic systems can be readily interfaced with microprocessor-based controllers. A hydraulic actuator can be equipped with electronic sensors to provide digital or analog feedback on its position, direction, and speed.

Disadvantages of Hydraulic Systems

Many disadvantages counteract the advantages of hydraulic systems. The disadvantages are listed below:

- **Energy level:** Hydraulic systems are high-pressure, high-power systems that require careful design and all safety precautions. High-speed oil jets can cause hydraulic lines to burst under high pressure, injuring personnel.

- **Contamination Control:** Hydraulic systems need effective contamination control as they are sensitive to dirt, moisture, and corrosion. The components in hydraulic systems tend to wear when subjected to high forces or torques.

- **Temperature Dependence:** The working of hydraulic systems in extreme temperatures deteriorates their performance.

- **Leakage:** A hydraulic system tends to leak because of sealing defects. The leakage pollutes the environment and makes the surroundings a mess.

- **Risk of Fire:** A hydraulic system with a mineral-based fluid medium also risks fire if fluid leakage occurs in a hot environment.

- **Energy losses:** Pressure drops in valves, lines, and fittings are the primary causes of losses in hydraulic systems.

- **Maintenance:** Precision parts of hydraulic systems require proper maintenance, as they are exposed to abuse, extreme climates, and polluted atmospheres. A fluid medium used in a system also requires proper filtration to maintain its quality.

- **Cost:** Hydraulic system components, such as pumps, valves, and filters, are expensive. For precise position-sensing applications, expensive electro-hydraulic valves are required.

- **Noise:** The excessive noise generated by hydraulic systems can have a disturbing effect on the workforce.

- **Waste disposal:** Disposing of used hydraulic fluids is problematic and costly.

Objective-type Questions

1. Which statement is not related to Pascal's law?
   a) Fluid velocity increases as pressure decreases and vice versa
   b) The pressure at any point in a static fluid is the same in all directions
   c) The pressure exerted on a confined fluid is transmitted equally in all directions
   d) The pressure in a confined fluid acts with equal force on equal areas

2. The fundamental law governing the operation of a hydraulic press is:
   a) Boyle's law
   b) Pascal's law
   c) Laminar flow
   d) Constant flow

3. Which is not a unit of pressure?
   a) Pascal
   b) Poise
   c) psi
   d) Kg/cm^2

4. 1 bar is equal to:
   a) 10 MPa
   b) 14.5 lb/in^2
   c) 100 Pascal
   d) 10 Kgf/cm^2

5. A hydraulic press is designed with an output piston three times the area of the input piston. Technician A says that the output piston develops a force three times that of the input piston, and Technician B says that the output piston moves one-third as far as the input piston. Which of the following statements is true?
   a) Technician A only
   b) Technician B only
   c) Both Technician A and Technician B
   d) Neither Technician A nor Technician B

6. Kinematic viscosity is a measure of the fluid viscosity under:
   a) Pressure
   b) Inertia force
   c) External load
   d) The force of gravity

7. Stoke is a unit of:
   a) Viscosity index
   b) Absolute viscosity
   c) Kinematic viscosity
   d) None

8. Which is not a unit of hydraulic fluid viscosity?
   a) Poise
   b) Stoke
   c) Pascal
   d) Saybolt Universal Seconds

9. The glass capillary viscometer measures:
   a) Viscosity index
   b) Viscosity grade
   c) Absolute viscosity
   d) Kinematic viscosity

10. Some ISO viscosity grades include:
    a) 4, 6, 14, 21, 38
    b) 20, 30, 40, 60, 80
    c) 5, 15, 25, 50, 100
    d) 22, 32, 46, 68, 100

11. As the temperature increases in a hydraulic system,
    (a) the fluid flow rate increases
    (b) the fluid viscosity decreases
    (c) the system pressure decreases.
    (d) the system parameters remain insensitive

12. High viscosity index points to:
    a) Low viscosity
    b) High viscosity
    c) Stable viscosity
    d) Variable viscosity

13. Flow rate in a hydraulic system can be measured in terms of:
    a) lpm
    b) m/s
    c) ft/s
    d) All of the above

14. In hydraulic systems, the flow velocity is:
    a) Same everywhere
    b) Greatest in return lines
    c) Greatest in suction lines
    d) Greatest in pressure lines

15. Mark the incorrect statement:
    a) The study of fluids at rest is called hydrostatics
    b) Fluid viscosity is affected by variations in temperature
    c) The flow rate in a hydraulic system determines the speed of hydraulic actuators
    d) With a constant flow rate, the velocity of the fluid passing through the narrow section decreases

16. Mark the correct statement:
    a) Fluid flow tends to be turbulent with high fluid viscosity
    b) Fluid flow tends to be laminar when the inertia forces in the fluid dominate
    c) Fluid flow tends to be laminar when the viscosity forces in the fluid dominate
    d) Turbulent flow can be avoided in hydraulic systems by increasing the velocity of fluid flow

17. What causes turbulent fluid flow in a hydraulic system?
    a) Lower fluid density
    b) Higher fluid velocity
    c) Higher fluid viscosity
    d) Smaller pipe section

18. Reynolds number (Re) is defined as the ratio of:
    a) Inertial forces to viscous forces
    b) Elastic forces to pressure forces
    c) Viscous forces to gravity forces
    d) Viscous forces to inertial forces

19. A fluid meant for a hydraulic system should have:
    a) low lubricity
    b) high viscosity
    c) high bulk modulus
    d) high compressibility

Review Questions
1) Briefly explain the operation of a simple hydrostatic system with a neat sketch.
2) Give a broad definition of the term 'hydraulics'.
3) What are the disadvantages of using water as a power transfer medium in hydraulic systems?
4) Why is oil, rather than water, an energy transmission medium in industrial hydraulic systems?
5) What is Pascal's law? State its application.
6) List three famous industrial pressure units and give the relevant conversion factors.
7) Mention the typical distinguishing maximum pressure levels for classifying hydraulic systems.
8) Describe the concept of the mechanical advantage pertaining to hydraulic power systems.
9) What is viscosity concerning hydraulic fluids?
10) Why is viscosity so relevant to fluid power applications?
11) What do you mean by the absolute viscosity of hydraulic fluid?
12) Give two units of absolute viscosity.
13) What is meant by the kinematic viscosity?
14) Differentiate between absolute viscosity and kinematic viscosity.
15) What are the units of kinematic viscosity?
16) Describe different ways to measure viscosity.
17) What is Saybolt Universal Second (SUS)?
18) Describe the effects of temperature variation on fluid viscosity.
19) Give a brief note on the viscosity classification system.
20) Describe the fundamental principles of viscometers.
21) Describe the viscosity index.
22) What is the purpose of specifying the viscosity index of a fluid?

23) What does a low viscosity index fluid indicate?
24) Differentiate viscosity and viscosity index.
25) Give the full expansion of ISO, ASTM, ANSI, and SAE.
26) Draw the characteristics of fluid viscosity against temperature for thick and thin fluids.
27) How do you differentiate between the flow rate and the velocity of fluid flow?
28) Explain the characteristic features of laminar flow.
29) Describe the characteristic features of turbulent flow.
30) Differentiate between the laminar and turbulent flows.
31) Why is turbulent flow avoided in hydraulic systems?
32) What design measures can be taken to avoid turbulence in hydraulic systems?
33) What is the Reynolds number?
34) What is the significance of the Reynolds number in deciding the type of fluid flow?
35) Explain the term 'bulk modulus' concerning hydraulic fluids.
36) Explain Bernoulli's Equation.
37) Mention different types of energies involved in a flowing fluid.
38) Explain potential energy due to elevation.
39) Explain potential energy due to static pressure.
40) List some advantages of hydraulic systems.
41) List some disadvantages of hydraulic systems.

Numerical Problems

1) What pressure does a hydraulic cylinder require with an active area of 16 in^2 for lifting 165347 lb of load? {Ans: 10334 psi}
2) Calculate the approximate force a hydraulic cylinder applies if it has a diameter of 2 in and is connected to a 2900 psi circuit. {Ans: 9106 lb}
3) Calculate the mechanical advantage required to lift a 1000 lb object with a 10 lb. {Ans: 100}
4) Calculate the approximate force in pounds required to raise a 225 lb load if the mechanical advantage of the system is 5. Design a possible hydraulic lift that would produce such a mechanical advantage. {Ans: 45 lb}
5) A hydraulic car lift used in a service station has an input pump piston and an output plunger to support a loading platform. The pump piston has a radius of 0.47 inches, and the plunger piston has a radius of 0.59 inches. The car and the plunger have a combined weight of 5620 lb. If the bottom surfaces of the piston and plunger are at the same level, what input force is required to lift the car and the output plunger? What pressure produces this force? {Ans: 35.65 lb, 51.37 psi}
6) Determine the Reynolds number, if an incompressible fluid of viscosity 0.255 lb.s/ft^2 and relative density of 1.75 slug/ft^3 flows through a 0.066 ft pipe with a velocity of 8.2 ft/s. Is the flow laminar or turbulent?
7) The diameter of the cylindrical pipe for a newly designed hydraulic system is estimated at 0.5 inches. A fluid with a viscosity of 68 cSt should be used for the system, and the fluid flow velocity is taken as 19.69 ft/s. Determine the Reynolds number for the fluid flow system and indicate if the flow regime is turbulent or laminar.
8) Determine the size of the suction line of a hydraulic system with a 15 gpm positive-displacement pump. The recommended flow velocity through the line is 3.61 feet per second. {Ans: 1.2 inches}

Objective-type questions - answer key: 1-a, 2-b, 3-b, 4-c, 5-d, 6-d, 7-c, 8-c, 9-a, 10-d, 11-d, 12-b, 13-c, 14-c, 15-c

Chapter 3 | Hydraulic Fluids – Functions, Characteristics, & Categories

Modern hydraulic applications demand compact machines with tighter tolerances and faster cycle times. They operate at higher pressures, temperatures, and speeds, and are designed to handle small amounts of fluid. Under these circumstances, the fluid used in a hydraulic system is subjected to severe stresses. Therefore, a hydraulic professional must know the functions, characteristics, and types of hydraulic fluids. This chapter describes these in detail.

Functions of Hydraulic Fluids
The fluid in a hydraulic system must perform many essential functions. The essential functions of the fluid are to:
- transmit power,
- provide lubrication to the internal moving parts in the system,
- provide sealing between the clearances in moving parts, and
- assist in the removal of contaminants and heat from the system.

Preparation of Hydraulic Fluids
Hydraulic fluid is prepared from a base stock and additives. The base stock possesses all the essential characteristics to perform well in a particular class of hydraulic systems. Examples of base stock include petroleum fluids, high-water-based fluids (HWBF), synthetic fluids, and vegetable fluids.

Many types of hydraulic fluids can be formulated by adding the base stock with various additives to meet the exacting requirements of complex systems. Blending the base fluid with suitable additives can improve its physical and chemical properties, making it more stable even in the presence of heat, oxygen, and water.

Table A5.1 of Appendix 5 gives some hydraulic fluid additives.

Fluid Characteristics
The following sections briefly outline some essential properties of hydraulic fluids.

Viscosity
Let us begin with viscosity. If the fluid medium in a hydraulic system is exposed to cold temperatures, it will thicken and its viscosity will tend to increase. More energy is required to pump the thick fluid, resulting in a higher pressure drop and excessive heat. This may lead to sluggish operation, higher power consumption, and lower mechanical efficiency. It may also produce cavitation and damage filters in the system.

If the fluid is heated, it will become thinner and have lower viscosity. A fluid that is too thin tends to rupture the fluid film between sliding surfaces in the system components, leading to leaks and higher friction. It also produces a higher rate of oxidation, reducing its service life.

The fluid used in a system must be thin enough to flow smoothly but thick enough to maintain a sufficient lubricating film between sliding surfaces and to provide proper sealing. The viscosity may be kept between 16 and 36 cSt at the usual operating temperature for most hydraulic applications.

Viscosity Index

Certain hydraulic systems are subjected to wide temperature variations. Mobile hydraulic systems are exposed to the outside environment. A high-precision hydraulic system is also sensitive to changes in the viscosity of its fluid medium at low temperatures. A hydraulic system exposed to wide temperature variations requires a fluid with a high viscosity index (VI) to minimize viscosity changes across a wide temperature range. Some fluids have large VIs to begin with. Some other fluids have their VIs reinforced through VI improvers. Such fluids are sometimes called multiviscosity fluids or multigrade fluids.

Fluid Compressibility

A good-quality hydraulic fluid should have very low compressibility (that is, a high bulk modulus) so that it remains stiff, providing a fast response from the associated system. However, the fluid's compressibility increases with temperature and applied pressure. A typical mineral-based fluid undergoes about a 0.5% reduction in volume for every 1000 psi of pressure applied, up to 4350 psi. Water-based and synthetic fluids have higher bulk moduli than mineral-based fluids.

The effect of compressibility shows up as a loss of fluid volume. This volume loss represents a loss of power, as no downstream actuator can recapture the compressive energy. Therefore, fluids with low compressibility must be selected for specialized systems such as precision machines or servo systems.

Lubricity

A fluid provides a load-carrying film in the clearance between two relatively moving adjacent surfaces in a hydraulic component. The film prevents metal-to-metal contact, thereby minimizing friction. Under modest load conditions, petroleum fluids can satisfy the system's lubrication requirements.

Wear Resistance

A fluid intended for use under modest load conditions should be formulated with anti-wear additives to improve its wear resistance. An anti-wear additive is stabilized zinc dithiophosphate (ZDP). Under highly stressed conditions, the ZDP may produce undesirable ash. Fluid manufacturers are seeking environmentally safe, zinc-free, ashless additives rather than zinc-based additives.

Under high loads, it is difficult to maintain a sufficiently thick fluid film within the clearances of moving parts in a hydraulic system component. To enhance its load-carrying property, the fluid used for this type of application should be formulated with an Extreme Pressure (EP) additive.

Oxidation Resistance

Over time, fluid passes through various components and naturally oxidizes, forming reaction products, such as acids, sludge, varnish, and gum. Exposure to heat, water, air, and metal catalysts accelerates the natural oxidation process. The signs of oxidation appear as changes in the fluid's color, odor, and acidity. Oxidation in hydraulic fluids and their by-products, such as sludge, can cause issues. For instance, sludge may block valves that control work processes in hydraulic systems, restricting their functionality.

A superior hydraulic fluid should resist any reaction with oxygen. A base fluid with excellent chemical stability can achieve better oxidation resistance. Additionally, antioxidants offer excellent oxidation resistance and effective acid neutralization.

Corrosion Resistance

Corrosion occurs when moisture and oxygen in the fluid react with metal surfaces. This leads to abrasive wear of the parts and increases leakage by opening up the tolerances of close-fitting parts.

System rusting occurs when moisture and oxygen attack ferrous parts. A suitable rust inhibitor added to the fluid in a system protects the fluid against system rusting and chemical corrosion.

Foam Resistance

Foam is a mass of air bubbles that collects at the air-liquid surface in a reservoir. It may be generated by the excessive churning or impingement of return fluid at the bottom of the reservoir. If the foaming process is excessive, the foam will likely be drawn again into the fluid. Therefore, the hydraulic fluid should have low foaming.

Stability

It refers to the fluid's resistance to degradation under extreme temperatures, increased chemical activity, or high water content. The fluid should have excellent thermal, chemical, and hydrolytic stabilities.

- Thermal Stability refers to the fluid's ability to resist degradation when subjected to high temperatures and extreme shear.

- Chemical Stability refers to the ability to resist degradation when subjected to increased chemical activity.

- Hydrolytic stability refers to the ability to resist chemical decomposition in the presence of water.

Fire Resistance

The basic parameters of a fire-resistant fluid are its resistance to ignition and to the propagation of flame from its source of ignition. Flash point and fire point are the two parameters used in this context.

Flash Point

Flash point is the lowest temperature at which a fluid gives off enough vapors to form an ignitable mixture. When this mixture comes into contact with a heated matter, it will likely generate flashes.

Fire Point

Fire Point refers to the lowest temperature at which a fluid gives off adequate vapors to its surrounding air. This vapor can sustain continuous combustion after ignition.

Pour Point

Pour Point refers to the lowest temperature at which a fluid can flow when cooled under the specified test conditions.

To wrap up, hydraulic fluids should have a long lifespan and be cost-effective. It is important to note that no single fluid possesses all the necessary properties for a given application. However, a wide range of hydraulic fluids is designed to meet the required properties for specific operating conditions.

The Categories of Hydraulic Fluids

As modern hydraulic systems require high-performance hydraulic fluids to meet stringent requirements, manufacturers prepare a variety of hydraulic fluids (Figure 3.1).

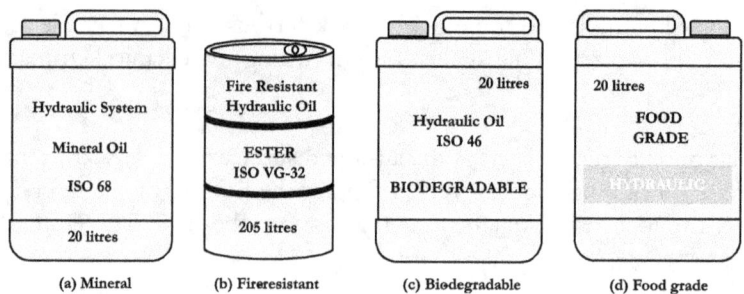

Figure 3.1 | Categories of hydraulic fluids

Petroleum-based Fluids

Petroleum fluids have been the preferred energy-transfer medium for hydraulic systems for many years. They have excellent lubricating and corrosion-inhibiting properties, are low-cost, and are available in a wide range of viscosities. However, they are flammable, toxic, and not very biodegradable. They must also be compatible with construction materials, such as seals.

Fire-resistant Fluids

They are required for high-temperature, hazardous hydraulic applications. Two basic types of fire-resistant fluids are: (1) High-water-based fluid (HWBF) and (2) Synthetic fluids. HWBFs are very much fire-resistant due to their high water content. Synthetic fluids have exceptional fire-resistant properties, but they are costly.

ISO 6743 divides fire-resistant fluids into HFA, HFB, HFC, and HFD.
- HFA: Fluid in water emulsions with a combustible proportion of 20% maximum
- HFB: Water in fluid emulsions with a combustible proportion of 60% maximum
- HFC: Water glycol solutions with a water proportion of at least 35%
- HFD: Water-free fluid on a synthetic base

Synthetic Fluids (HFD type)

Synthetic fluids are prepared by blending alkaline compounds with additives. Synthetic fire-resistant fluids are prepared from (1) Phosphate esters, (2) Polyol esters, and (3) Halogenated hydrocarbons. They have good fire resistance and excellent lubrication characteristics. However, they are expensive and often incompatible with many seal materials. They may also give off toxic vapors. Further, they require a special disposal plan.

Biodegradable Fluids

There is demand for ecologically safe green fluids for environmentally sensitive applications such as agriculture, forestry, construction, offshore drilling, and mining, where fluid leakage could harm the environment. A biodegradable fluid is the best choice for such applications. When spillage occurs, a readily biodegradable fluid breaks down 60% of the fluid into harmless products through reaction with naturally occurring bacteria when exposed to the atmosphere for 28 days in a standard test. The most important base fluids of biodegradable hydraulic fluid are: (1) Synthetic esters and (2) Vegetable oils.

Food-grade Fluids

Devices used for the manufacturing and packaging of beverages, food, cosmetics, and medicines must be hygiene-specific. These devices must be designed to use food-grade fluids to protect against the risk of incidental contact between products and the fluids. Fluid manufacturers prepare a food-grade fluid from a highly refined, non-toxic Polyalphaolefin (PAO) synthetic base fluid, a highly specialized, non-toxic, food-grade additive package, and a food-grade anti-microbicide. Food-grade fluids must be maintained to a high level of cleanliness, sanitation, and quality.

Typical Hydraulic Fluid Specifications

Sample specifications of some hydraulic fluids are given in Table 3.1.

Table 3.1 | Hydraulic fluid specifications

Type of fluid (Hypothetical)	Viscosity (ISO Grade)	Viscosity Index
Type F1	32	98
Type F2	46	98
Type F3	68	98
Type F4	32	164
Type F5	46	141
Type F6	68	370

Section 5B in Appendix 5 gives the properties of some hydraulic fluids.

Objective-type Questions

1. An ideal hydraulic fluid is one that
    a) behaves like a perfect gas
    b) possesses a low bulk modulus
    c) can flow through pipes with high inertia
    d) has zero viscosity and remains perfectly stiff

2. Which is not a function of hydraulic fluids?
    a) Store energy
    b) Seal clearances
    c) Transmit power
    d) Lubricate system parts

3. What is the typical range of viscosities selected for hydraulic fluids, measured at 40°C (100°F)?
    a) 01 to 10 cSt
    b) 10 to 400 cSt
    c) 100 to 400 cSt
    d) 100 to 1000 cSt

4. Viscosity Improver (VI) can be used in hydraulic fluid to:
    a) improve the viscosity of the fluid
    b) prevent fluid breakdown against high shear stress
    c) stabilize the viscosity against variations in temperature
    d) protect the integrity of the fluid against harsh working conditions

5. An extreme pressure additive is used in a hydraulic system fluid to:
   a) improve the load-carrying capacity of the fluid
   b) prepare the fluid for high-pressure applications
   c) prevent the metal-to-metal contact of sliding surfaces
   d) stabilize the viscosity of the fluid against high pressure

6. Multi-grade hydraulic fluids are:
   a) fluids with VI improvers
   b) fluids without VI improvers
   c) a mixture of different grades of fluids
   d) a class of fluids with high ISO Viscosity Grades

7. An anti-wear additive added to most of the hydraulic fluids is the:
   a) Silicone oil
   b) Polyalphaolefins
   c) Polymethacrylates
   d) Zinc dithiophosphate

8. Oxidation of hydraulic fluid causes:
   a) the breakdown of fluid constituents
   b) the decrease in the lubricity of the fluid
   c) the formation of acids, sludge, and varnish
   d) All of the above

9. The demulsibility property of a hydraulic fluid allows:
   the heat to be quickly dissipated from the fluid
   the water to be readily separated from the fluid
   the air to be released from the fluid
   None of the above

10. Hydrolytic stability of hydraulic fluid indicates the resistance against:
    a) heat
    b) pressure
    c) shear stress
    d) None of the above

Review Questions
1) What are the primary and supporting functions of hydraulic fluids?
2) How are hydraulic fluids prepared?
3) Mention five additives used in the preparation of hydraulic fluids.
4) List five properties of hydraulic fluids.
5) Briefly explain the viscometric characteristics required of hydraulic fluids.
6) What are the adverse effects of using low-viscosity fluids in hydraulic systems?
7) What are the adverse effects of using highly viscous fluids in hydraulic systems?
8) Why is determining the correct fluid viscosity significant for a hydraulic application?
9) Explain what the viscosity index of fluid means.
10) What is the primary function of VI improvers?

11) How does varying the temperature affect the fluid's viscosity in a hydraulic system?
12) Describe the lubricity property of hydraulic fluids.
13) What are extreme-pressure (EP) additives used in hydraulic fluids?
14) How does the extreme-pressure (EP) additive in a hydraulic fluid act while in use?
15) What is the purpose of formulating hydraulic fluids with anti-wear additives?
16) Differentiate between the following: EP additives and anti-wear additives
17) Explain the need for good oxidation resistance for hydraulic fluids.
18) What are the adverse effects of oxidation in a hydraulic fluid?
19) What essential measures can be taken to protect a hydraulic fluid against oxidation?
20) Briefly explain the corrosion problems in hydraulic systems.
21) Explain the foaming process and the foam's effects on hydraulic fluids.
22) Why is it important to have the right air release property for a hydraulic fluid?
23) What are the adverse effects of air present in hydraulic fluids?
24) What are the effects of water present in hydraulic fluids?
25) What are the adverse effects of operating hydraulic fluids at higher temperatures?
26) State the reasons for the heat build-up in the fluid medium of a hydraulic system.
27) What are the factors affecting the overall stability of hydraulic fluids? Explain.
28) What are the adverse effects of using hydraulic fluids at higher temperatures?
29) What actions may you take to improve the thermal stability of hydraulic fluids?
30) Give a brief account of the chemical stability of hydraulic fluids.
31) Give a brief account of the hydrolytic stability of hydraulic fluids.
32) Briefly explain, with examples, the compatibility issue of seal materials with hydraulic fluids.
33) Give a brief note on the filterability of hydraulic fluids.
34) Explain the fluid property terms: (1) Demulsibility and (2) foaming.
35) Briefly explain the following fluid property terms: (1) Pour point, (2) flash point, and (3) fire point
36) What are the essential properties of hydraulic fluids? Discuss any four of them in detail.
37) What are the main classes of hydraulic fluids? List their distinguishing features.
38) Explain the advantages and disadvantages of petroleum-based fluids used in hydraulic systems.
39) Briefly explain the two basic types of fire-resistant fluids.
40) What are the different types of water-based fluids used in hydraulic systems?
41) What are the essential characteristics of water-based fluids?
42) What precautions must be taken while maintaining water-based fluids used in hydraulic applications?
43) What are high-water-based fluids (HWBF)?
44) Differentiate oil-in-water emulsions and water-in-oil emulsions.
45) Give a brief account of synthetic hydraulic fluids.
46) Differentiate between the following: (1) Water-based and (2) Synthetic fire-resistant fluids
47) Give a brief note on eco-friendly hydraulic fluids.
48) What are the areas of application for biodegradable fluids? Explain.
49) How do ecologically safe hydraulic fluids protect the environment?
50) How do fluid manufacturers typically formulate biodegradable fluids?
51) Give a brief account of food-grade hydraulic fluids.
52) What are the general requirements for hydraulic fluids?

Objective-type questions - answer key: *1-d, 2-a, 3-b, 4-c, 5-a, 6-a, 7-d, 8-d, 9-b, 10-d*

Chapter 4 | Hydraulic Fluid Contamination & its Control

Hydraulic fluids are subjected to various types of contamination. Contaminants can affect their physical and chemical properties. If the contaminants are not monitored and controlled, the fluid is likely to fail. These failures include deterioration of the fluid properties and consequent fluid breakdown. At the system level, contamination can affect component performance and service life, leading to erratic operation, increased heat generation, frequent fluid replacement, catastrophic system failure, and higher costs.

Three critical issues concerned with the proper maintenance of hydraulic fluids are: (1) knowing the type of contaminants, (2) controlling contamination, and (3) assessing the health of the fluids. Analysis of the fluid in a system can help detect an emerging problem.

Contaminations in Hydraulic Fluids

Contaminants are the natural enemy of hydraulic components and systems. About 70-80% of system failures are caused by contaminants. Even minute particles can damage hydraulic components due to the minuscule clearances in today's hydraulic components.

Solid Particles

It includes hard particles, such as dust, dirt, silica, and wear metals, as well as soft particles, such as elastomers and fibers. Silts are particles less than 5 microns, and chips are particles greater than 5 microns. Silt particles of a size corresponding to the typical clearance of components are more dangerous to a hydraulic system than larger chip particles. The system itself can generate metal particles from component wear. Abrasive particles can scrape metal from component surfaces. The freely circulating particles can cause premature wear of parts.

Water

Water can significantly impact the aging process when introduced into the system fluid. Unprotected reservoir openings, leaking seals, and ineffective heat exchangers are potential sources of water entry in hydraulic systems. Fluid can dissolve water up to its saturation point. Above the saturation point, water remains in the free or emulsified state. A mineral fluid can contain up to 100 ppm of water, that is, up to 0.01%. Moisture can provide oxygen for chemical reactions. Further, excessive water contamination in a hydraulic system can accelerate the aging of the system fluid, a risk you should be aware of.

Chemical Contaminants

They are formed by the breakdown of additives due to chemical reactions. The reaction products can generate additional contaminants, such as acids or oxidants, in the presence of water and heat. They can cause physical and chemical changes in the additive elements. These changes can lead to the deterioration of additives and subsequent fluid breakdown.

Air

Air can enter the fluid medium through system leaks, pump aeration, or excessive fluid turbulence in the reservoir. Next, air can exist either in the 'free state' or an entrained state. An air pocket trapped in a part of the system is an example of free air. Air bubbles less than one millimeter in diameter dispersed in the fluid medium are entrained air. The entrained air can cause cavitation and foaming as it cycles through the system. It also tends to make the system operation spongy and weaken the system response.

Heat

A hydraulic fluid can deteriorate at higher temperatures and may lose its lubricating properties. Therefore, it is required to maintain the fluid's temperature within satisfactory limits. Typically, the maximum temperature specified for fluids used in general hydraulic systems is 176°F, mainly because of rubber parts.

Contamination Control

Fluid contamination control involves removing particles, water, air, sludge, acids, and chemicals from the system fluid. The particles can be removed by installing correctly-sized filters at appropriate locations in the system. Ferrous particles and rust matter can be removed by installing magnets. Removal of acids, sludge, gums, varnishes, and other oxidation products requires an adsorbent filter with active elements. Water in the fluid can be removed by installing a water-removal filter or a vacuum dehydrator. Heat can be removed by installing a heat exchanger. Air contamination can be eliminated by providing air bleeds and diffusers.

Fluid Cleanliness Standards

The cleanliness of hydraulic fluids must be monitored to ensure satisfactory operation of hydraulic system components. Many national and international organizations have developed standards for specifying the particle size classification and contamination concentration levels. The particle size classification standard is ISO 11171. The standard for specifying contamination concentration levels is ISO 4406. The SAE aerospace standard AS4059, for specifying particulate contamination in hydraulic fluids in different classes, was developed in 1988 as a replacement for the NAS 1638 standard. The details of the standard AS4059 are given in Annexure 7.

Particle Size Classification Standard

The standard ISO 11171 specifies three particle sizes — 4, 6, and 14 microns — to represent the concentrations of fine and coarse particles.

Fluid Cleanliness Level Standard

The standard ISO 4406 specifies the cleanliness level of a given sample of fluid using a three-number range code representation, such as 18 / 16 / 14, based on the number of particles of size greater than 4, 6, and 14 microns, respectively, present in one milliliter (ml) of the sample fluid.

Table 4.1 | Range codes based on the number of particles

Range code	Number of particles per ml	
	>	<=
21	10 000	20 000
20	5 000	10 000
19	2 500	5 000
18	1 300	2 500
17	640	1 300
16	320	640
15	160	320
14	80	160
13	40	80
12	20	40
11	10	20

Assume that the number of particles of sizes larger than 4µ, 6µ, and 14µ found in one ml of the hydraulic fluid sample are 1510, 406, and 55, respectively. Then, from Table 4.1, the fluid cleanliness range code can be derived as 18/16/13. Table A6.1 in Appendix 6 gives the complete contamination code rating system as per ISO 4406.

Cleanliness Level Targets
Hydraulic equipment manufacturers, fluid suppliers, and fluid power associations have established target cleanliness levels for fluids used in general types of hydraulic components. For initial learning, Table 4.2 lists the typical target cleanliness levels for fluids used in some hydraulic components. Also, Table A6.2 of Appendix 6 gives detailed recommended fluid cleanliness levels.

Table 4.2 | Cleanliness level targets for hydraulic components

Component	Target cleanliness
Gear pumps and motors	20 / 18 / 15
Vane and piston pumps/motors and valves	19 / 17 / 14
Proportional valves	17 / 15 / 12
Servo valves	16 / 14 / 11

Hydraulic Fluid Analysis
Fluid analysis can be performed to assess the health of a fluid medium. The fluid analysis essentially counts contaminant particles, detects oxidation levels, identifies component wear, assesses additive condition, establishes the overall contamination level, and verifies the fluid composition. If the analysis meets the specified cleanliness target, only filters need to be maintained and the fluid retested periodically. If not, appropriate corrective actions must be taken to rectify the problems.

Contamination can be measured using portable laser particle counters. Calibrated and evaluated to ISO standards, the portable particle counter identifies and reports the range code for the number of particles greater than a specified size in 1 mL of the sample fluid.

A proper fluid analysis first establishes the target cleanliness level, sampling location, and testing frequency. Ensure that the sample taken from the system is representative of the system fluid. Analysis can be conducted on a fluid sample by: (1) Patch test, (2) Laboratory analysis, and (3) Online fluid monitoring.

Patch Test
It is a simple visual analysis of the fluid sample extracted from a hydraulic system using a fluid analysis kit. It comprises a microscope, filter test patches, a vacuum pump, sample bottles, and visual correlation charts or photographs.

A 100 mL sample of the fluid is passed through the test patch's filter media. The patch is then dried and analyzed for color and content under the microscope. The reference photographs of known particle concentration levels are compared to determine the approximate ISO cleanliness code and the type of particles captured on the patch.

Laboratory Analysis
Laboratory analysis is a comprehensive examination of the fluid sample. Most of the laboratories offer the following essential fluid parameters:
- Particle counts,
- Water content,
- Viscosity,
- Total Acid Number, TAN, and
- Spectrometric analysis for finding the wear metals and trending graphs

The Total Acid Number (TAN) indicates the fluid's age and can be used to decide when to replace it. It is a measure of potassium hydroxide (KOH), in milligrams, required to neutralize the acid present in one gram of the fluid. Typically, it should be less than 1.4 milligrams of potassium hydroxide per gram. A borderline value is between 1.4 and 2.6 mg (KOH)/gram. A TAN value above 2.6 mg (KOH)/gram is unsatisfactory.

Online Fluid Monitoring
With advances in computer technology and the introduction of sophisticated online fluid-monitoring instruments, fluid analysis can be consistently performed on-site while the system operates. Today's online contamination monitoring instruments can detect changes in the quality, contamination level, and chemical composition of the fluid medium in a hydraulic system. They can also measure the number of wear metals present in the fluid.

Objective-type Questions
1. Three critical issues concerned with the proper maintenance of hydraulic fluids are:
    a) Knowing the type of contaminants
    b) Means for controlling contamination
    c) Assessing the health of the fluids
    d) All of the above

2. Mark the wrong statement
    a) Silt particles of a size corresponding to the component clearance are more dangerous to a hydraulic system than larger chip particles
    b) Fluid can dissolve water up to its saturation point
    c) The entrained air can cause cavitation and foaming
    d) An adsorbent filter can remove ferrous particles

3. The ISO standard for specifying contamination concentration levels in hydraulic fluids is the:
    a) 4402
    b) ISO 4406
    c) ISO 5599
    d) ISO 1209

4. An appropriate fluid cleanliness specification for a servo application as per ISO 4406 could be:
    a) 21/18/15
    b) 20/17/14
    c) 19/16/13
    d) 17/14/11

5. For the target cleanliness level of 17/15/12, as per ISO 4406, which of the following cleanliness levels is unacceptable
   a) 15/13/11
   b) 16/14/11
   c) 17/15/12
   d) 18/16/13

Review Questions
1) How does the fluid in a hydraulic system get contaminated?
2) What are the different types of contaminants present in hydraulic fluids? Briefly explain each type.
3) Describe the effects of particle contamination in hydraulic systems.
4) Explain the effects of water present in hydraulic systems.
5) Explain the effects of oxidative reactions in hydraulic fluids.
6) Describe the effects of excessive heat on the fluid medium of a hydraulic system.
7) Why is the cost of contamination very high in hydraulic systems? Explain.
8) What are the reasons for the fluid breakdown in hydraulic systems?
9) How does fluid contamination destroy hydraulic components?
10) What are the sources of contamination in hydraulic fluids? Briefly explain.
11) Briefly explain a typical approach to reasonable contamination control in hydraulic systems.
12) Mention some of the symptoms of excessive contamination in hydraulic systems.
13) Enlist some necessary measures to avoid contamination in hydraulic systems.
14) What are the ways to remove particle contaminants from a hydraulic system?
15) What are the ways to remove water from a hydraulic system?
16) What are the ways to remove air from a hydraulic system?
17) What are the ways to remove heat from a hydraulic system?
18) Explain various ways of conditioning hydraulic fluids.
19) What measures can be taken to maximize the life of hydraulic systems?
20) Explain the ISO cleanliness standards for evaluating the contamination level of hydraulic fluids.
21) Explain the meaning of the fluid cleanliness level 18/16/13 as per the ISO 4406 standard.
22) Why is it essential to conduct fluid analysis in a hydraulic system?
23) Explain the laboratory analysis of hydraulic fluids.
24) Why do maintenance technicians take fluid samples during preventive maintenance of hydraulic systems?
25) Explain the procedure to conduct the fluid analysis of the fluid in a hydraulic system.
26) What are the two crucial maintenance actions in a healthy hydraulic system?
27) What corrective measures are necessary if the health of the fluid used in a hydraulic system is not meeting the target cleanliness levels?
28) Give two important methods for finding the fluid cleanliness level in a hydraulic system. Briefly explain any one method.
29) What are the expected results of the fluid analysis of hydraulic fluid?
30) What are the advantages of the online monitoring of hydraulic fluids?
31) Give a brief note on each of the following: (a) Hydraulic fluid cleanliness, (b) The disposal of used hydraulic fluids, and (c) The recycling and reclaiming of hydraulic fluids.

Objective-type questions - answer key: *1-d, 2-d, 3-b, 4-d, 5-d*

Chapter 5 | Hydraulic Filters

A modern hydraulic system is susceptible to many types of contamination in its fluid medium. Contaminants can cause premature wear of internal surfaces, promote leakage, and clog flow paths. Water in the system fluid can lead to corrosion and accelerated component wear. Therefore, an efficient filtration system should be an integral part of every hydraulic system to separate particulate matter and water from the system fluid. Filters are installed at appropriate locations in a hydraulic circuit to effectively control contamination. The following sections present the constructional features, types, and specification parameters of filters.

Function of Filters
Filters are necessary devices for removing particulate contamination from hydraulic systems. When fluid flows through a filter, it traps contaminants while allowing the fluid to flow freely. As the filter clogs, the differential pressure across it increases until it bursts or collapses.

Parts of Filters
Filters consist of many essential parts and some optional parts. Figure 5.1 shows an exploded view of a hydraulic filter. A filter consists of the following:
- Filter head
- Filter bowl or housing
- Filter element

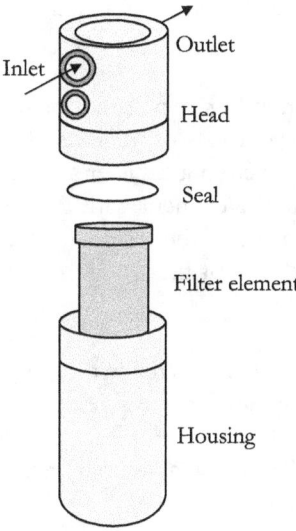

Figure 5.1 | Exploded view of a hydraulic filter

The filter element is housed in the filter bowl. The bowl and filter element (cartridge) can remain separate and be assembled onto the filter head as a cartridge-type filter unit. Alternatively, the filter element and the housing can be configured as an integrated unit and screwed onto the filter head as a spin-on unit. Hydraulic filters can be arranged with other useful constituents, such as bypass valves, clogging indicators, etc. They can also be organized in duplex, in-line, or in-tank configurations to realize many useful functions. Remember, various internal parts of a filter must be compatible with the type of fluid used. Let us now further elaborate on the filter parts.

Filter Head

Figure 5.2(a) shows that a filter head holds the filter element and its housing. It consists of ports for the inlet and outlet. It may also consist of optional ports for the pressure gauge and visual and electrical indicators. It is made of cast aluminum, a standard material, or ductile iron for high-pressure applications.

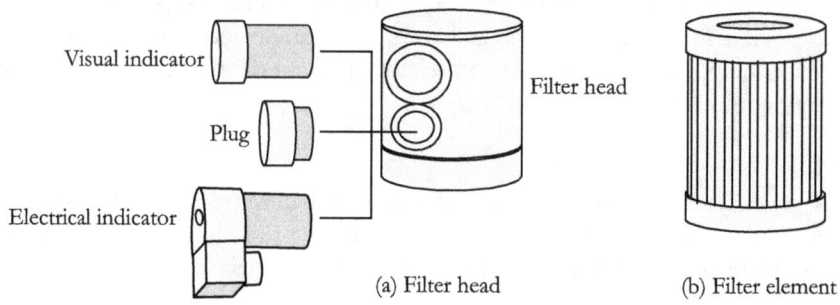

(a) Filter head   (b) Filter element

Figure 5.2 | Hydraulic filter parts

Filter Housing

A filter housing encloses the filter element and confines the system fluid within the unit. It must also withstand the pressure within the unit. Filter housings are usually made of ductile iron or stainless steel. Filter housing styles are cartridge, spin-on, duplex, in-tank, and in-line.

Filter Element

A filter element (or cartridge), as shown in Figure 5.2(b), usually comprises a steel wire screen, cellulose media, or synthetic glass fiber media. It consists of millions of tiny micron-sized pores. A piece of the filter media is usually pleated and assembled in a canister as a disposable element. The pleats are made deeper and more numerous to provide the element with greater filtration surface area. The fluid can flow from the outside of the filter element to the inside as in an 'out-to-in' type of filter or from the inside to the outside as in an 'in-to-out' type of filter.

Wire Mesh Filter Media

Figure 5.3(a) shows that the wire-mesh media are made of epoxy-coated stainless steel wire fabric. The filter captures contaminants in a fluid stream on the side of the wire screen facing the fluid flow. This type of filtration is regarded as surface filtration. A piece of wire-mesh filter media is used to make a coarse filter, usually called a strainer. Typically, a strainer traps very large, abrasive particulate matter that could tear a standard filter.

Typical filtration ratings for wire mesh media include 25 μ, 44 μ, 50 μ, 60 μ, 74 μ, 90 μ, 100 μ, 150 μ, 200 μ, etc.

Cellulose Filter Media

A piece of cellulose (or paper) media, as shown in Figure 5.3(b), is made from plant fibers held together by resins. The pores in the cellulose media are microscopic. The multi-layered, thick-walled media absorbs contaminants throughout its depth as the fluid flows through it. Therefore, this type of media is known as depth media. Typical filtration ratings for cellulose media in hydraulic systems include 10 μ and 25 μ.

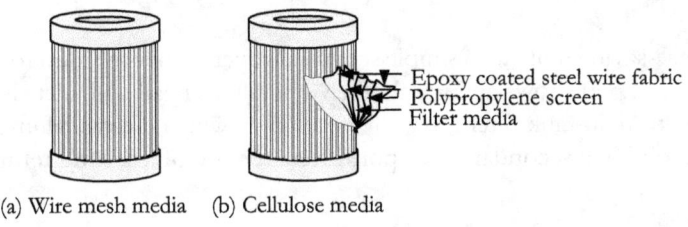

(a) Wire mesh media (b) Cellulose media

Figure 5.3 | Hydraulic filter element types

Glass Fiber Synthetic Filter Media
Synthetic glass fiber media are man-made, consistent, and rounded to provide the least resistance to flow. It is made of inorganic micro-fine glass fibers. They are randomly laid into a multi-layered web with tapered pore geometry. This type of construction ensures that larger pores are on the upstream surface and finer and finer pores are towards the downstream part of the media. The thick-walled media captures contaminants throughout its depth. Therefore, a filter with synthetic media can produce high filtration efficiency. Typical filtration ratings for glass fiber media in hydraulic applications include 3 μ, 5 μ, 6 μ, 10 μ, 20 μ, 16 μ, and 25 μ.

Surface Media vs. Depth Media
- As stated earlier, surface media are made from woven wire. The filter captures contaminants in a fluid stream on the side of the wire screen facing the fluid flow.
- The depth media are thick-walled filter elements with absorbent materials. The filter usually absorbs contaminants throughout the depth of the filter element as the fluid flows through the media.

Cartridge Type Filters
Earlier filters were cartridge-type filters with replaceable cartridges in permanent housings. Figure 5.4(a) shows a cartridge-type hydraulic filter.

Spin-on Type Filters
Figure 5.4(b) shows a spin-on filter. It is a self-contained filter element and housing assembly. The ineffective spin-on unit can be unscrewed from its mount and discarded. A new unit can then be screwed onto the mount. This design has made filter replacements more convenient. However, spin-on type filters generate more wastage with each filter unit replacement.

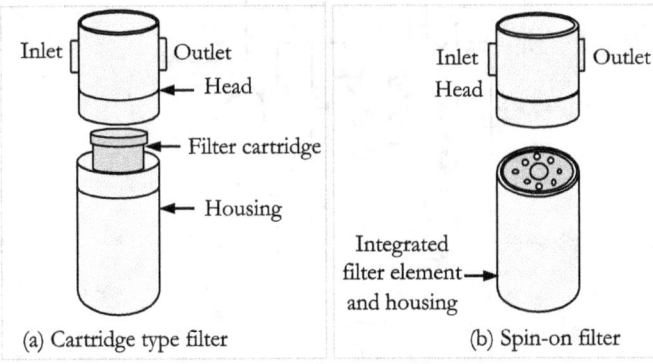

(a) Cartridge type filter   (b) Spin-on filter

Figure 5.4 | Hydraulic filters

In-tank Filter Type

In-tank filters are space-saving units with simple screw-on covers. They are ideal for low pressure return-line applications. Figure 5.5(a) shows an in-tank filter. The filter's head and inlet sit above the tank, with the housing in the tank. An in-tank filter is usually heavy-duty with a die-cast aluminum head and a steel or nylon canister. An optional secondary inlet port offers the use of a second return line.

In-line Filter Type

The in-line filter assembly, shown in Figure 5.5(b), features a heavy-duty steel canister with an inlet and outlet, a cartridge-type filter element, and a top cover. The steel housing design provides a strong, durable, and dependable unit. The filter cartridge can be replaced easily by removing the top cover.

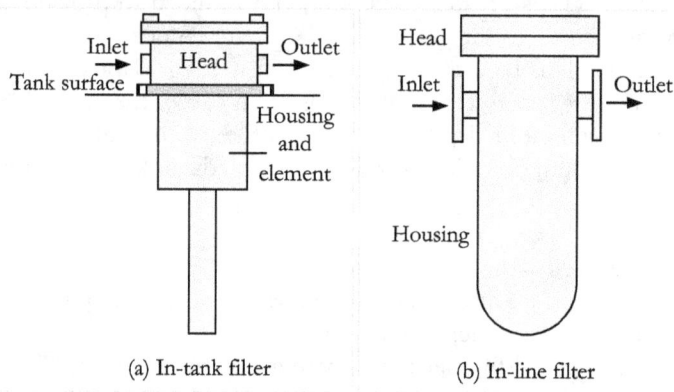

Figure 5.5 | Types of hydraulic filters

Filter with By-pass Valve

As shown in Figure 5.6, a bypass valve can be included in a filter to protect against a filter burst. This filter assembly is a type of relief valve that diverts all or part of the fluid flow back to the system reservoir (tank) when the permissible maximum pressure differential across the filter is reached.

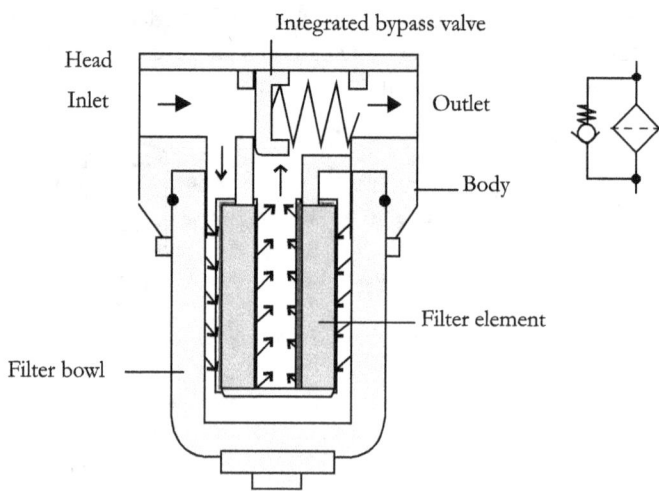

Figure 5.6 | A cross-sectional view of a filter with a by-pass valve

By-pass Valve Setting: Bypass valves have cracking pressures typically in the range from 2 psid to 100 psid. The choice of cracking pressure for a filter depends on its location. Suction and return-line filters have lower settings than those of pressure-line filters. The filter element collapse/burst pressure should be at least 1.5 times the full-flow pressure drop across the bypass valve and may range from 50 to 150 psid.

Duplex Type Filter

A duplex filter, as shown in Figure 5.7, is an assembly of two (or more) filters with a selector valve to select one filter at a time for active filtration. When the filter element in the selected filter becomes dirty and needs servicing, the valve is shifted without shutting down the system, thereby blocking flow through the ineffective filter and allowing flow through the second filter. The clogged filter element can then be cleaned or replaced while flow continues to pass through the currently active filter.

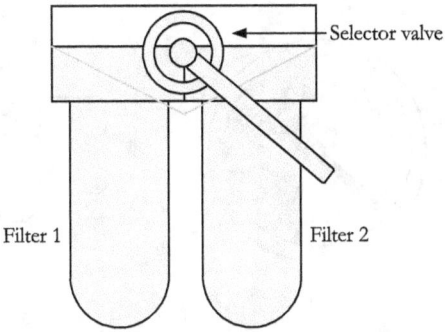

Figure 5.7 | A duplex-type hydraulic filter

Full-flow Filtration

In full-flow filtration, the entire fluid volume passes through the filter element during each cycle, eliminating the possibility of contaminated fluid reentering the system.

Proportional-flow Filtration

In proportional or partial-flow filters, a portion of the fluid entering the filter passes through the filter element, while the remaining portion bypasses the filter element through the venturi without filtration.

Service / Contamination Indicators

A hydraulic filter element is efficient only if its particle-capture efficiency is fully exploited. For this purpose, a contamination indicator or clogging indicator is fitted to the filter. Vacuum indicators and switches can be used with strainers to indicate when the vacuum reaches a specific level.

A service indicator allows the filter element to be replaced only when its full dirt-holding capacity has been reached, rather than periodically replacing the filter elements on a time-based schedule. The contamination indicators for a filter usually include a pressure gauge and visual and electrical indicators that signal the need to replace the filter element.

Contamination indicators are especially important with a suction filter to detect excess pressure drop in the suction line.

Pressure Gauge

Figure 5.8(a) shows a pressure gauge. It is used to measure pressure differential directly across the filter element. It consists of a mechanical bourdon tube with an elastic chamber, a link, a geared sector, a pinion, a pointer, and a calibrated scale. Pressure applied to the elastic chamber causes the Bourdon tube to move outwards. The higher the pressure, the greater the tube's deflection radius. The deflection converts the applied pressure into the corresponding movement of the attached pointer across the scale via links, levers, and gearing. The pressure gauge is a safety measure that monitors overpressures in the system and assists with troubleshooting.

Visual Indicator

Figure 5.8(b) shows two types of visual indicators. A visual indicator can be a dial indicator or a visual pop-up indicator with a red button to indicate that the filter element is clogged and the set differential pressure across the filter is exceeded.

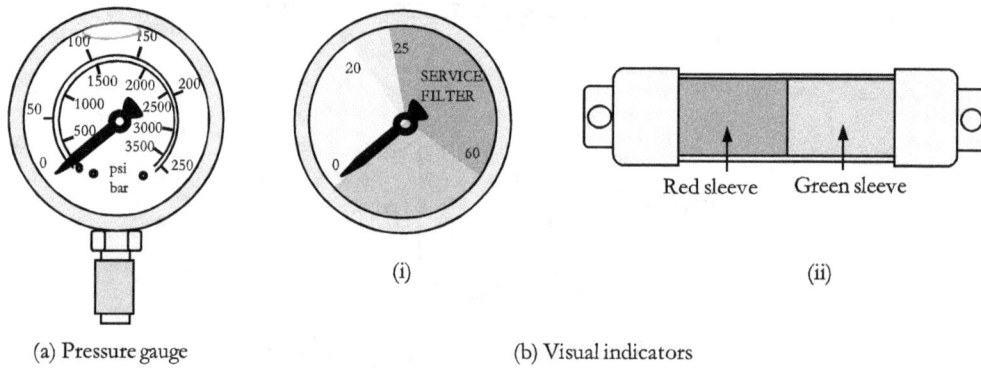

Figure 5.8 | Service / Contamination indicators for hydraulic filters

Electrical Indicator

An electrical switch can be provided in a filter when a connection to a remote sensing device, such as an alarm, horn, or light, is desired. The switch can respond to the differential pressure across the filter element.

Symbols of Service Indicators

The self-explanatory Figure 5.9 gives symbols of service indicators for hydraulic filters.

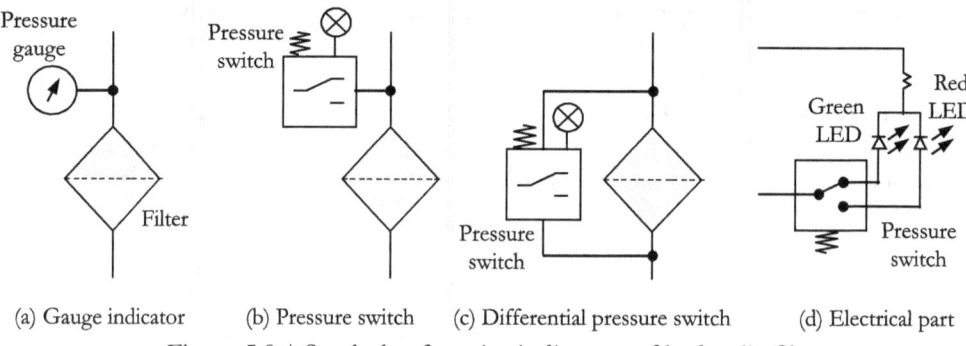

Figure 5.9 | Symbols of service indicators of hydraulic filters

Seals
Filters with Buna-N seals are suitable for most petroleum-based fluid applications. Filter seals made of fluorocarbon, such as Viton, are used for applications with operating temperatures above 180°F and fluids such as phosphate ester, water-glycol, water/oil emulsions, and HWCF.

Magnet
The filter unit can be provided with a magnet, as shown in Figure 5.10(a), to attract and hold ferrous particles down to even smaller than 1 micron. Fluid flow through the filter element is generally from the outside to the inside (out-to-in). For the effective filtration of metallic particles in systems with high contamination load, a filter designed for the flow from the inside of the filter element to the outside (in-to-out) through the filter element can be used.

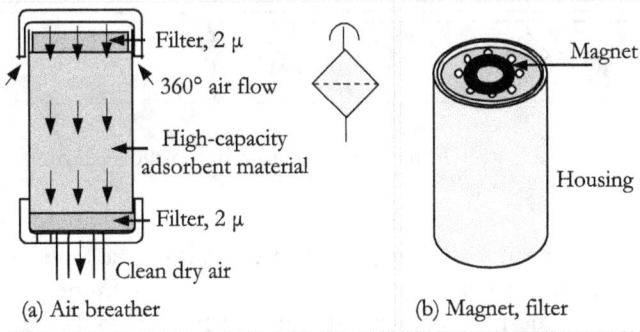

Figure 5.10 | Accessories for hydraulic filters

Air Breather
As shown in Figure 5.10(b), an air breather unit usually consists of air filters and a quantity of desiccant material. It is fitted to the reservoir in a hydraulic system to protect the system fluid in the reservoir against particulate contamination that enters through the reservoir opening. The filter removes solid particles down to 3 μm (0.000118 inches) at 97% efficiency, and the desiccant material prevents moisture from entering the reservoir in a hydraulic power pack.

.Filters - Installation Locations
According to their locations in a hydraulic system, filters are categorized into four types, as shown in Figure 5.11: (1) Strainer/ Suction filter, (2) Pressure filter, (3) Return-line filter, and (4) Offline filter. An air filter is also used in a hydraulic system fitted on the top of the reservoir.

Strainer
A strainer is installed at the pump suction side within the hydraulic reservoir. It comprises a zinc-plated housing, stainless steel wire-mesh screens, and rugged steel core centers. It is a coarse filter, usually having a mesh width ≥ 149 μ (0.005867 inches). It protects the pump from coarse particles at the lowest cost. However, a blocked strainer can starve the pump, which, in turn, can cause cavitation.

A strainer used in a hydraulic system is cleanable and reusable. But it is difficult to clean because it is submerged in the reservoir within the system. Another drawback of the strainer is that it promotes cavitation in the pump intake line when clogged. They offer the least benefit.

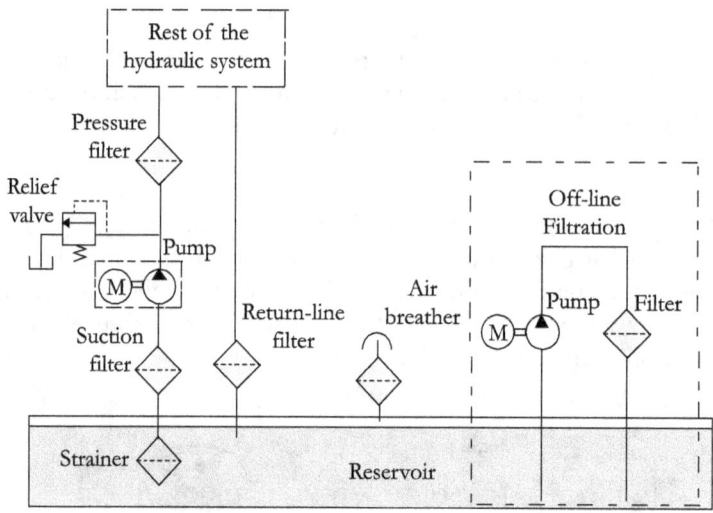

Figure 5.11 | Filter locations in a hydraulic system

Suction Filter

A conventional suction filter is mounted outside the reservoir, above the fluid level, in a service-friendly manner. The space-saving in-tank suction filter can be mounted semi-immersed in the fluid. A suction filter is a coarse filter with a mesh width typically ranging from 5 to 149 µ. It protects the pump economically from coarse particles. A bypass valve with low cracking pressure can be incorporated into a suction filter to prevent the starvation of the associated pump. Suction filters are less expensive and easier to service than other types of filters. However, they offer only medium benefits.

Pressure Filter

A pressure filter is installed downstream of the pump. It usually sees the full pressure as set by the relief valve. It is constructed with a rugged casing to withstand the maximum system and shock pressures. It can be smaller and finer (10 – 20 µ). It must be sized for the specific flow rate. The filter should have an internal bypass valve with a cracking pressure of about 100 psi to prevent collapse or burst. Pressure filters are available in medium-pressure (up to 2000 psi) and high-pressure (up to 6500 psi) ratings. They are available in cartridge, spin-on, in-line, and/or duplex configurations.

The fluid from the pump is usually contaminated with internally generated contamination. The primary function of the pressure filter is to keep the contaminated fluid clean. It also protects expensive and dirt-sensitive downstream components. Therefore, pressure filters offer great benefits. However, they are more expensive because their housing is under pressure. Servicing is also difficult because of their heavy-duty construction.

Return-line Filter

A return filter is installed in the return line, which may see no more than 300 to 435 psi. It is usually a low-pressure housing. It can tolerate a higher pressure differential across its filter element. It removes particles typically≥ 25 µ in mineral-based fluids and ≥ 10 µ in synthetic fluids. The purpose of the return-line filter is to trap dirt from the system's working components and particles entering the system through any worn seals on those components. Return-line filters are available in the in-tank, in-line, and duplex configurations.

Return-line filters offer high benefits. However, they may be subject to high-flow surges during operation. A relief valve may be placed across the filter to provide an additional path for the excess return flow.

Offline Filtration / Kidney-loop Filter

As shown in Figure 5.12, the offline filtration system consists primarily of a separate pump, filter unit, and hoses. The filter unit uses a fine filter element and a low-pressure housing that are easily serviced. The components of the offline filtration system can be arranged on a mobile cart. An offline filtration system with quick-disconnect couplers can be temporarily retrofitted to an existing hydraulic system. The unit can serve many hydraulic systems. Alternatively, an offline filtration system can be permanently integrated as a dedicated unit.

In the offline filtration system, fluid is continuously pumped out of the reservoir, passed through the filter, and allowed to return to the reservoir. This subsystem operates independently of the main hydraulic system and is used to achieve the best possible filtration results in hydraulic systems. However, this type of filtration system requires an additional pump-motor unit.

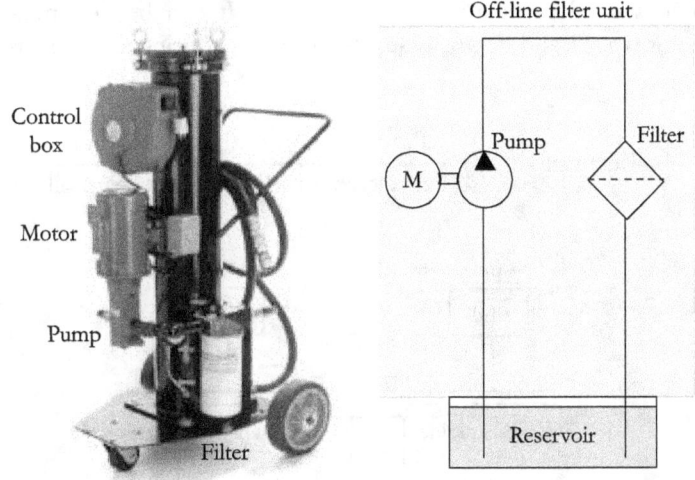

Figure 5.12 | An offline hydraulic filter unit
Courtesy: Engineered Filtration, Inc., USA

Performance Ratings of Filters

The cleanliness of a hydraulic system's fluid is linked to the performance of the filter elements used in the system. These filter elements are rated based on their ability to separate contaminants of particular sizes from the system fluid under specific test conditions.

Filter manufacturers publish various performance data for filters. Two basic parameters specified by the manufacturers are the mesh number (sieve number) and the micron ratings.

The following sections describe these two parameters. Also, there is an industry standard called the 'multi-pass test' to measure filter performance ratings, such as the Beta (ß) Ratio and filter efficiency.

Mesh Number/Sieve Number

The mesh size, or fineness, of a wire-mesh filter can be expressed as its mesh number or sieve number. It is the number of openings in a wire mesh from the center of any wire to the center of a parallel wire one inch away, as shown in Figure 5.13. Table A8.1 of Appendix 8 gives the mesh-to-micron conversion chart.

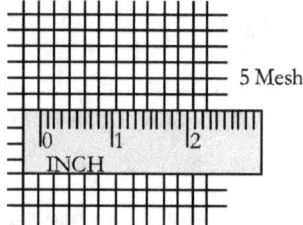

Figure 5.13 | Illustrating the measurement of mesh width

Beta Ratio

It is a measure of a filter element's filtration efficiency. It can be determined by monitoring oil contamination levels upstream and downstream of a test filter during a single pass, as shown in Figure 5.14. That is,

$$\text{Beta ratio, } \beta_{(x)} = \frac{\text{Particle count in the upstream oil}}{\text{Particle count in the downstream oil}}$$

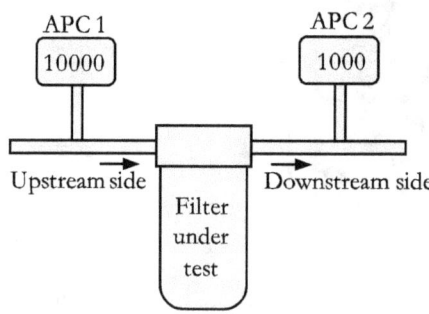

Figure 5.14 | A single-pass test setup for measuring the Beta ratio

Filter Efficiency

Filter efficiency has the same meaning as the Beta ratio. It is given by:

$$\text{Efficiency}_{(x)} = (1 - \frac{1}{\beta}) \times 100$$

It is important to be cautious when using the Beta ratio because it does not account for real-world operating conditions, such as pressure surges and temperature variations, which can affect the filter's performance. Also, a filter's Beta ratio does not show how much dirt it can hold. Despite these limitations, the Beta ratio is an important parameter for assessing a filter's expected performance.

Table 5.1 shows Beta ratios and filter efficiencies for specific upstream and downstream particles.

Table 5.1 | Beta ratios and filter efficiencies

Upstream particles ($\geq x$ μm)	Downstream particles ($\geq x$ μm)	Beta (ß) ratio	Efficiency $_{(x)}$
1,00,000	50,000	2	50.0%
1,00,000	5,000	20	95.0%
1,00,000	1,333	75	98.7%
1,00,000	1,000	100	99.0%
1,00,000	500	200	99.5%
1,00,000	100	1000	99.9%

Micron Ratings
A filter's micron rating indicates its ability to remove a specified percentage of particles of a specific size.

- The absolute micron rating of a filter is the smallest size of particles it can capture more than 98.6% on the first pass through it.

The nominal micron rating of a filter is the smallest micron size of particles it can capture in a specified quantity, ranging from 50% to 95% on the first pass through it.

For example, the filtration efficiencies of samples of filters manufactured in similar batches are measured at different micron sizes and recorded as shown in Table 5.2. From the Table, the absolute micron rating is 5 μ, and the nominal micron rating is 2, 3, or 4 μ.

It may be noted that paper filters are usually rated nominally, and synthetic filters are usually rated in absolute terms.

Table 5.2 | Filtration efficiencies

Micron size	Filtration efficiency
20 μ	99.9%
10 μ	99.1%
5 μ	98.61%
4 μ	90%
2 μ	50%

Differential Pressure (ΔP)
Differential pressure (ΔP) across the filter indicates the difference between its inlet and outlet pressures.

Particle Capture Efficiency
It is the weight of the specified artificial contaminant (ISO medium test dust) that must be added to the fluid upstream of the filter to produce a given pressure differential across the filter under the specified conditions. It indicates the amount of solid that the filter element can hold before it needs to be replaced.

Example 5.1

In a single-pass test for finding the performance of a hydraulic test filter, 100000 particles of size ≥ 6 µm are counted upstream using a laser particle counter, and 1000 particles of the same size range are counted downstream of the filter. Calculate the Beta ratio and the filter's efficiency.

Solution

$$\begin{aligned}
\text{Upstream particle count, } C_u &= 100000 \\
\text{Downstream Particle count, } C_d &= 1000 \\
\text{Beta ratio, } \beta_{(x)} &= C_u / C_d \\
\text{Beta ratio, } \beta_{(6)} &= (100000 / 1000) = 100 \\
\text{Efficiency}_{(6)} &= (1 - \tfrac{1}{\beta}) \times 100 \\
&= [1 - (1/100)] \times 100\% = 99\%
\end{aligned}$$

The Multipass Test

The arrangement for a multipass test is shown in Figure 5.15. The multi-pass test gives reproducible test data for assessing the performance of hydraulic filter elements. This test provides an accurate, universally accepted method for evaluating the efficiency of a filter element in removing particles of standard test dust from the sample fluid over a wide range of particle sizes under controlled laboratory conditions. This test is standardized by ISO 16889, SAE J1858, ANSI, and NFPA.

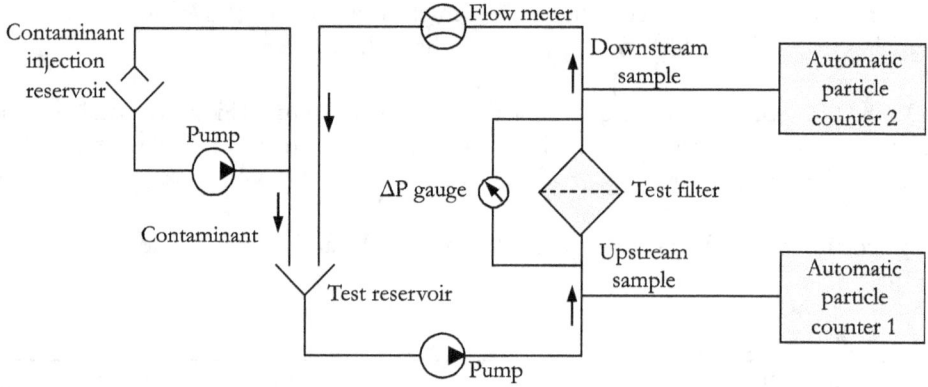

Figure 5.15 | A test setup for a multipass test

Filter Selection

Filters are essential for maintaining hydraulic systems at a satisfactory level of cleanliness. Therefore, proper filter selection is crucial. The following factors must be considered for the selection of hydraulic filters:
- Know the system characteristics, such as the flow rate, the working pressure, the viscosity of the system fluid, and the amount of expected system contamination
- Consider the cleanliness level requirements of the system
- Consider the performance specifications of each filter/filter element, such as its micron rating, Beta ratio/ filter efficiency, dirt-holding capacity, and the maximum permissible pressure differential across the filter
- Decide the pressure rating, reliability, service life, and the cost of the filters
- Consider the compatibility of the filter materials with the fluid used in the system

Objective-type Questions

1. Which type of hydraulic filter can be used to prevent the collapse of a filter due to the clogging of its filter element?
   a) Filter with bypass valve
   b) Duplex filter
   c) Proportional filter
   d) Full-flow filter

2. A strainer employs:
   a) Mechanical type, surface filtration
   b) Mechanical type, depth filtration
   c) Absorbent type, surface filtration
   d) Absorbent type, depth filtration

3. Which type of hydraulic filter do you use for its easy maintenance without shutting down the process line?
   a) Filter with bypass valve
   b) Duplex type of filter
   c) Proportional filter
   d) Full-flow filter

4. A return-line filter generally employs:
   a) Mechanical type, surface filtration
   b) Mechanical type, depth filtration
   c) Absorbent type, surface filtration
   d) Absorbent type, depth filtration

5. Which hydraulic filter do you use to protect valves and cylinders in a hydraulic system directly?
   a) Suction filter
   b) Pressure filter
   c) Return-line filter
   d) Offline filter

6. A strainer employs:
   a) Cellulose media
   b) Wire-mesh media
   c) Synthetic media
   d) Depth media

7. Which type of hydraulic filter do you use for filtering fluid independent of the main hydraulic system?
   a) Suction filter
   b) Pressure filter
   c) Return-line filter
   d) Offline filter

8. Synthetic filter media are made from:
   a) Cellulose
   b) Glass fibers
   c) Steel wires
   d) Metal disks

9. Which type of hydraulic filter can be used to trap dirt from the working components of a hydraulic system?
   a) Suction filter
   b) Pressure filter
   c) Return-line filter
   d) Offline filter

10. Which of the following types of filters offers a high benefit to hydraulic systems?
    a) Strainer
    b) Suction filter
    c) Return-line filter
    d) None

11. What is the efficiency of the hydraulic filter with a beta ratio of 20 for 4μ particles?
    a) 50%
    b) 80%
    c) 95%
    d) 99%

12. When a hydraulic filter performance is expressed as $ß_4 = 75$, what does this rating mean?
    a) A beta ratio of 4, 75% efficient at removing 75-micron particles
    b) A beta ratio of 75, 98% efficient at removing 4-micron particles
    c) A beta ratio of 4, 75% efficient at removing 4-micron particles
    d) A beta ratio of 75, 75% efficient at removing 4-micron particles

13. Which of the Beta ratios given below provides the better wear protection at 10 microns?
    a) $ß_5 = 50$
    b) $ß_{10} = 200$
    c) $ß_{25} = 200$
    d) $ß_{200} = 1000$

14. The pressure differential across a hydraulic filter unit increases due to:
    a) The use of higher-viscosity fluid
    b) The clogging of the filter media
    c) The friction of the fast-moving fluid
    d) All of the above

Review Questions
1) Why is it necessary to install an efficient filtration system in a hydraulic system?
2) Classify hydraulic filters according to their nature of filtration.
3) Differentiate the surface and depth filter media used in hydraulic systems.

4) What materials are used to make the filter media for hydraulic filters? Briefly explain.
5) What are the advantages of synthetic filter media?
6) Differentiate the cellulose filter media and the synthetic filter media.
7) Explain the working principle of a hydraulic filter with a bypass valve.
8) Explain the function of a duplex-type hydraulic filter
9) How do you differentiate between full-flow filtration and proportional-flow filtration?
10) Explain the significance of the location-based hydraulic filtration.
11) Explain the functional and constructional aspects of strainers used in hydraulic systems.
12) What are the differences between a strainer and a filter concerning a hydraulic system?
13) What are the typical locations in hydraulic systems where filters can be installed?
14) Explain the essential function and installation features of suction filters in hydraulic systems.
15) Explain the essential function and requirements of pressure filters in hydraulic systems.
16) What are the disadvantages of suction filters?
17) Explain the primary function of the return-line filters used in hydraulic systems.
18) Enumerate the advantages and disadvantages of pressure filters in hydraulic systems.
19) What are the advantages of return-line filters?
20) Explain the hydraulic offline filtration technique with a neat sketch.
21) What are the advantages of the offline filters used in hydraulic systems?
22) What are the performance specifications of hydraulic filters?
23) Define the two types of micron ratings for hydraulic filters.
24) What is the difference between the absolute and nominal micron ratings?
25) Explain the multi-pass test for a hydraulic filter with the schematic diagram of the test setup.
26) What parameters can be determined through the multi-pass test for a hydraulic filter?
27) How is the efficiency of the hydraulic filter evaluated?
28) What is the Beta ratio of a hydraulic filter? Explain with a numerical example.
29) What is filter efficiency, and what is its significance in hydraulic systems?
30) How is the Beta ratio of a hydraulic filter related to its efficiency?
31) Explain the meaning of Beta ratio, $ß_{4(c)} = 2$, of a hydraulic filter.
32) Write short notes on the following terms related to hydraulic filters: (1) Synthetic filter media, (2) Proportional flow filtration, (3) Multi-pass test, and (4) Dirt holding capacity.
33) Write short notes on the following terms of hydraulic filters: (1) Depth filter, (2) Duplex filter, (3) Offline filtration, and (4) Beta ratio.
34) What system characteristics do you consider when specifying a hydraulic filter?
35) What factors do you consider while selecting hydraulic filters?

Numerical Problems

1) Find out the Beta ratio of the filter when 10000 particles of $>= 20$ μm enter the filter, and 9950 of these particles are trapped by the filter. Also, calculate the filtration efficiency for the given particulate size. {Ans: $ß_{(20)} = 200$, $\eta_{(20)} = 99.5\%$}
2) What is the meaning of $ß_4(c) = 1000$, of a hydraulic filter? Also, calculate the corresponding filter efficiency.
3) In a multi-pass test, 30000 and 1000 particles of size 4 microns or greater were recorded in automatic particle counters upstream and downstream of the test filter, respectively. Determine the filter's efficiency for a given particulate size. {Ans: $\eta_{(4)} = 96.67\%$}

Objective-type questions - answer key: *1-a, 2-a, 3-b, 4-d, 5-b, 6-b, 7-d, 8-b, 9-c, 10-c, 11-c, 12-b, 13-b, 14-d*

Chapter 6 | Hydraulic Reservoirs

Hydraulic Power Pack
A hydraulic system requires sufficient high-quality fluid at all times for efficient operation. As shown in Figure 6.1, a power pack supplies the required fluid to the system.

It is a compact, portable, custom-designed, or pre-engineered assembly with essential and optional components. The essential components are a fluid-filled reservoir (tank), a close-coupled pump-motor unit, a pressure relief valve, and a pressure gauge, and the optional components include a heat exchanger, a temperature controller, directional control valves, filters, etc.

It also includes necessary instrumentation and other accessories, such as accumulators, hoses, level gauges, thermostats, and quick-disconnect couplings. The modern way to configure a power pack is through standardized sub-assemblies.

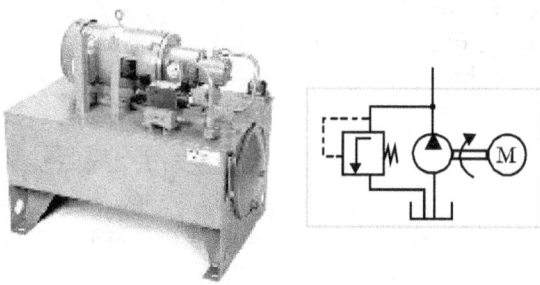

Figure 6.1 | A hydraulic power pack
Courtesy: Advance Motion Control, USA

A pump in the power pack converts the power from its prime mover into hydraulic power at the pressure and flow rate required by all system actuators. The fluid is continuously passed through various intermediate components to the system actuators and then returned to the associated reservoir.

The circulating fluid can accumulate contaminants and absorb heat from the system. Therefore, the fluid needs to be serviced before it is circulated again. For this purpose, the fluid is stored in the reservoir and conditioned before it is circulated again.

Hydraulic Reservoirs
A reservoir stores and conditions a given quantity of hydraulic fluid. A well-designed reservoir:
- allows a reasonable dwell time for the fluid
- allows most of the contaminants to drop out
- assists in dissipating the heat
- allows air bubbles to come to the surface and dissipate
- compensates for the fluid volume changes
- provides a convenient mounting place for the pump-motor unit and valves

The reservoir is properly designed, correctly sized, and equipped with numerous accessories to achieve the objectives mentioned above.

Constructional Features

A well-designed hydraulic reservoir should be completely enclosed and self-contained. Figure 6.2 shows the cross-sectional view of a reservoir. It is provided with the following parts: (1) Baffle plate, (2) Suction line, (3) Return line, (4) Filler-cum-breather, (5) Drain plug, (6) Strainer, (7) Fluid level indicator, (8) Pressure gauge, (9) Removable cover, (10) Diffuser, and (11) Magnetic tank cleaner. Hydraulic reservoirs are modelled in vertical and horizontal designs. Gaskets should be used under all covers and clean-out plates for tight sealing. Sealing flanges should be used around all pipes fitted to the reservoir. The bottom part of the reservoir is usually inclined from side to side or 'V'-shaped.

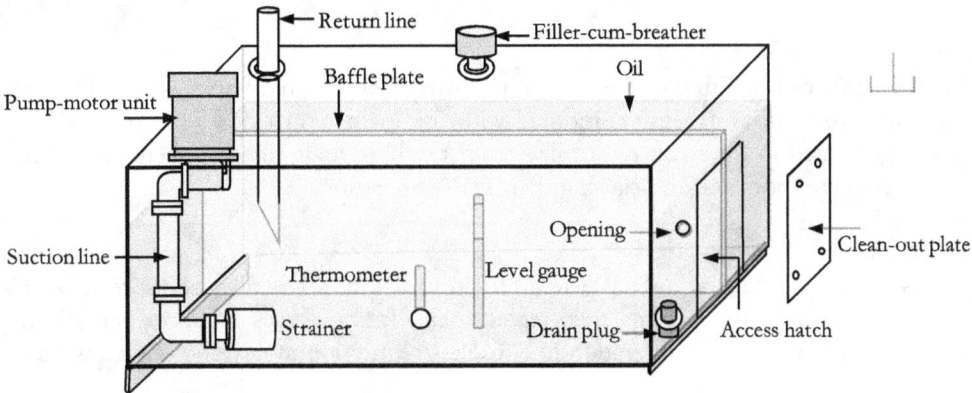

Figure 6.2 | A cross-sectional view of a hydraulic reservoir

The reservoir must be located in an area with good airflow for rapid heat dissipation. The servicing parts, such as sight glasses, filters, filler breather, and drain cock, must be easily accessible.

Baffle Plate

As shown in Figure 6.2, a baffle plate is fitted lengthwise through the middle of the reservoir. Its purpose is to separate the suction chamber from the return chamber. The pump should draw the fluid from the suction chamber, and the return flow should be discharged into the return chamber.

The baffle plate should extend slightly above the reservoir's maximum fluid level. A few tiny openings are drilled in the baffle plate at one end, far away from the suction and return lines. This arrangement tends to equalize the fluid levels on both sides of the baffle plate, as the fluid is directed from the return line to the suction line. It also ensures that the return fluid takes a circuitous path through the reservoir to the suction line. In this way, the fluid gets more dwell time within the reservoir. The extended dwell time allows contaminants to settle within the reservoir and helps it dissipate the fluid's heat as quickly as possible. Further, it assists in deaerating the fluid.

Suction Line

The suction line carries fluid from the reservoir to the pump's inlet. Its bottom end should be located some distance above the reservoir's bottom to prevent settled contaminants from re-entering the pump. The suction line is usually fitted with a strainer, suction filter, or both. It may be noted that an increased vacuum will develop in the suction line if the suction strainer and filter become clogged with contamination or if the fluid remains too cold at start-up.

Return Line

The return line carries the fluid from the system back to the reservoir. The suction and return lines should be located on the same side of the reservoir, but on either side of the baffle plate. The return line must terminate below the fluid level and extend up to two to four times the return line pipe diameter above the base plate of the reservoir to reduce turbulence and foaming. The open end of the return line may be cut at 45° to prevent fluid from being blocked if the line is pushed to the bottom of the reservoir. Further, it is advantageous to orient the opening towards the reservoir sidewall to maximize fluid-contacting surface area for faster heat transfer and fluid cooling. The return line may be slightly oversized and fitted with a diffuser to reduce the velocity of the return fluid.

Drain Line

A separate drain pipe can be fitted to the reservoir to carry leakage fluid directly from the components through external lines, without being combined with the main return flow. This provision prevents back-pressure generated in the main return line from reaching drain-line-connected sections, such as spring chambers in components, and causing their faulty operation.

Filler-cum-Breather

An opening is usually provided at the top plate of the reservoir to act as filler-cum-breather. It is a filler that fills the reservoir with fluid during fluid replacement. It also serves as a breather, allowing air to enter and exit the reservoir during operation to equalize the interior and exterior air pressures.

Figure 6.3 shows a breather unit that can be fitted to the breather opening. An air filter of 5 microns (0.00019685 inches) (or better), incorporated in the breather unit, prevents the ingress of airborne contaminants into the reservoir. It protects the fluid from contamination found in the harsh environment surrounding it. It is, however, essential to install a filter of sufficient capacity to allow the rapid discharge of the air displaced by the large volume of fluid returning to the reservoir. The breather unit may include sufficient desiccant material, such as silica gel, to dehumidify the air entering the unit. It may, however, be noted that the reservoir may also be provided with a separate filler opening with a cap and an air breather opening with a filter. The filler cap should be chained to the reservoir to keep it secured.

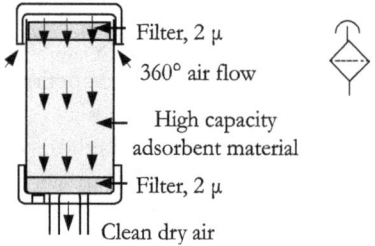

Figure 6.3 | An air breather unit

Drain Plug

As mentioned earlier, the bottom plate of the reservoir is provided with a downward gradient or a V-shape. A drain plug is provided at the lowest point of the bottom plate so that the system fluid can be drained entirely without any difficulty during fluid replacement. The deepest end of the reservoir is on the opposite side of the suction and return lines. A drain plug is shown in Figure 6.2.

Strainer

A strainer, suction filter, or both are fitted to the suction line to prevent dirt, grit, sludge, rust, and other contaminants from entering the associated pump. The strainer is usually fitted inside the reservoir, submerged in the fluid. A suction strainer is shown in Figure 6.2. It is usually a pleated stainless steel wire screen. Vacuum indicators and switches can be used with strainers to provide a signal when the vacuum reaches a specific level.

Fluid Level Indicator

The fluid level in the reservoir must be monitored to avoid potential pump starvation. This monitoring is assisted by incorporating a sight window or a fluid level indicator in the reservoir or using a level gauge. The sight window or fluid level indicator provides a quick visual inspection of the reservoir's fluid level. The transparent fluid level indicator is made of polyamide resin and protected with a steel guard.

Figure 6.4(a) and (b) show two models of fluid level indicators. Integrated level gauges, such as a dipstick, can also be used to monitor the reservoir's fluid level. The material used for the indicator should be compatible with the reservoir's fluid.

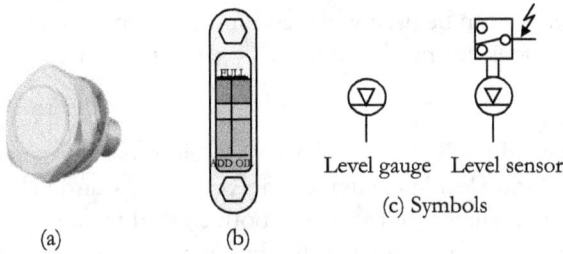

Figure 6.4 | Fluid level gauges

Bourdon Tube Pressure Gauge

As shown in Figure 6.5, a pressure gauge measures the system's pressure level. It is also a safety measure, an overpressure monitoring system, and assists with troubleshooting. A vacuum gauge installed at the pump inlet must be used to check the pump inlet condition.

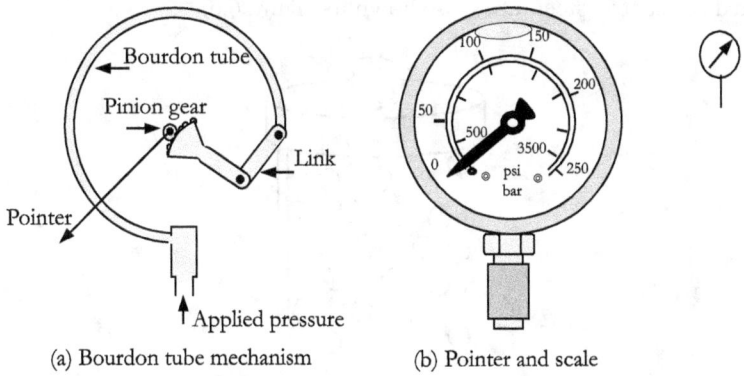

Figure 6.5 | Pressure Gauge

A pressure gauge consists of a Bourdon tube with an elastic chamber, a link, a geared sector, a pinion, a pointer, a calibrated scale, and a snubber. The bourdon tube is constructed from brass and bronze alloys. Typically, the pressure gauge has a stainless steel bezel and case filled with glycerin to reduce the flutter of its pointer and avoid internal damage.

Pressure applied to the elastic chamber causes the Bourdon tube to move outwards. The higher the pressure, the greater the tube's deflection radius. The deflection converts the applied pressure into the corresponding movement of the attached pointer across the calibrated scale via the links, levers, and gearing.

A fixed orifice snubber is used in the gauge's pressure connection to dampen pointer oscillation and protect the gauge from damage caused by pressure surges.

The service life of a permanently fitted pressure gauge can be prolonged by isolating the gauge from the associated system using a shut-off valve or gauge isolator, except when taking a reading. Another purpose of using the gauge isolator is that the technician can maintain the gauge without interrupting the system.

A spring-loaded pressure gauge can be used when continuous monitoring of system pressure is desired. This type of pressure gauge does not need a snubber or gauge isolator.

Removable Cover
The reservoir must be designed for easy internal access to clean out any accumulated residues and rust particles. Periodic servicing and cleaning activities can be carried out quickly if a large opening with a removable cover or clean-out plate is fitted at one or both ends of the reservoir. The clean-out openings should be placed to give access to both sides of the baffle plate. A clean-out cover is shown in Figure 6.2.

Diffuser
A diffuser is connected directly to the return line and installed below the minimum fluid level. The diffuser's closed surface should face towards the pump intake. Its purpose is to reduce the velocity of the fluid returning to the reservoir, thereby preventing foaming and resuspension of deposited dirt. Additionally, it helps reduce turbulence and noise in the system. This type of diffuser is usually made of two specially crafted concentric steel tubes, as shown in Figure 6.6.

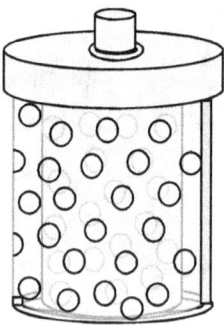

Figure 6.6 | A diffuser

Baffle Screen

As shown in Figure 6.7, a baffle screen is meant to dissipate air bubbles from the fluid. A screen can be installed inside the reservoir at an angle to collect bubbles that can then be pushed to the top surface and dissipated.

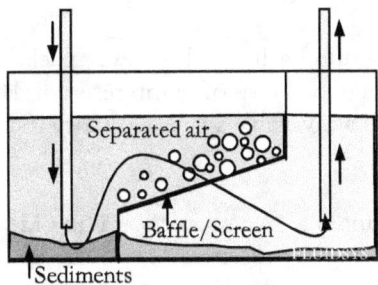

Figure 6.7 | Baffle/Screen

Magnetic Tank Cleaner

Abrasive ferrous particles appear in the fluid medium of a hydraulic system due to the constant flaking effect of its moving parts, chipping due to subsurface casting flaws, and particles generated by boring and machining operations. Tank cleaners with permanent magnets can be incorporated to attract and hold the ferrous particles in the fluid that ordinary filters might miss. This feature helps prevent the recirculation of metal particles through the system and excessive wear of its components.

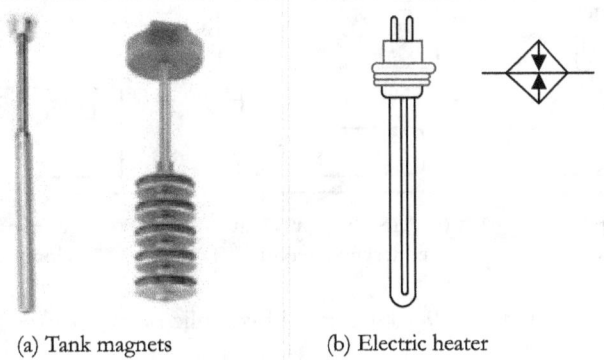

(a) Tank magnets   (b) Electric heater

Figure 6.8 | Reservoir accessories

Figure 6.8(a) shows two designs of tank magnets: the standard slim-line model and the heavy-duty multi-magnet model. The standard model with a small magnetic unit and a short support rod is suitable for everyday use. A multi-magnet unit with an extended support rod can be used where large-scale abuse is expected during system operation.

Tank Heater

If a hydraulic system is exposed to freezing temperatures, the fluid in the reservoir becomes so thick that the pump is at risk of cavitation damage. An electric immersion heater, shown in Figure 6.8(b), can keep the fluid warm.

Standard Sizes of Reservoirs

The standard sizes of hydraulic reservoirs range from 0.25 to 80 gallons, with custom-made models available up to 500 gallons. Reservoirs made of thermoplastic with a capacity of up to 2.5 gallons are available for mobile applications.

Pump-Reservoir Layouts

There are many methods of configuring a hydraulic power pack. To maximize a hydraulic system's performance, it is essential to have a good grasp of pump reservoir layouts. When choosing a layout for a particular application, factors such as available space, tolerance to cavitation, and ease of maintenance should be considered.

The pump-motor unit can be mounted above, under, or alongside the reservoir, as shown in Figure 6.9(a), (b), and (c), respectively. Each pump-reservoir layout has advantages and disadvantages. Various factors, such as the ease of maintaining the power pack and replacing its components, must be considered when deciding on the pump reservoir layout.

Pump-above-Reservoir

The top surface of the reservoir makes a convenient location for mounting the pump and its drive motor. Figure 6.9(a) shows the pump-reservoir layout with the pump above the reservoir. However, the main disadvantage of the layout is that the pump must generate sufficient suction at its suction side to draw the fluid into the pump and accelerate it. The length of the piping must be as short as possible. The piping arrangement must have few or no bends.

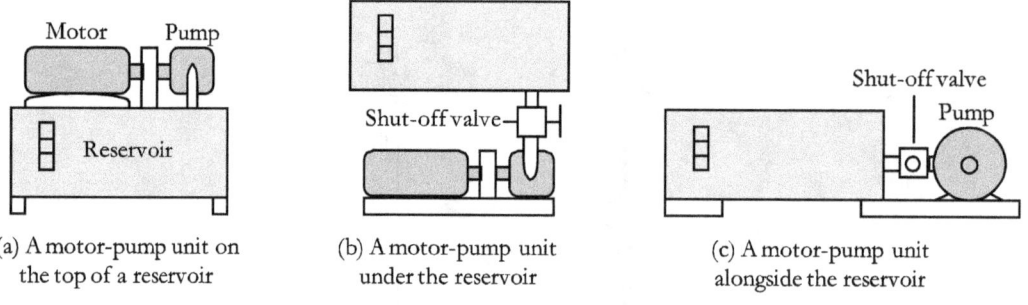

Figure 6.9 | Layouts of hydraulic power packs

Pump-under-Reservoir

Figure 6.9(b) shows the pump-reservoir layout with the pump unit arranged below the reservoir. This layout takes advantage of the reservoir's static head pressure. With the reservoir above, the pump inlet is always flooded with fluid. This design therefore improves suction conditions and reduces the risk of whirlpools and cavitation.

Pump-alongside-Reservoir

Figure 6.9(c) shows the pump-reservoir layout with the pump alongside the reservoir. In this arrangement, the pump inlet is always flooded with the fluid. A shutoff valve in the inlet line allows maintenance or repair work on the pump unit without draining the reservoir. The reservoir configuration can also help fulfill application requirements or restrictions, such as flooded suction, spatial limitations, noise and vibration control, and contamination elimination.

Sizing of Reservoirs

Reservoirs used in hydraulic systems differ in size (capacity). A well-designed reservoir for a hydraulic system must be correctly sized to supply the required fluid quantity to the system at all times and to accommodate the thermal expansion of the system fluid. Moreover, the reservoir must be sized to ensure that the circulating fluid has a reasonable dwell time to allow it to settle out particulate contamination, dissipate heat as quickly as possible, suppress turbulence in the return fluid, and release the entrained air.

An undersized reservoir can produce higher fluid temperatures than the design temperature, while an oversized reservoir incurs higher capital costs.

The size of a hydraulic reservoir depends on the applications concerned. Determining its size is not an exact science. Remember, reservoirs are generally oversized.

Typical reservoir sizes include 10, 15, 20, 30, 40, 50, 60, 80, 100, 120, 150, 200, 300, 400, 500, 600, and 700 gallons.

There are generally two approaches to finding reservoir sizes. One approach is an approximate method based on the volumetric flow rate involved in the associated system. Another approach to holistically finding the most appropriate reservoir size is to consider the heat balance feasible in the system.

Reservoir Size, Based on the System Flow Rate

For a hydraulic system, where the mineral fluids are used, and medium-to-high frequency demands are expected, a reservoir with a capacity ([gallons) of three to five times the volume flow rate (gpm) of the system fluid is adequate. That is,

Reservoir size, gallons = (3 to 5) x Pump flow rate, gpm

Under this general rule, the fluid returned to the reservoir has three to five minutes of dwell time before it circulates again.

Including a 10% air cushion is essential to ensure proper pumping action.

Exceptions: However, the recommended reservoir size for a hydraulic system should be larger than usual under certain exceptional situations. For example, a larger reservoir should be used if the system is liable to be exposed to high ambient temperatures or if some special fluids are intended to be used. A reservoir size of 5 to 8 times the pump flow rate per minute is recommended for a hydraulic system using a water-glycol or Polyol ester fluid medium.

Further, the reservoir must be large if the associated system has accumulators. In mobile and aerospace hydraulic systems, space constraints often limit the size of reservoirs that can be used. For example, in a mobile application, 1.5 to 2 times the pump flow rate is often a more realistic target.

However, remember, using a heat exchanger, the reservoir size could be as small as required to accommodate the fluid corresponding to the pump flow rate in one minute.

Example 6.1: Estimate the size of a hydraulic reservoir for industrial use with high-frequency operations that meets the following specifications. The reservoir should be filled with a petroleum-based fluid, and the application will likely be exposed to high temperatures. The fluid flow rate in the system is 13.2 gpm. Assume that no heat exchanger is used in the system.

Solution
    Pump flow rate, Q   = 13.2 gpm

The reservoir will be filled with petroleum oil and used in high-frequency operations in a hot environment. Therefore, it is appropriate to assume:

    Reservoir size   = 5xQ
                            = 5x13.2 = 66 gallons

Reservoir Size, Based on the Heat Balance

To determine the heat balance in an existing system or one under design, it is necessary to quantify the heat generated by energy losses. Then, it is necessary to determine how much heat can be dissipated from the reservoir's heat-dissipating area. From these calculations, the optimal reservoir size and heat exchanger size can be determined. The following sections outline the heat load in an existing reservoir and in a new system under design.

Heat Load of an Existing Reservoir

Consider an existing hydraulic system with a reservoir of known volume. The fluid in the system heats up due to inefficiencies. The fluid's temperature initially increases as the system is switched on, then levels off at a steady temperature. The heat load of the reservoir with a fluid can easily be determined using the procedure given below:
1. Measure the volume of the reservoir (V, gallons)
2. Measure the fluid temperature (T1) at the start-up
3. Measure the fluid temperature (T2) after the system has been in operation for a specified duration (Δt, in minutes)
4. Determine the differential temperature, ΔT (°F) = T2 − T1

$$\text{Heat load (hp)} = \frac{V \text{ (gallons)} \times C_p \text{ (btu/lb)} \times \varrho \text{ (lb/ft}^3\text{)} \times \Delta T \text{ (°F)}}{317.3 \times \Delta t \text{ (min)}}$$

Where,
C_p is the specific heat of the fluid in BTU/lb (°F)
ϱ is the density of the fluid in lb/ft^3
1 hp = 2544 btu/hour

Heat Load of a New Reservoir

The heat load of a new reservoir being designed can be estimated using the system's input power as a guide. It can be obtained from the drive motor or engine nameplate. The losses of the system's components can be taken from the manufacturer's published data. Generally, losses from pumps and actuators can be taken as 5 to 15% of the component's input power, and from valves and flow restrictions as 10 to 20% of the component's input power. The total heat losses in a hydraulic system are about 30-50% of the drive motor or engine's input power.

Heat dissipation by Hydraulic Reservoirs

The heat dissipation factor, an important consideration in deciding the reservoir capacity, is explained below.

Assume that a reservoir with a heat-dissipating surface area A dissipates heat H through conduction and radiation. The reservoir base can be excluded from the heat-dissipating surface if it is not elevated by at least 6 inches above the ground. Let ΔT be the temperature difference between the reservoir walls and the ambient air. The formula for the heat dissipated by the reservoir is given by:

The general formula: $H = k \times \Delta T \times A$, where k is a constant
In the English system: $H\ (hp) = 0.001 \times \Delta T\ (°F) \times A\ (ft^2)$

Example 6.2
If the fluid temperature in a hydraulic reservoir is 140°F, and the ambient air temperature is 75°F, how much heat will the reservoir with 20 ft² of surface area dissipate?

Solution

Fluid temperature	= 140°F
Ambient air temperature	= 75°F
Temperature differential	= 140 - 75 = 65°F
Surface area of the reservoir	= 20 ft²
Heat dissipation, H	= 0.001 x ΔT (°F) x A (ft²)
	= 0.001 x 65 x 20 = 1.3 hp

Example 6.3
A steel tank with dimensions of **19.69 in (L) x 15.75 in (W) x 16.54 in (H)**, designed for use as a hydraulic reservoir, dissipates heat from the hydraulic fluid at **158°F**. The ambient temperature is **77°F**. Calculate the tank's heat-dissipation rate.

Solution

Tank dimension	= 19.69 in x 15.75 in x 16.54 in
Fluid temperature	= 158°F
Ambient temperature	= 77°F
ΔT	=158-77 = 81°F
A	=650 + 520 + 310
	= 1480 in²
	=10 ft²
H	= 0.001 x ΔT (°F) x A (ft²)
	= 0.001 x 81 x 10
	= 0.81 hp

Heat Exchangers /Coolers

Heat is usually generated in a hydraulic system due to inefficiency and poor design. Devices used in the hydraulic system, such as flow control valves, sequence valves, pressure-reducing valves, and undersized directional control valves, can also contribute to heat development in the system.

If the cooling effect from the reservoir is insufficient, a heat exchanger (or cooler) must be fitted to increase the system's heat dissipation rate. However, heat exchangers are expensive, and their maintenance can run high.

Two main types of heat exchangers are used in hydraulic systems. They are: (1) air-cooled heat exchangers and (2) water-cooled heat exchangers.

Air-cooled Heat Exchanger

When the heat to be removed in a hydraulic system is comparatively small, an air-cooled heat exchanger can be employed.

Figure 6.10(a) shows a schematic diagram of a typical air-cooled heat exchanger. It primarily consists of a blower and radiator. Air-cooled heat exchangers usually consist of finned tubes through which hot fluid passes, transferring heat to the fins. Then, blowers force air around the fins, removing heat through forced convection.

Airborne contaminants in the blower's surroundings, such as heavy dust, water, and coolant vapors, can quickly reduce the efficiency of the air-cooled heat exchanger. The water-cooled type of heat exchanger is much more efficient, as explained in the following section.

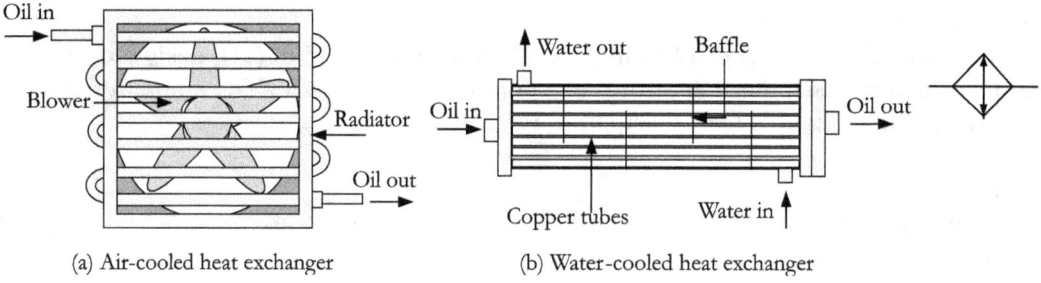

Figure 6.10 | Heat exchangers for hydraulic systems

Water-cooled Heat Exchanger

When the heat to be removed is relatively high or the surrounding atmosphere is likely to be hot, a water-cooled heat exchanger should be used in a hydraulic system.

Figure 6.10(b) shows a schematic diagram of the commonly used shell-and-tube type of water-cooled heat exchanger. It has a series of small tubes inside the shell. Fluid flows through the tubes, and cold water flows through the shell surrounding the tubes. The relatively cold water carries away the heat from the fluid.

Water-cooled heat exchangers are more efficient at cooling fluids to lower temperatures than air-cooled heat exchangers.

Noise in Hydraulic Systems

Industries are increasingly alarmed about the effects of plant noise on their workforce. Therefore, the need for a quiet work environment is growing. Like any other system, hydraulic systems may be a source of great noise.

Effect of Noise on Health

Excessive noise can destroy workers' hearing. Workers exposed to factory noise often complain of nervousness, sleeplessness, and fatigue. The noise can also cause physical and psychological stress. Usually, workplace noise first affects workers' ability to hear high-frequency (high-pitched) sounds.

Noise Reduction Techniques

It is necessary to design hydraulic systems, especially power units, with appropriate noise-reduction techniques to mitigate the damaging effects of noise. The following activities are recommended to reduce the noise and vibration in a hydraulic machine/system.
- Select the proper frame and base,
- Enclose noisy machine parts,
- Use resilient mounts for the machine, whenever possible,
- Use densely perforated plates in the machine,
- Select pumps and drive motors with low noise level ratings,
- Align the pump-motor unit in the system properly,
- Overhaul/replace the worn or damaged pump unit in the system,
- Replace any broad belt in the system with a set of narrow belts separated by spacers,
- Correctly route the hoses,
- Avoid long and unsupported conductor runs,
- Reduce cavitation,
- Prevent entry of air into the system fluid.

Objective-type Questions

1. Which of the following is <u>not</u> the purpose of a hydraulic reservoir?
    a) Dissipate the heat from the hydraulic system
    b) Drop contaminants out of the fluid in the system
    c) Suppress the shock pressures in the hydraulic system
    d) Compensate for the volume changes of the fluid

2. A baffle plate in a hydraulic reservoir is to:
    a) Prevent the development of excess pressure
    b) Prevent the ingress of contaminants into the reservoir
    c) Reduce the velocity of return fluid
    d) Separate the suction chamber from the return chamber

3. A filler-cum-breather is used in a hydraulic system to:
    a) Fill the fluid into the reservoir
    b) Breathe the air in and out of the reservoir
    c) Prevent the entry of airborne contaminants into the reservoir
    d) All of the above.

4. Which part of the hydraulic reservoir is used to reduce the velocity of the return fluids?
   a) Baffle
   b) Diffuser
   c) Magnets
   d) Strainer

5. Noise in hydraulic systems can be reduced by:
   a) Using the right type of filters
   b) Using sound-absorbing materials
   c) Controlling contamination
   d) Sizing the reservoirs correctly

6. A hydraulic power pack:
   a) Stores fluid
   b) Conditions fluid
   c) Delivers fluid
   d) All of the above

Questions
1) What is a hydraulic power pack?
2) Draw a simple sketch of a hydraulic reservoir and explain its main parts.
3) What are the primary functions of hydraulic power packs?
4) What are the essential and optional components of hydraulic reservoirs?
5) What are the features included in the power pack of a hydraulic system to control the contamination and the temperature in the system?
6) Give five factors you consider in the design of hydraulic reservoirs.
7) Mention the purpose of the following elements provided in hydraulic reservoirs: (1) Baffle plate, (2) Filler breather, and (3) Diffuser.
8) What is the standard rule for finding the size of a hydraulic reservoir?
9) What are the circumstances under which the selection of a hydraulic reservoir is larger than usual?
10) Briefly explain the various pump-tank layouts in hydraulic systems.
11) Describe the function of heat exchangers in hydraulic systems.
12) What are the different types of heat exchangers used in hydraulic systems? Explain briefly.
13) What are the probable sources of noise in hydraulic systems?
14) What are the effects of exposure to noise on the health of the industrial workforce?
15) List some techniques for reducing the noise that hydraulic power sources produce.

Numerical Problems
1) A hydraulic reservoir is to be filled with a petroleum-based fluid. The fluid flow rate in the system where the reservoir should be employed is 5.28 gpm. Assume that no heat exchanger is used in the system. What is the reservoir size for industrial hydraulic use with medium-frequency work operations?
2) Calculate the surface area required for a hydraulic reservoir to obtain a heat dissipation capacity of 0.118 hp if the ambient temperature is 68°F and the maximum desired temperature is 154°F.

Objective-type questions - answer key: 1-c, 2-d, 3-d, 4-b, 5-b, 6-d

Chapter 7 | Hydraulic Pumps

A pump in a hydraulic system is usually a part of a power pack. It is a positive displacement device coupled to a prime mover. It converts mechanical energy from its prime mover into hydraulic energy using an incompressible fluid. It draws fluid from the associated reservoir when driven by its prime mover, then pushes it into the system. This pumping action creates flow. The system develops pressure due to the resistance to flow. The resistance to flow arises from fluid viscosity, flow restrictions, and/or load on actuators.

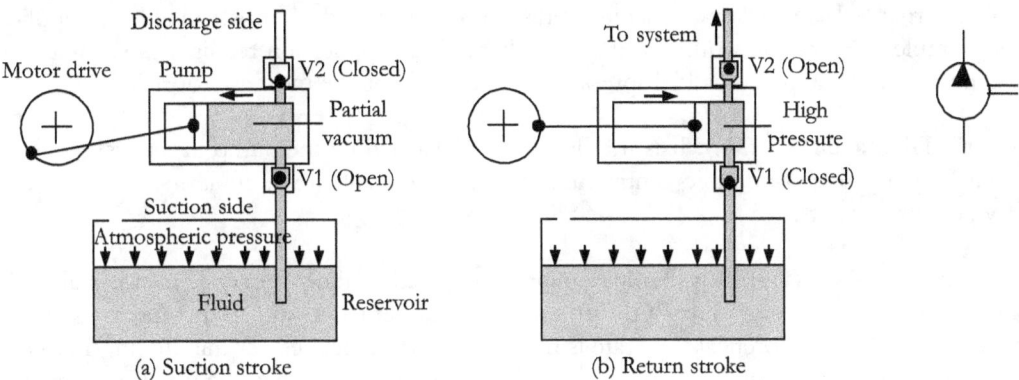

Figure 7.1 | Schematic diagrams showing two critical positions of a hydraulic pump system

The pumping action is illustrated in Figure 7.1. The pump piston moves back and forth when driven by a motor at a constant speed, serving as the pumping element. The pump incorporates two valves, V1 and V2, for drawing fluid into and containing it within the system. The suction stroke creates a partial vacuum on the suction side of the pump, drawing fluid from the associated tank into the pump chamber through the open valve V1. On the return stroke, valve V1 is closed, and the fluid is trapped in the chamber. The closed valve V1 provides tight sealing and prevents the fluid from flowing back to the tank. The motion of the pumping element induces a force on the pumped fluid, which is then discharged through the opened valve V2. In this way, the pump creates a flow. In short, the pump draws, traps, seals, and discharges the fluid during the pumping cycle. In a practical pump, the trapped fluid is also moved from the suction port to the discharge port.

As stated earlier, a partial vacuum is created at the suction side of a pump. If a high vacuum is developed at the pump inlet, it can cause excessive wear on the internal surfaces due to cavitation. Therefore, the pump should be located at an elevation of no more than 6 feet above the fluid surface in the reservoir. However, the height should be limited to 3 feet, as additional restrictions in the inlet line will increase the vacuum.

Further, the design goal must be to construct a pump that runs quietly and produces no vibrations. The use of special seals and the provision of close tolerances characterize the construction of the positive displacement pump. Under normal pressure, these features assist in transferring a fixed volume of fluid during each cycle, with a small amount of leakage, thus giving rise to the term 'positive displacement.' However, the pump is liable to promote leakage due to increased friction in the pumping elements. A pump should be constructed to withstand high pressures, generate low pressure and flow pulsations, and produce low noise levels.

Terms and Definitions – Hydraulic Pumps

Positive-displacement pumps come in many varieties, sizes, flow rates, and power ratings. Some essential parameters for their operation include pressure rating, priming, slippage, displacement, flow rate, torque, input power, output power, and efficiency. The following sections provide definitions of hydraulic pump terms.

Pressure Rating

It is the pressure that overcomes all resistances in the system, including useful work and losses. Alternatively, it is the maximum pressure the pump can withstand without damaging its parts or causing excessive internal leakage. A low-to-medium-pressure pump would be most suitable for applications involving simple or moderate work. On the other hand, if an application requires substantial work, as in large construction equipment, a high-pressure pump would be the most appropriate.

Volumetric Displacement (V_D): It is the fluid volume carried by the pump in one revolution of its driveshaft. It is expressed in cubic centimeters per revolution (cc/rev), cubic inches per revolution (in3/rev), or similar units.

Theoretical Flow Rate (Q_T): It is the fluid volume displaced by the pump at its inlet per unit of time. It can be determined by the product of the volumetric displacement of the pump and the speed of the pump's driveshaft. The theoretical flow rate is measured in cubic inches per minute (in^3/min) or gpm. The mathematical equation for the theoretical flow rate (Q_T) of the pump in the English system of units is as follows:

$$Q_T \, (in^3/min) = V_D \, (in^3/rev) \times N \, (rpm)$$

$$Q_T \, (gpm) = Q_T \, (in^3/min) \, / \, 231$$

Pump Slippage (Q_s): It represents the internal leakage of fluid from the pump's discharge port to its suction port. The internal leakage is due to some unavoidable small clearance between the pump's internal parts. The slippage is a function of the pump speed, the differential pressure across the pump, the degree of wear of its interior surfaces, and the viscosity of the fluid passing through the pump. Any increase in slippage reduces pump efficiency.

Actual Flow Rate (Q_A), Pump

It is the actual volume of fluid discharged by the pump per unit of time. It is given by the theoretical flow rate minus the pump slippage. That is,

$$\text{Actual flow rate} = \text{Theoretical flow rate} - \text{Slippage}$$

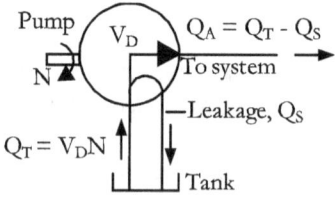

Figure 7.2 | A symbolic representation of a hydraulic pump showing its flow rate parameters

The actual flow rate is measured in cubic-inch per minute (in³/min) or gpm. Figure 7.2 depicts the relationship between the theoretical and actual flow rates in the pump drawing fluid from the tank and delivering it to the system.

Example 7.1 | What is the theoretical flow rate of a fixed-displacement hydraulic pump with a volumetric displacement of 0.0046 ft³/rev operating at 2000 rpm?

Solution

Volumetric displacement, V_D = 0.0046 ft³/rev
Pump speed, N = 2000 rpm = 33.33 rps

Theoretical flow rate, Q_T = $V_D \times N$
= 0.0046×2000 = 9.2 ft³/min

Actual Torque (T_A): It is the actual torque delivered to the pump by its prime mover and is given by:

$$\text{Actual torque, } T_A(\text{in.lb}) = \frac{\text{Actual power delivered to the pump (hp)} \times 63025}{N \text{ (rpm)}}$$

$$\text{Actual torque, } T_A(\text{ft.lb}) = \frac{\text{Actual power delivered to the pump (hp)} \times 5252}{N \text{ (rpm)}}$$

$$\text{Actual torque, } T_A(\text{ft.lb}) = \frac{\text{Actual power delivered to the pump (hp)} \times 550}{N \text{ (rps)}}$$

Theoretical Torque (T_T): The pump's theoretical torque is a function of its volumetric displacement and the system pressure. It equals the actual torque minus the torque losses due to friction in the pump.

$$\text{Theoretical Torque, } T_T(\text{in.lb}) = \frac{V_D(\text{in}^3/\text{rev}) \times P \text{ (psi)}}{2\Pi}$$

Power Relationships, Hydraulic Pump
Figure 7.3 shows the pump block diagram, with the power relationships at the input and output. Assume that a prime mover, such as an electric motor, drives the pump.

Input Power, Pump
It is the power delivered to the pump by its prime mover. The speed of the driveshaft and torque imparted by the motor determine the pump's input power.

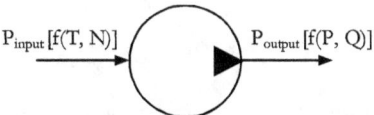

Figure 7.3 | The parameters for the power relationships of a pump

$$\text{Pump input power (hp)} = \frac{T_A(\text{in.lb}) \times N(\text{rpm})}{63025} = \frac{T_A(\text{ft.lb}) \times N(\text{rpm})}{5252} = \frac{T_A(\text{ft.lb}) \times n(\text{rps})}{550}$$

Pump Output Power: It is the power delivered by the pump. The pressure and the pump's actual flow rate determine the output power.

$$\text{Pump output power (hp)} = \frac{P(\text{psi}) \times Q_A(\text{gpm})}{1714}$$

Example 7.2 | A hydraulic pump delivers 10.6 gpm at 2176 psi for a work operation. Calculate the hydraulic power developed by it.

Solution
With usual notations,
Flow, Q_A = 10.6 gpm
Pressure, P = 2176 psi

Output power, P_{out} = P (psi) x Q_A (gpm) / 1714
= 2176 x 10.6/1714 = 13.46 hp

Efficiencies of Hydraulic Pumps

An ideal hydraulic pump has no fluid leakage or frictional loss. Such a hypothetical pump is said to be 100% efficient. However, in a practical hydraulic pump, both fluid leakage and frictional losses occur to some extent, and therefore, its efficiency is always less than 100%. Two types of efficiencies are identified to account for the two types of losses in the pump. They are: (1) Volumetric efficiency and (2) Mechanical efficiency. The volumetric efficiency indicates the extent of leakage in a frictionless pump, while the mechanical efficiency indicates the extent of frictional losses in a leak-free pump. The combined effect of the leakage and the frictional losses in the pump can be gauged from its overall efficiency.

Volumetric Efficiency (η_v)

Figure 7.4 presents a symbolic diagram of a hydraulic pump, showing the parameters for its volumetric efficiency. The volumetric efficiency is the ratio of the actual pump flow rate at a given pressure to the theoretical flow rate determined by the pump's geometric displacement, assuming no frictional losses in the pump. It indicates the extent of leakage within the pump. It is given by:

$$\text{Volumetric Efficiency }(\eta_v) = \frac{\text{Actual flow rate}}{\text{Theoretical flow rate}} = \frac{Q_A}{Q_T}$$

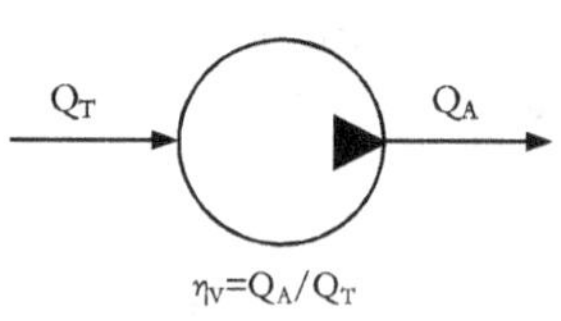

$Q_T \rightarrow \bigcirc \rightarrow Q_A$

$\eta_V = Q_A/Q_T$

Figure 7.4 | A diagram of a hydraulic pump showing the parameters for its volumetric efficiency

Assume that a new gear pump has a volumetric efficiency of about 94%. That means, for every 5 gpm of fluid drawn through the pump's suction port, 4.7 gpm is delivered through the pump's discharge port, and the remaining 0.3 gpm of leakage fluid slips past the gears on its way back to the suction port.

Mechanical Efficiency (η_m)

Figure 7.5 presents a symbolic diagram of a hydraulic pump, showing the parameters for its mechanical efficiency. Mechanical efficiency is the ratio of the power delivered by the pump to the power delivered to the pump, assuming no leakage. It indicates the extent of energy losses in the pump due to friction. It is given by:

$$\text{Mechanical Efficiency } (\eta_m) = \frac{\text{Pump output power, assuming no leakage}}{\text{Actual power delivered to the pump}} = \frac{P \times Q_T}{T_A \times N}$$

The mechanical efficiency of the pump can also be calculated in terms of its torque units. That is,

$$\text{Mechanical Efficiency } (\eta_m) = \frac{T_T}{T_A}$$

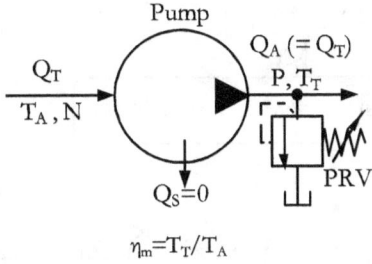

Figure 7.5 | A symbolic diagram of a hydraulic pump showing the parameters for its mechanical efficiency

Overall Efficiency (η_o)

Figure 7.6 presents a symbolic diagram of a hydraulic pump, showing the parameters that determine its overall efficiency. The overall efficiency is the ratio of the actual power delivered by the pump to the actual power delivered to the pump. It is given by:

$$\text{Overall Efficiency } (\eta_o) = \frac{\text{Actual power delivered by the pump}}{\text{Actual power delivered to the pump}} = \frac{P \times Q_A}{T_A \times N}$$

The overall efficiency of the pump is also given by the product of its volumetric efficiency (η_v) and its mechanical efficiency (η_m). That is,

$$\eta_o = \eta_v \times \eta_m$$

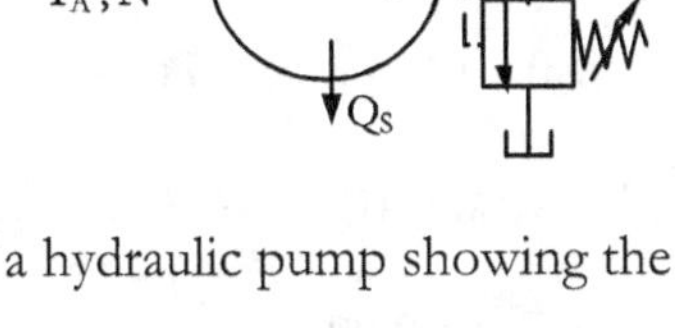

Figure 7.6 | A symbolic diagram of a hydraulic pump showing the parameters for its overall efficiency

Summary of Relations for Hydraulic Pumps

Figure 7.7 summarizes important relations of hydraulic pumps in English units.

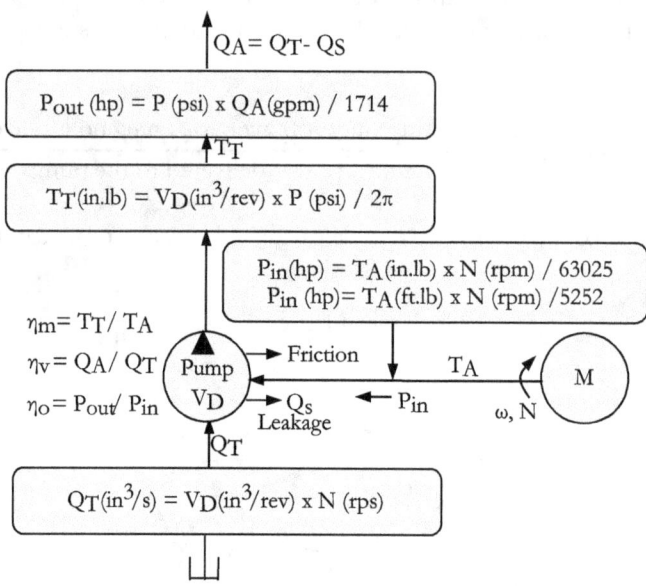

Figure 7.7 | Summary of relations for hydraulic pumps

Example 7.3 | A hydraulic pump with a displacement of 6.35 in³/rev delivers 25.4 gpm operating at a pressure of 1450 psi when driven by a prime mover with a torque of 1593 in.lb at 1000 rpm. Find out: (1) the volumetric efficiency, (2) the mechanical efficiency, and (3) the overall efficiency of the pump.

Solution

Pump displacement, V_D	= 6.35 in³/rev
Actual flow rate, Q_A	= 25.36 gpm
Speed, N	= 1000 rpm
Pressure, P	= 1450 psi
Actual torque, T_A	= 1593 in.lb

Theoretical flow rate, Q_T = V_D (in³/rev) x N (rpm) = 6.35 x (1000) = 6350 in³/min
= 6350/231 gpm = 27.49 gpm

Volumetric efficiency, η_v = (Q_A/Q_T) x 100% = (25.36/27.49) x 100% = 92%

Input power, P_{in} = T_A (in.lb) x N (rpm)/63025 = 1593 x 1000/63025 = 25.3 hp

Output power, P_{out} = P (psi) x Q_T (gpm)/1714 = 1450 x 27.49/1714 = 23.25 hp

Mechanical efficiency, η_m = (P_{out}/P_{in}) x 100% = (23.25/25.3) x 100% = 92%

Overall efficiency, η_o = $\eta v \times \eta m$ = 0.92 x 0.92 = 0.84 = 84%

Classification of Hydraulic Pumps

Hydraulic pumps are classified in many different ways. In general, they are broadly classified into the following two types. They are: (1) Positive displacement (hydrostatic) pumps and (2) Non-positive displacement (hydrodynamic) pumps.

The positive-displacement pump delivers a definite amount of fluid to the system. In contrast, the non-positive-displacement pump's internal leakage increases with downstream pressure, and the flow output is not constant. Remember, positive-displacement pumps are more important in industrial and mobile hydraulic systems.

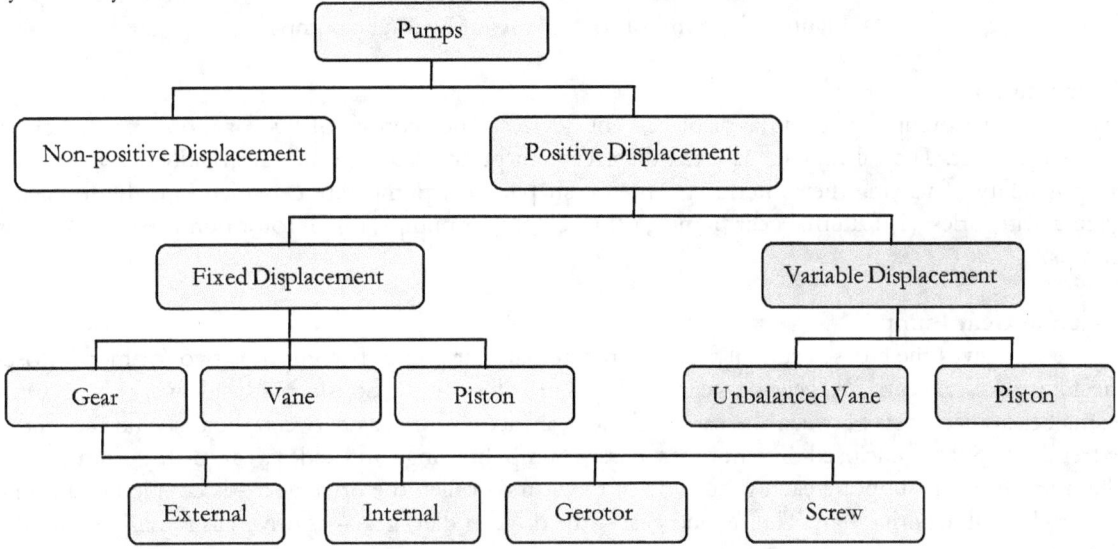

Figure 7.8 | Classification chart for hydraulic pumps

Figure 7.8 shows the classification of hydraulic pumps. Positive-displacement pumps can be classified into fixed-displacement and variable-displacement types.

The fixed-displacement pump delivers a fixed amount of fluid for each revolution of its driveshaft. In contrast, the variable-displacement pump is designed to deliver a variable volume of fluid per cycle when running at a given speed. In this case, the pump output can be varied by changing the physical relationship of the internal pump elements. However, variable-displacement pumps are more complex to design.

Positive-displacement pumps can be further classified by the type of pumping element used, such as gears, vanes, or pistons. Accordingly, there are three main types of positive-displacement pumps. They are (1) Gear pumps, (2) Vane pumps, and (3) Piston pumps. The pumping elements generate the flow. There are many variations in the design of the leading pump types. All these types can be developed as fixed-delivery types, but only vane and piston pumps can be designed as variable-delivery types.

Alternatively, the positive-displacement pumps can be classified into the following types. They are (1) reciprocating type and (2) rotary type. In a reciprocating pump, the pumping element reciprocates; in a rotary pump, it rotates.

Symbolic Representation of Hydraulic Pumps

Figure 7.9 gives the symbols of various types of hydraulic pumps.

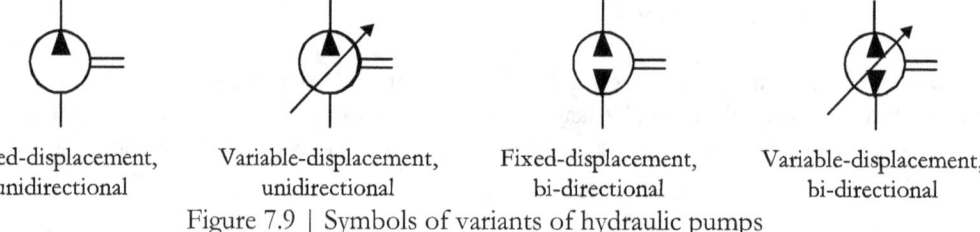

Fixed-displacement, unidirectional   Variable-displacement, unidirectional   Fixed-displacement, bi-directional   Variable-displacement, bi-directional

Figure 7.9 | Symbols of variants of hydraulic pumps

Gear Pumps

Hydraulic gear pumps are positive-displacement devices. They consist of two or more gears meshing with each other. They always act as fixed-displacement pumps because they are rigidly designed, with no possibility of varying their internal geometric shape. Gear pumps are classified into the following general categories: (1) External gear pumps, (2) Internal gear pumps, (3) Gerotor pumps, and (4) Screw pumps.

External Gear Pump

Figure 7.10 gives the cross-sectional view of an external gear pump. It comprises two (or more) close-meshing identical gears (i.e., gear-on-gear) enclosed in a close-fitting housing with side wear plates. Only a small clearance exists between the case and the gear faces, as well as between the case and the teeth's extremities. Space, enclosed by the gear teeth, pump housing, and side wear plates, forms fluid chambers. A shaft supports each gear. One of the gears – called the drive gear - is coupled to a prime mover through its driveshaft. The second gear—the driven (idler) gear—is driven as it meshes with the drive gear.

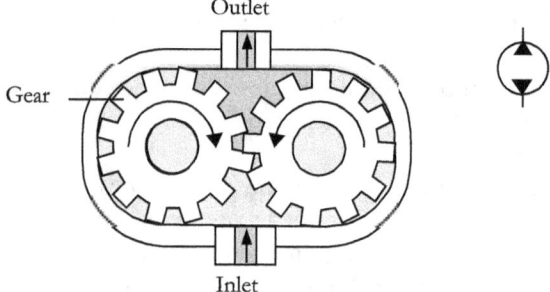

Figure 7.10 | A schematic diagram illustrating the operation of the external gear pump

The gears rotate in opposite directions when driven by the prime mover. They mesh within the housing at a point between the inlet and outlet ports to provide tight sealing. As the gears rotate, the diverging teeth create an expanding volume at the pump inlet. This expanding volume creates a partial vacuum on the inlet side, drawing fluid from the associated reservoir into the chamber. The fluid is trapped in the chambers on either side. The trapped fluid then moves around the periphery of the rotating gears as two streams. Remember, the pump has a positive internal seal against leakage. The two streams recombine and are positively ejected from the pump's delivery port. Therefore, when driven by the drive motor, the intermeshing gears displace a fixed volume of fluid across the pump in one revolution of its driveshaft, creating flow.

Characteristic Features of External Gear Pumps

In general, gear pumps are simple and the least expensive of the hydraulic pump types. The rigid design of gears and their housing allows for the development of high pressures and the pumping of highly viscous fluids. Also, gear pumps are more tolerant of harmful fluid contamination. Therefore, they are most commonly found in mobile equipment.

A gear pump designed with fewer teeth tends to produce greater fluid pulsations. A large number of teeth for each gear can make the discharge relatively smooth and continuous. Gear pumps are considered unbalanced because their bearings experience greater pressure on one side. Thus, the gear pumps tend to wear disproportionately on one side over time. A gear pump with larger clearances between its mating surfaces, operating at higher speeds, and using a low-viscosity fluid can produce more leakage.

Volumetric Displacement of External Gear Pumps

Figure 7.11 shows the cross-sectional view of an external gear pump of length 'L' with two gears. Each gear has an outside diameter 'D_o' and an inside diameter 'D_i'. The volumetric displacement of the pump can be found by considering the volume of a hollow cylinder of outside diameter 'D_o' and inside diameter 'D_i.' There are two such volumes in the pump. Therefore, the amount of space available for the fluid to occupy is one-half of the total volume. Thus, the volumetric displacement 'V_D' can be expressed by:

$$V_D = \frac{\Pi}{4}(D_o^2 - D_i^2)L$$

Where,
  D_o, D_i and L in ft
  V_D in ft³ per revolution [Note: 1 ft³ = 1728 in³]

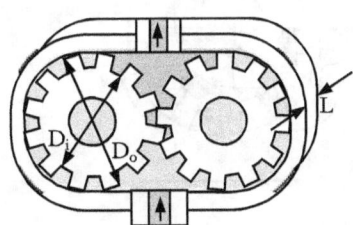

Figure 7.11 | The cross-sectional view of an external gear pump

Example 7.4 | Calculate the volumetric displacement of an external gear hydraulic pump with two gears. Each gear has an OD of 3 in, an ID of 2 in, and an axial length of 1 in. Find the pump's volumetric efficiency if the actual pump flow at 1800 rpm is 0.0586 ft³/s.

Solution

D_o = 3 in
D_i = 2 in
L = 1 in
N = 1800 rpm
Q_A = 0.0586 ft³/s

$$V_D = \frac{\pi}{4}(3^2 - 2^2) \times 1 = 3.925 \text{ in}^3/\text{rev} = 0.00227 \text{ ft}^3/\text{rev}$$

Theoretical flow rate, Q_T = 0.00227 × 1800/60 = 0.0681 ft³/s
Volumetric efficiency = (0.0586/0.0681) × 100 = 86%

Advantages and Disadvantages of External Gear Pumps: Table 7.1 gives the advantages and disadvantages of external gear pumps.

Table 7.1 | Advantages and disadvantages of external gear pumps

Advantages	Disadvantages
- Good suction performance	- Short life expectancy
- Mid-volume, mid-pressure delivery	- Do not support variable displacement delivery
- Highly tolerant of contamination	- High pulsation delivery
- Least expensive pump designs	- Operation is noisy due to the meshing of gears
- Work equally well with watery type fluids and highly viscous fluids	- Promote disproportionate wear on one side due to pressure imbalance
- Support bi-directional operation	- Least efficient pump designs

Internal Gear Pump

An internal gear pump is a positive-displacement pump. Figure 7.12 gives the cross-sectional view of an internal gear (gear-within-a-gear) pump. The pump consists of an outer rotor gear, an inner spur gear, and a crescent-shaped spacer, all enclosed in a housing. The gears are set eccentrically to each other. The gear with fewer teeth operates inside the rotor gear. The stationary crescent spacer is machined into the space between these gears and separates them. It acts as a seal between the suction and discharge ports. It also creates two pathways for the fluid to flow.

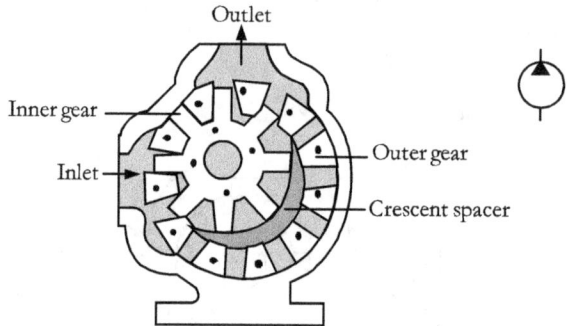

Figure 7.12 | A schematic diagram illustrating the operation of an internal gear pump

The prime mover can drive any one of the gears via its shaft, which is supported on bearings. The prime-mover-coupled gear drives the other gear in the same direction. The rotation of gears causes the teeth to unmesh near the inlet port, creating a partial vacuum in the pump's inlet chamber, which then draws fluid from the system reservoir. The chambers formed between the inner and outer gear teeth on both sides of the crescent spacer trap the fluid. The trapped fluid is then moved through the pump chambers. The transported fluid is positively displaced out of the delivery port, as the pump has a positive internal seal that prevents leakage.

Advantages and Disadvantages of Internal Gear Pumps: Internal gear pumps are compact and provide smooth and non-pulsating flow. They have good suction ability and produce less cavitation.

However, their design is more complicated and expensive to manufacture. They have the overhung load on their shaft bearings.

Gerotor Pump

A Gerotor pump is a fixed-displacement pump. It is similar to the internal gear pump. However, the crescent-shaped spacer is absent in the Gerotor pump. Figure 7.13 gives the cross-sectional view of a Gerotor pump. It comprises an off-center inner gear rotor (Gerotor element) and an outer female gear rotor (idler) enclosed within the pump's housing. There is precisely one tooth difference between the outer and inner gear rotors. The tips of the teeth on the inner gear are always in contact with those on the outer gear. The driven inner gear rotor draws the outer gear rotor along as they mesh.

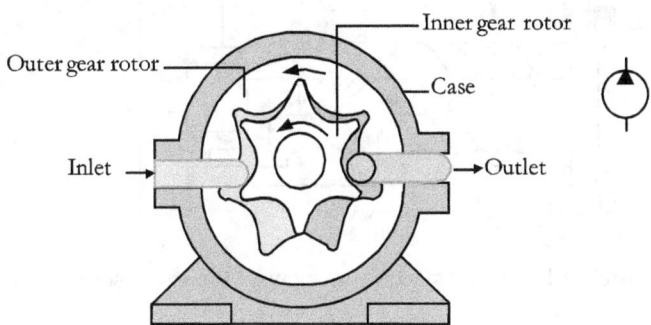

Figure 7.13 | A schematic diagram with the cross-sectional view of a Gerotor pump

The teeth diverge near the inlet port as the gears rotate in the close-fitting housing. This expanding volume creates a partial vacuum in the pump's inlet chamber, drawing fluid from the system reservoir into the chamber. The fluid is trapped in the chambers between the gears and then moved through the pump by the rotating gears. The fluid is positively ejected from its delivery port as the pump has a positive internal seal against leakage.

Advantages and Disadvantages of Gerotor Pumps: Design flexibility is one of the greatest attributes of Gerotor pumps. Table 7.2 gives their advantages and disadvantages.

Table 7.2 | Advantages and disadvantages of gerotor pumps

Advantages	Disadvantages
- Simple concept and compact design	- Limited to medium pressures
- Provides smooth pumping action	- Excessive fluid leakage
- Suitable for a wide range of viscosities	- Provides only fixed-displacement
- Inexpensive and highly efficient	- Overhung load on the shaft bearing

Screw Pumps

Screw pumps are also positive-displacement pumps. Next, many variants of the screw pump are available in the market. The two basic types are: (1) single-screw pumps and (2) multiple-screw pumps.

The single-screw hydraulic pump consists of a screw fitted in the pump housing, and the screw threads are eccentric to the axis of rotation. The spiraled rotor in the pump rotates eccentrically in the stator.

The multiple-screw pumps have multiple external screw threads. These pumps are further classified as two-screw pumps and three-screw pumps. The two-screw pump is constructed with two parallel, intermeshing rotors that rotate within a housing machined to close tolerances. The three-screw pump consists of a central drive rotor with two meshing idler rotors rotating in a housing machined to close tolerances. These pumps may employ timing gears to maintain the clearances between the screws as they rotate. Further, the multiple-screw pumps may be single- or double-suction. The following paragraphs explain the operation of a three-screw pump.

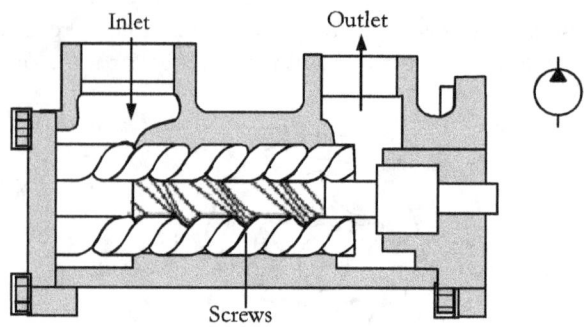

Figure 7.14 | A cross-sectional view of the three-screw pump

Three-Screw Pumps: In this type of pump, a set of three screws – one drive screw and two driven screws (idlers)- is installed in the pump housing, as shown in Figure 7.14. The center screw of the pump drives the other two screws that are located on either side of the center screw. As the screws rotate, the fluid is carried axially through the sealed chambers formed between the screws and the housing. The fluid is pushed along its axis by rotor rotation and smoothly forced out of the other end. The absence of metal-to-metal contact and the reduction of fluid pulsations result in quiet operation.

Three-screw pumps have low noise levels, high reliability, and long life. They are the largest class of multiple-screw pumps in service today.

Advantages and Disadvantages of Screw Pumps: Table 7.3 gives the advantages and disadvantages of screw pumps.

Table 7.3 | Advantages and disadvantages of screw pumps

Advantages	Disadvantages
- Good suction characteristics	- Performance is viscosity-dependent
- Suitable for a wide range of fluids	- Complicated screw shape
- High tolerance to contamination	- The efficiency of screw pumps is relatively low
- Produce non-pulsating delivery	- Relatively high cost because of close tolerances

Vane Pumps

Vane pumps are positive-displacement pumps used in many hydraulic applications. They are high-performance pumps but are inherently more complex than gear pumps. Based on design type, vane pumps can be classified into two types. They are: (1) Unbalanced vane pumps and (2) Balanced vane pumps. They can also be classified into fixed-displacement and variable-displacement types.

The vane pump has many parts that are subjected to wear. As long as the wear is uniform, the vane tips establish firm contact with the cam surface. This property minimizes fluid leakage from the pump, thereby improving its efficiency. However, increased fluid contamination can lead to uneven vane wear.

Unbalanced Vane Pump

Figure 7.15 shows the cross-sectional view of an unbalanced vane pump. It consists of a prime-mover-driven rotor with sliding vanes in close-fitting radial slots. Further, the rotor moves within a larger circular cavity. The centers of the rotor and the cavity are offset by a certain distance, causing an eccentricity. The vane tips bear against the casing and form an adequate seal. Side plates are used to keep the fluid confined to the fluid chambers.

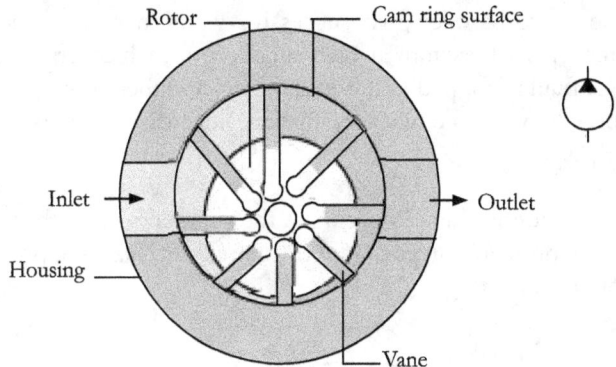

Figure 7.15 | A cross-sectional view of an unbalanced vane pump

As the rotor rotates, the space between two successive vanes increases near the inlet port. This expanding volume creates a partial vacuum on the suction side, drawing fluid into the chambers formed by the vanes. The fluid is trapped in these chambers and then moved through the pump by the rotating vanes. The fluid is positively ejected from its delivery port as the pump has a positive internal seal against leakage.

As shown, the pumping occurs in the chambers on one side of the rotor and the driveshaft. Therefore, this design imposes considerable side loads on these components.

Variable Displacement Vane Pump

The displacement of a hydraulic pump depends on the geometry and working relationship of its internal operating parts. When running at a constant speed, the displacement can be varied by altering the physical relationship of its internal working elements. That is, the displacement can be varied by modifying the degree of eccentricity between the rotor and the pump casing from zero to maximum.

The variable displacement vane pump has a movable outer casing ring surrounding the rotor. There is no pumping action if the rotor is dead center with the casing.

However, if the center of the rotor is offset to the maximum extent from the center of the ring, then the eccentricity between the rotor and the ring reaches the maximum, and the pump flow tends to approach its maximum value.

The pump flow can infinitely vary by positioning the rotor between these extremes.

Balanced Vane Pump

Figure 7.16 shows the cross-sectional view of a balanced vane pump. It consists of a prime-mover-driven rotor with sliding vanes in close-fitting radial slots. Next, the rotor moves concentrically inside an elliptical cam ring (casing). The pump also consists of a set of diagonally opposite suction ports and a set of diagonally opposite discharge ports. This design creates two separate pumping areas on opposite sides of the rotor.

As the rotor rotates, the space between two successive vanes at each suction side increases. This expanding volume creates a partial vacuum at each suction port, which draws fluid into the chambers formed by the vanes. The fluid is trapped in these chambers and then moved through the pump by the rotating vanes. As the space between the two rotating vanes decreases, the fluid is squeezed through the corresponding discharge port.

As shown in the Figure, the pumping action occurs in the chambers on both sides of the rotor and the shaft equally. Therefore, the pressure forces on the rotor are fully balanced, preventing the side-loading of the drive shaft and the rotor bearing.

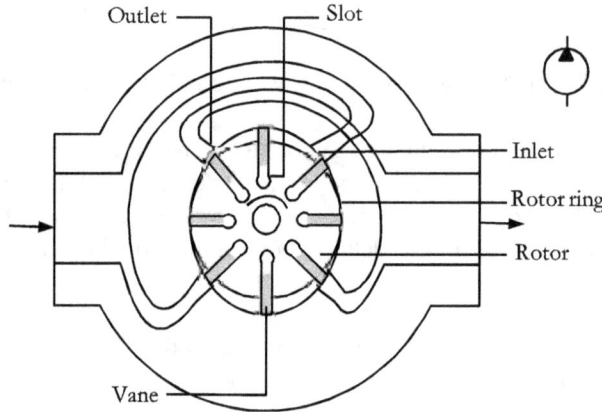

Figure 7.16 | A cross-sectional view of a balanced vane pump

Pressure-compensated Vane Pump

In some hydraulic applications, pressure-compensated vane pumps are necessary. The pressure-compensated vane pump stops pumping the fluid when a predetermined system pressure is reached. A variable-displacement vane pump used in a hydraulic system can be equipped with an additional spring to automatically adjust the pump's eccentricity. The spring initially sets the cam ring to its maximum eccentricity, allowing the pump to deliver the highest volume of system fluid.

Due to a preload on the spring, the cam ring does not move initially until the applied force is sufficient to counter the preload. The range of pressure over which the cam ring does not move is the pump's constant-flow region.

Once the preload force is exceeded, the cam ring shifts. As the spring's pressure increases, the cam ring becomes concentric with the rotor, and the pump flow becomes zero. The pump's flow profile appears smooth because the cam ring can move over an infinite range of positions between the minimum and maximum eccentricities.

Advantages and Disadvantages of Vane Pumps: Vane pumps deliver fluid with low pulsation, operate at low noise levels, and work well with low-viscosity fluids.

However, they are sensitive to contamination, require fluids with excellent anti-wear properties, have a complex structure, and are unsuitable for high pressures.

Volumetric Displacement of Vane Pumps

Figure 7.17 shows the cross-sectional view of an unbalanced vane pump with its rotor having a diameter 'D_R' rotating inside a movable cam ring with a diameter 'D_C.' Let 'L' be the rotor width and 'e' be the eccentricity of the rotor's center line with the cam ring's center line.

As explained earlier, moving the cam ring can adjust the degree of eccentricity. That is, if the cam ring is concentric with the rotor ring, the eccentricity is zero, and there would be no fluid flow. On the other hand, if the cam ring is moved to its maximum eccentricity, the maximum volume of fluid would be pumped. The maximum eccentricity, e_{max} can be found in Figure 7.17 and is given as:

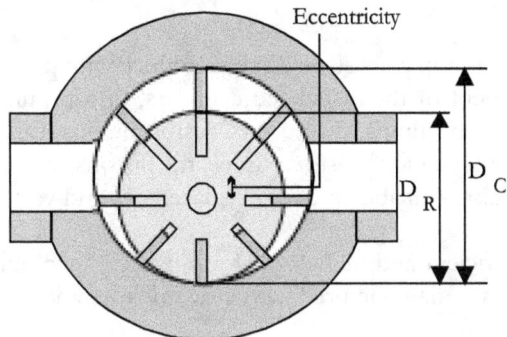

Figure 7.17 | A schematic diagram showing the parameters for determining the volumetric displacement of a vane pump

$$e_{max} = \frac{D_C - D_R}{2}$$

The maximum volumetric displacement $V_{D,\,max}$ is given by:

$$V_{D,\,max} = \frac{\pi}{4}(D_C^2 - D_R^2)\,L = \frac{\pi}{4}(D_C + D_R)(D_C - D_R)\,L$$

$$= \frac{\pi}{4}(D_C + D_R)(2e_{max})\,L = \frac{\pi}{2}(D_C + D_R)(e_{max})\,L$$

The actual displacement V_D is obtained by replacing the maximum eccentricity (e_{max}) with the eccentricity (e). That is:

$$V_D = \frac{\pi}{2}(D_C + D_R)(e)\,L$$

Example 7.5 | Find the volumetric displacement of an unbalanced hydraulic vane pump having a cam ring diameter of 3.15 in, a rotor diameter of 2 in, and a width of 2.4 in. Assume the eccentricity of the center line of the rotor with the center line of the cam ring of the motor as 0.24 inches.

Solution

With usual notations,
$D_C = 3.15$ in
$D_R = 2$ in
$L = 2.4$ in
$e = 0.24$ in

Volumetric Displacement, $V_D = \dfrac{\pi}{2}(D_C + D_R)(e)L$

$= \dfrac{\pi}{2}(3.5+2)(0.24) \times 2.4 \text{ in}^3 = 4.97 \text{ in/rev}$

Piston Pumps

A positive displacement piston pump consists of a cylinder block with pistons attached to the driveshaft. Based on the spatial arrangement of the associated cylinders, piston pumps can be broadly classified into axial-piston and radial-piston pumps. In the axial-piston pumps, the pistons are arranged parallel to the cylinder block, whereas in the radial-piston pumps, the pistons are arranged radially in the cylinder block. The piston pumps are also available in fixed-displacement and variable-displacement designs.

Piston pumps are the most efficient and highest-performing hydraulic pumps. They are very sensitive to abrasive wear because they are manufactured with very fine tolerances.

Axial-piston Pumps

The axial piston pump is a positive-displacement pump with many pistons arranged in a circular array within the associated cylinder block. A prime mover drives the cylinder block, which rotates with the driveshaft.

Axial piston pumps can be classified into in-line and bent-axis types. In the inline axial piston pump, the cylinder block's centerline is aligned with the driveshaft's centerline. In the bent-axis axial piston pump, the center line of its cylinder block is at an angle with the center line of its driveshaft.

In-line Axial-piston Pumps

Figure 7.18 shows the cross-sectional view of an in-line axial-piston pump. It mainly consists of: (1) a group of cylinders (typically an odd number of cylinders) with pistons, (2) a cam plate (swashplate), (3) a stationary valve plate, and (4) a driveshaft. The cylinders are arranged in parallel and formed into a round block about an axis. The axis of the cylinder block is aligned in-line with the axis of the driveshaft. The cylinder pistons are fitted to the cam plate through ball joint shoes. If the cam plate is not angled, the rotating pistons remain stationary within their bores. However, if the cam plate is angled, the rotating pistons move back and forth in their respective cylinder bores. The valve plate contains two kidney-shaped openings. The large opening is the suction port, and the small opening is the discharge port.

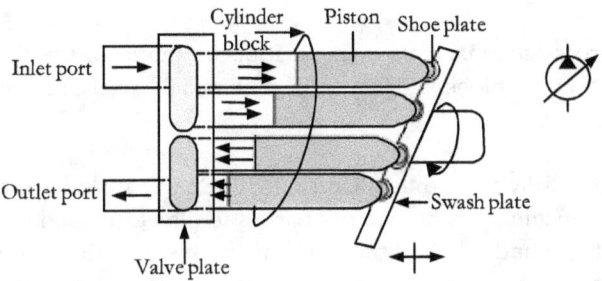

Figure 7.18 | A cut-section view of an in-line axial-piston type hydraulic pump

As the cylinder block rotates, each cylinder alternates between the low-pressure suction side and the high-pressure discharge side through the valve plate. In other words, as any bore in the rotating cylinder block turns past the suction port, the piston is pulled out from its bore. The outward movement of the piston creates a partial vacuum at the inlet port. The development of the partial vacuum draws the fluid from the associated reservoir into the bore. The piston is pushed into its bore as it passes the discharge port. The inward movement of the pistons forces the fluid out of the discharge port, thus creating the flow.

The size and number of pistons of a piston pump and the length of each piston stroke determine its displacement. A pump with a fixed swashplate angle is a constant-displacement pump. In the variable-displacement type, a mechanism varies the swashplate angle over its full range.

Bent-axis Piston Pump

Figure 7.19 shows the cross-sectional view of a bent-axis piston pump. It mainly consists of a: (1) cylinder block, (2) driveshaft, and (3) valve plate. The cylinder block contains many cylinders arranged in parallel and formed into a round block about an axis. The ball-ended cylinder pistons are fixed to the driveshaft. The axis of the cylinder block is set at an angle to the driveshaft axis via a jointing mechanism that tends to reduce side loads. When driven, the pistons reciprocate in their respective bores. The valve plate contains two semi-circular openings for the suction and discharge ports. The bores in the cylinder block are connected, successively, to the low-pressure suction port and the high-pressure discharge port by the valve plate as it rotates. This action creates the flow.

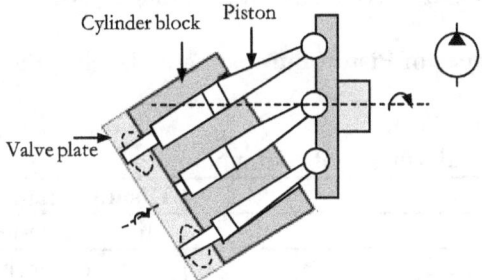

Figure 7.19 | A cut-section of a bent-axis piston-type hydraulic pump

Bent-axis piston pumps can be either fixed-displacement or variable-displacement. In the fixed-displacement type, the offset angle is fixed. In the variable-displacement type, a mechanism varies the offset angle from zero to an absolute maximum.

Radial Piston Pumps

Figure 7.20 shows the cross-sectional view of a radial piston pump of the rotating cylinder type. This design consists of: (1) a cylinder block with seven or nine radial barrels, (2) a reaction ring, (3) a pintle, and (4) a driveshaft.

The pistons are arranged radially around the driveshaft. The pintle includes suction and discharge ports that connect to the inner openings of the cylinder barrels to direct the fluid in and out of each cylinder. Due to the centrifugal force and the backpressure on the pistons, the pistons in the cylinder block remain in constant contact with the reaction ring. The rotor moves eccentrically to the cylinder block, and the pistons move in and out as the cylinder block rotates.

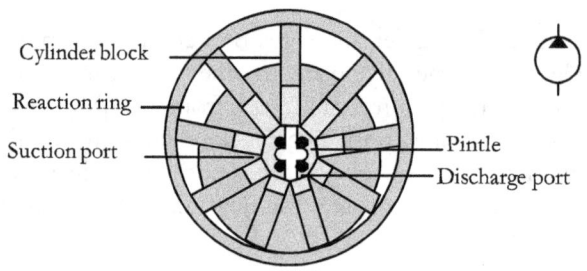

Figure 7.20 | A simplified schematic diagram of a radial piston pump

The pump's drive shaft rotates the cylinder block. When the cylinder bores on one side pass near the inlet port in the pintle, the pistons in those bores are forced outward by the reaction ring. The outward movement of the pistons creates a partial vacuum at the inlet port. The development of the partial vacuum draws the fluid from the associated reservoir into the respective cylinder bores. When the cylinder bores on the other side move past the point of maximum eccentricity, their pistons are forced inward by the reaction ring. This inward movement of the pistons forces the fluid out of the discharge port.

Radial piston pumps are available in fixed-displacement and variable-displacement designs. In the fixed-displacement design, the physical relationship of the pump elements cannot be altered. Hence, the pump can deliver only a fixed volume of fluid per revolution. In the variable-displacement design, the pump delivery can be varied by altering the eccentricity of the cylinder block.

Advantages and Disadvantages of Piston Pumps: Table 7.4 gives the advantages and disadvantages of piston pumps.

Table 7.4 | Advantages and disadvantages of piston pumps

Advantages	Disadvantages
- High power-to-weight ratio	- Intricate design
- Highest efficiency, highly reliable	- Most expensive
- Operate at higher pressures	- Higher maintenance costs
- Good serviceability	- Sensitive to contamination
- Provide long life	- Limited flow rates

Example 7.6 | A fixed-displacement axial-piston pump is designed with nine bores. Each bore has a diameter of 0.59 inches and a maximum stroke of 0.98 inches. If the pump is driven at 1400 rpm, what is the theoretical flow rate of the pump?

Solution

No. of bores = 9
Bore diameter, D = 0.59 in
Maximum stroke, L = 0.98 in
Speed, N = 1400 rpm

Volume of each bore = $(\pi \times D^2/4) \times L = [\pi \times (0.59)^2/4] \times 0.98 = 0.27$ in^3

Volumetric displacement, $V_D = 9 \times 0.27 = 2.43$ in^3/rev

Theoretical flow rate = V_D (in^3/rev) x n (rps) = $2.43 \times (1400/60) = 56.7$ in^3/s
= 56.7 / 3.85 = 14.73 gpm

[Note: 1 gpm = 3.85 in^3/s or 1 in^3/s = 0.25974 gpm]

Comparison of Positive-displacement Pumps

Selecting a suitable pump for a given hydraulic application depends on many factors. Generalizing is extremely difficult, but some significant features can be identified by comparing gear, vane, and piston pumps. Table 7.5 provides a comparison of various types of positive-displacement pumps.

Table 7.5 | Comparison of positive displacement hydraulic pumps

Parameter	Gear pumps	Vane pumps	Piston pumps
Design	Simple, rugged	Slightly complex	Complex
Displacement type	Fixed	Fixed or variable	Fixed or variable
Pulsation	High	Low	High
Fluid sensitivity	Least sensitive	Sensitive	Sensitive
Effect of contaminants	Tolerant	Less tolerant	Sensitive
Leakage	Prone to leakage	Less prone	Prone to leakage
Noise level	High	Low	High
Size	Small to medium	Small to medium	Medium to large
Power-to-weight ratio	Low	Low	High
Efficiency	Least	Medium	Highest (>90%)
Cost	Cheapest	Medium	Costly
Maintenance costs	High, due to wear	High, due to wear	Very high
Service life	Longest	Long	Very long
Application	Light-, medium-duty	Light-, medium-duty	Heavy-duty
Pressure level	Medium	Lowest	Highest
Displacement	Medium	Lowest	Highest
Viscosity range	highest	Low	Low

Materials of Construction, Pumps

Hydraulic pumps are constructed with a wide variety of materials. Gears, vanes, pistons, casing, and brackets are made of cast iron, ductile iron, steel, stainless steel, and/or higher alloys. Shaft seals are made with elastomers, plastics, and Teflon. Bushings are made of carbon graphite, bronze, silicon carbide, ceramic, or other unique materials. By precisely matching the system fluid to the materials of construction of the pump parts that come into contact with it, a pump's life-cycle performance can be improved.

Mounting of Hydraulic Pumps

Hydraulic pumps are generally mounted horizontally on the top plates of the reservoir. They often use foot- or flange-mounting designs. The mounting arrangements for a pump are often provided at its end caps. Flanges are provided for stationary mounting, and clevis-type fittings for swing-type mounting. The mounting flange should remain perpendicular to the driveshaft.

The prime mover and pump shafts must be properly aligned during their assembly. The maximum permissible parallel misalignment is usually less than 0.004 inches, and the maximum permissible angular misalignment is usually less than $0.2°$, or according to the recommendations of pump manufacturers. Avoid any stress on the shaft against the bending or thrust loads.

Employ a flexible coupling whenever possible to ensure optimal damping. The mounting bolts should be properly seated and torqued.

The direction of rotation of a unidirectional hydraulic pump must be the same as indicated on the pump.

Side Loads on Hydraulic Pumps

In certain hydraulic pumps, such as gear pumps and unbalanced vane pumps, all of the high-pressure fluid's pumping action takes place in the chambers on one side of the rotors and driveshafts. This design imposes considerable side loads on the rotors and shafts. A misaligned pump shaft can also impose large side loads on its driveshaft. Side loads on a pump can cause bearing failure, seal damage, and eventual shaft breakage.

Characteristic Curves of Hydraulic Pumps

The performance of hydraulic pumps can be assessed by using the following characteristic curves: (1) Flow vs. speed, (2) Flow vs. operating pressure, (3) Drive power vs. operating pressure, and (4) Efficiency vs. operating pressure.

The flow rate curve in Figure 7.21(a) shows the relation between a hydraulic pump's flow rate and the speed of its driveshaft at a specified pressure.

Figure 7.21(b) shows a graph of the pressure versus flow rate. As discussed earlier, ideally, there is no internal slippage in a positive-displacement pump, and the amount of fluid pumped remains constant for each revolution of its driveshaft, regardless of pressure. However, the mating components are not perfect, and small leaks occur past the pump's clearances. The leak increases with increasing pressure. The self-explanatory Figures 7.21(c) and (d) show the pressure-power and pressure-efficiency curves.

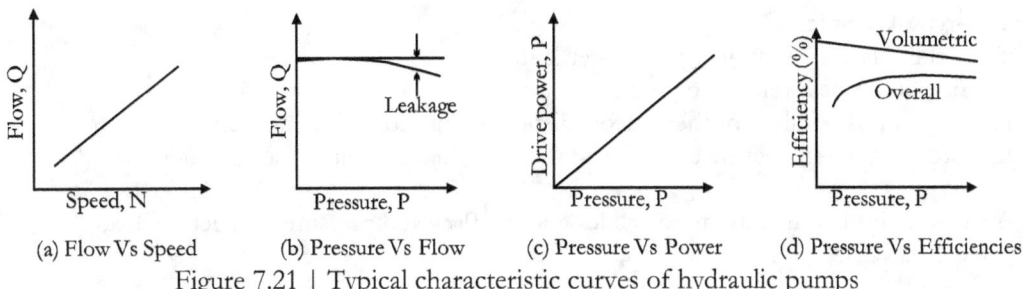

Figure 7.21 | Typical characteristic curves of hydraulic pumps

Requirements of Hydraulic Pumps

When selecting hydraulic pumps, the most critical customer requirements are their compact dimensions, minimal pressure and volume pulsations, high efficiency, low cost, long service life, reduced vibration, and low operating noise levels.

Selection of Hydraulic Pumps

Hydraulic pumps of varying designs and types are available in the market. The designer must satisfy many application and system requirements before selecting the types of pumps for the system. Some application requirements include expected load, type of fluid and its viscosity, operating temperature range, overall cost, and permissible noise level. Some system requirements include maximum pressure, flow rate, displacement, duty cycle, reliability, and ease of maintenance.

Application Notes, Hydraulic Pumps

Various types of hydraulic pumps, such as gear, vane, and piston pumps, and their variants are available to meet a wide range of application demands in industrial, mobile, aerospace, marine, mining, agriculture, and construction fields. The following paragraphs present some generalizations regarding hydraulic pump applications. However, remember that many exceptional factors must be considered when deciding the pump type for a given application.

For light- and medium-duty industrial applications, gear and vane pumps are most commonly used, whereas for power-intensive applications, piston and screw pumps are most suitable. External gear pumps are most appropriate for applications involving rough handling and dirty environments, such as mobile equipment and conveyor systems. Internal gear pumps are used in applications requiring low-speed, quiet operation, such as hydraulic presses, drilling machines, lifting devices, and marine and petrochemical applications. For energy-efficient applications, where space and weight are a premium, as in aircraft, Gerotor pumps can be used. Screw pumps are used in applications requiring quiet operation, such as machine tools, hydraulic presses, rolling mills, sheet metal machines, plastic molding machines, hydraulically driven propellers, submarines, and offline filtration systems.

Vane pumps are suitable for sophisticated applications requiring variable displacement and low noise. They are used in automotive power steering and transmission applications, marine and railway winches, oil and gas drilling equipment, earthmoving and construction equipment, plastic injection molding machines, sophisticated machine tools, and large presses.

Piston pumps find applications in aerospace, agricultural, automotive, mobile, and construction equipment; marine equipment; metal-forming and stamping machines; machine tools; oilfield equipment; and mining.

Objective-type Questions

1. Which of the following is an <u>incorrect</u> statement?
    a) A vane pump is less tolerant of contamination.
    b) A gear pump is suitable for the variable displacement mode of operation.
    c) A hydrostatic pump delivers a definite amount of fluid with little leakage under normal pressures.
    d) A variable-displacement pump provides enough flow and pressure to meet the load requirement.

2. Which of the following is (are) the reason(s) that cause(s) excessive leakage in the pump used in a hydraulic system?
    a) High-speed pump operation
    b) Use of a low-viscosity system fluid
    c) Significant gaps between its mating surfaces
    d) All of the above

3. Mark the pair of terms related to gear pump classification.
    a) Axial, radial
    b) Fixed, variable
    c) External, internal
    d) Unbalanced, balanced

4. Mark the attribute that is <u>not</u> true for hydraulic piston pumps.
    a) High power-to-weight ratio
    b) Least expensive
    c) Highest efficiency
    d) Limited flow rates

5. Mark the least expensive hydraulic pump.
    a) Internal gear pump
    b) Vane pump
    c) External gear pump
    d) Piston pump

6. Mark the correct statement.
    a) Adjusting the eccentricity between the rotor and cam ring can vary the volumetric displacement of an unbalanced vane pump.
    b) The balanced vane pump imposes considerable side loading on the driveshaft.
    c) In radial piston pumps, the centerline of the cylinder block is in-line with the centerline of the driveshaft.
    d) A hydraulic gear pump balances pressure forces on the gears.

7. Which parameter(s) affect the noise level in a hydraulic pump?
    a) Pressure
    b) Size of the pump
    c) Speed of the pump
    d) All of the above.

8. What is the theoretical flow rate of a fixed-displacement pump with a volumetric displacement of 0.0004 ft³/rev operating at 1450 rpm?
   a) 0.45 cfm
   b) 0.58 cfm
   c) 5 cfm
   d) 580 cfm

9. The volumetric displacement of a hydraulic pump represents the volume of the fluid that passes:
   a) through the pump in one second
   b) through the pump in one hour
   c) from the discharge side of the pump to its suction side
   d) through the pump in one revolution of its driveshaft

10. What fluid power is generated by a pump delivering 6.87 gpm at 2175 psi?
    a) 5.2 hp
    b) 6.7 hp
    c) 8 hp
    d) 8.7 hp

11. What is the volumetric efficiency of a hydraulic pump with a volumetric displacement of 30 in³/rev and an actual fluid flow rate of 427.17 in³/s at 1000 rpm? The system pressure is 2175 psi.
    a) 85.4%
    b) 70.2%
    c) 46.8%
    d) 95.1%

Review Questions
1) What is the primary function of a hydraulic pump?
2) How is the power developed in a pump related to the pump flow rate and the system pressure?
3) Briefly explain the terms of hydraulic pumps: (1) Positive displacement and (2) Slippage.
4) Briefly explain the following hydraulic terms: (1) Pump displacement, (2) Pump delivery, rate, and (3) Pump power.
5) Mention any five types of pumps used in hydraulic systems.
6) How are hydraulic pumps classified?
7) Classify positive displacement pumps based on construction.
8) Classify positive displacement pumps based on delivery.
9) Classify positive displacement pumps based on motion.
10) Classify positive displacement pumps based on displacement.
11) Classify positive displacement hydraulic pumps with suitable examples.
12) Why are centrifugal pumps rarely used in fluid power systems?
13) Describe the two basic types of hydraulic pumps, describing their essential differences.
14) Explain the term 'positive displacement' when specifying a hydraulic pump.
15) Explain the working of a positive-displacement pump.
16) Explain how a positive-displacement pump is selected for a hydraulic system.
17) Define volumetric efficiency.
18) Define mechanical efficiency.
19) Define overall efficiency.

20) What are the main reasons for the popularity of hydraulic gear pumps?
21) Differentiate between fixed-displacement and variable-displacement hydraulic pumps.
22) What are the reasons for excessive leakage and wear in gear pumps used in hydraulic systems?
23) How are hydraulic gear pumps classified?
24) Why can't gear pumps be used as variable displacement pumps?
25) Explain the essential constructional features of hydraulic gear pumps.
26) Explain the working principle of external gear hydraulic pumps.
27) Explain the difference between the external and internal gear pumps.
28) List the essential parts of an external gear pump.
29) Explain various designs of gears used in external gear pumps.
30) List some advantages and disadvantages of external gear pumps.
31) Mention some applications of external gear pumps.
32) State the differences between the helical and herringbone gears for hydraulic pumps.
33) State the factors that affect the efficiency of a gear pump when used in a hydraulic system.
34) Explain some critical performance measures for hydraulic gear pumps.
35) How is the displacement of the hydraulic gear pump determined? '
36) Explain the working principle of internal gear hydraulic pumps.
37) List some advantages and disadvantages of internal gear pumps.
38) Mention some applications of internal gear pumps.
39) Explain the working principle of a gerotor pump.
40) List some advantages and disadvantages of gerotor pumps.
41) Mention two applications of gerotor pumps.
42) Explain the working principle of a three-screw hydraulic pump.
43) List some advantages and disadvantages of hydraulic screw pumps.
44) Give two applications of hydraulic screw pumps.
45) Explain the working principle of an unbalanced rotary vane pump with a suitable diagram.
46) Explain the pumping action of a hydraulic vane pump.
47) Classify rotary vane hydraulic pumps.
48) List out the essential parts of a vane pump.
49) Explain the working principle of a balanced rotary vane pump.
50) Differentiate between the balanced and unbalanced vane pumps.
51) Explain the working principle of a variable displacement vane pump.
52) List some advantages and disadvantages of vane pumps.
53) Mention two important applications of vane pumps.
54) Write a short note on variable displacement vane pumps.
55) Briefly explain a vane pump's working principle and constructional details.
56) Explain the pumping action of a piston pump.
57) Describe the primary classification of piston pumps.
58) Explain the working principle of an in-line axial piston pump.
59) How can the displacement of an axial piston pump be varied?
60) Explain the working principle of a bent-axis piston pump.
61) Differentiate the following: in-line axial-piston pump and bent-axis piston pump.
62) Explain the working principle of radial piston hydraulic pumps.
63) Describe the classification and performance features of different types of hydraulic pumps.
64) List three advantages of piston pumps.
65) Draw the typical characteristic curves of hydraulic pumps and explain them.

Numerical Problems

1) Six equal-sized, single-acting hydraulic cylinders are simultaneously extended 5 inches in 2.2 seconds in an automated machining operation. Each cylinder has a bore diameter of 2.56 inches. Determine the minimum flow rate of a hydraulic pump for this application. [Ans: 70.19 in^3/s]

2) What is the displacement of a hydraulic pump required to produce 8 gpm while turning at a speed of 1750 rpm? [Ans: 0.00457 gallons/rev]

3) A 6.7 hp motor drives a hydraulic pump. The pump operates at 1450 psi. Assuming no losses, what is the flow rate through the system in units of gpm? [Ans: hp x 1714/psi=7.92 gpm]

4) What theoretical flow is produced by a fixed-displacement hydraulic pump running at 1750 rpm with a displacement of 3.78 in^3/rev? What is the actual flow generated if the pump's volumetric efficiency is 95%? [Ans: 6284 in^3/rev]

5) Determine the theoretical and actual torque requirements to run a hydraulic pump with a 2 in^3/rev displacement if the pressure rise is 2000 psi and its mechanical (torque) efficiency is 90%. [Ans: Theoretical torque=636.94 in.lb, Actual torque=573.25 in.lb]

6) A hydraulic pump delivers 9.25 gpm with a pressure rise of 1450 psi. The pump's overall efficiency is 87%. Calculate the pump shaft power. [Ans: 9 hp]

7) An electric-motor-driven hydraulic pump delivers 7.93 gpm at 3191 psi. What is the motor's power rating if the overall efficiency is 80%? {Ans: 14.76 hp]

8) What fluid power is generated by a pump delivering 15 gpm at 2000 psi? {Ans: 17.5 hp]

9) A fixed-displacement pump with a volumetric displacement of 8 in^3/rev operating at 2000 rpm has a 610 in^3/min slippage. Calculate the pump's volumetric efficiency. [Ans: 96%]

10) A fixed-displacement pump, driven at a speed of 1200 rpm, is supplied with an actual torque of 203567 in.lb. What is the mechanical efficiency of the pump operating with a theoretical flow rate of 12205 in^3/s at 2465 psi? Also, the volumetric and overall efficiency of the pump should be calculated if the pump slippage is 305 in^3/min. [Ans: η_m=85%, η_v=97.5%, η_o= 82.22%]

11) A hydraulic pump has a volumetric displacement of 5 in^3/rev. It develops a pressure of 972 psi and delivers 4625 in^3/min of fluid at 1000 rpm. What is the pump's overall efficiency if the input torque imparted is 885 in.lb? What is the theoretical torque required to operate the pump? [Ans: Ans: η_o=80.8%, T_T=773.89 in.lb]

12) The hydraulic pump has a nominal displacement of 3.05 in^3/rev. It delivers 4439 in^3/min at 1500 rpm, with a pressure rise of 1450 psi. The shaft power is 18.19 hp. Calculate its volumetric and overall efficiencies. [Ans: 97% and 89%]

13) Determine the theoretical flow rate, output power, and input power for a hydraulic pump generating an actual flow rate of 20 gpm with a pressure differential of 2000 psi if its volumetric and mechanical efficiencies are 95% and 90%, respectively. [Ans: 17.05 gpm, 23.3 hp, 27.25 hp]

14) An external gear hydraulic pump has 0.23 ft outside diameter, 0.13 ft inside diameter, and 0.066 ft width. What is the actual flow rate if the volumetric efficiency is 90%? The pump speed is 1400 rpm. [Ans: 0.03842 ft^3/s]

15) A variable displacement vane pump is used in the power steering system of an off-road vehicle. This pump is equipped with a control device for adjusting the cam ring's eccentricity. The cam ring diameter is 2.95 inches, and the rotor diameter is 2.17 inches. The width of the rotor is 1.97 in. Find (1) the maximum displacement possible and (2) the displacement at an eccentricity of 0.2 inches. {Ans: 12.36 in^3/rev, 6.18 in^3/rev]

Objective-type questions - answer key: 1-b, 2-d, 3-c, 4-b, 5-c, 6-a, 7-d, 8-b, 9-d, 10-d, 11-a

Chapter 8 | Cavitation

Cavitation is a harmful condition that occurs in hydraulic system components like pumps, valves, and actuators. It occurs when the system fluid does not fill the existing space in a hydraulic component.

Terms: Suction Head Pressure and Vapor Pressure

Suction head pressure and vapor pressure are key terms for understanding cavitation. The net positive suction head pressure is the pressure the fluid experiences at the pump inlet. The liquid's vapor pressure is the minimum pressure required to prevent it from vaporizing.

Temperature-pressure Characteristic (Vaporization Line) of a Fluid

Figure 8.1 shows the fluid's temperature-pressure characteristic (vaporization line), which acts as the dividing line between the liquid and vapor phases. That is, a measure of the fluid in the liquid form vaporizes as the temperature increases or the pressure decreases. Similarly, the entrained air in the fluid also comes out as bubbles when the system pressure falls below the air's saturation pressure.

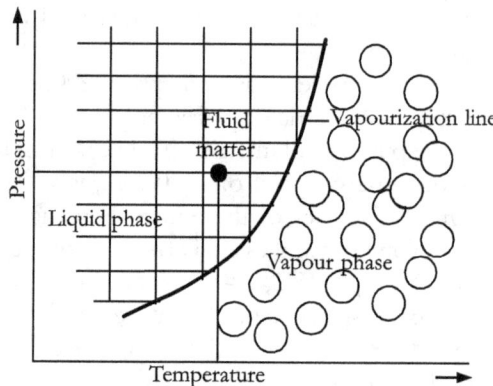

Figure 8.1 | A graphical representation of the vaporization line of fluid

Process of Cavitation in Various Stages

Cavitation occurs when the net positive suction head pressure acting on the fluid at the suction side of a pump falls below the fluid's vapor pressure or the saturation pressure of the entrained air. It may also happen when the local pressure in the venturi of a valve falls below the fluid's vapor pressure or the saturation pressure of the entrained air due to the accelerating fluid.

The schematic diagrams of Figure 8.2 illustrate the cavitation process in a simplified manner.

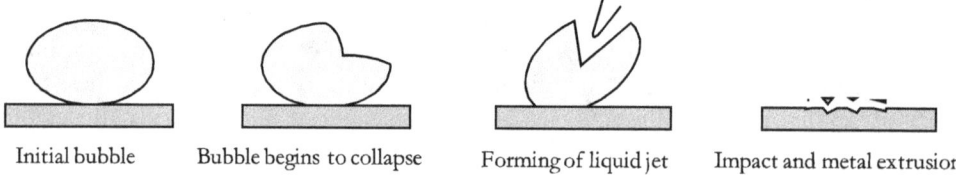

Figure 8.2 | Schematic diagrams illustrating various stages in the process of cavitation

The fall in pressure causes bubbles to form in the fluid. Soon, the bubbles collapse as the pressure on the fluid exceeds its vapor pressure (saturation pressure) when it migrates to a high-pressure area. The collapse of bubbles can cause localized pressures of up to 10000 psi. Next, the process of bubble formation and collapse is nothing but cavitation.

Harmful Effects of Cavitation

Bubble collapse is a violent event that affects fluid contact surfaces in hydraulic components and can cause pitting and erosion of their internal surfaces. Cavitation may cause flow or pressure fluctuations in the system. It can also cause excessive pump vibration, which could damage pump bearings, wear rings, and seals.

All these harmful effects of cavitation can significantly damage components and degrade the system's performance. Therefore, early detection of cavitation can minimize component damage, reduce downtime, and save money.

Detection of Cavitation

Cavitation in a pump can be distinguished by the loud popping noise of the collapsing bubbles. Along with the noise, they also create distinct vibration patterns in the pump, which help detect cavitation. Undetected intermittent cavitation in the pump can eventually cause significant damage. Therefore, a sophisticated hydraulic system employs an automated monitoring system to detect cavitation as it occurs. It continuously collects the pump's vibration data, automatically analyzes it, and sends a message when cavitation is detected.

Elimination of Cavitation

Cavitation in a hydraulic system pump can be prevented if the suction head pressure acting on the fluid at the pump's inlet is maintained above the fluid's vapor pressure and the saturation pressure of the entrained air. This condition can be achieved by regularly maintaining the pump's speed at its rated value, thereby retaining the pump's flow rate.

Further, it is required to maintain the fluid in the reservoir to the marked level and keep the fluid temperature below the maximum permissible limit.

In addition, using a fluid with the proper viscosity and maintaining low pressure drops across the strainer/suction filter are other simple rules to reduce cavitation.

Action needs to be taken during the design phase to reduce cavitation. The goal must be to keep the suction line as short as possible and at least one size larger than the discharge line.

Diesel Effect

Apart from cavitation, the diesel effect (hydraulic dieseling) can also be observed in hydraulic systems. A partial vacuum may develop at points of restriction in a hydraulic circuit, causing the air dissolved in the fluid to precipitate. When the pressure rises again, the air bubbles' temperature increases sharply. Consequently, the fluid bursts into gas bubbles, and the fluid vapor/air mixture may ignite spontaneously within milliseconds. This process is known as the diesel effect.

The diesel effect can cause piston rods and seals to burn or crack, and can age the fluid.

Objective-type Questions

1. The formation of vapor bubbles at the inlet of a hydraulic pump leads to:
    a) Aeration
    b) Oxidation
    c) Cavitation
    d) Over-pressurization

2. The occurrence of cavitation in a hydraulic pump can be prevented by:
    a) Pumping at high speeds
    b) Keeping the length of the pump's inlet line larger than the return line
    c) keeping the suction line at least one size smaller than that of the pump discharge line
    d) Maintaining pump suction head pressure at a higher level than the vapor pressure head of the fluid

3. Mark the incorrect statement regarding preventing cavitation in a hydraulic system.
    a) Keep the suction line as long as possible
    b) The suction line should be one size larger than the discharge line
    c) Maintain a low pressure drop across the strainer
    d) Maintain the pressure acting on the fluid medium at the inlet of the pump at a higher level than the vapor pressure of the fluid

4. Cavitation in a hydraulic system can cause all the following problems except
    a) flow fluctuation in the system
    b) pressure fluctuation in the system
    c) excessive pump vibration
    d) a reduction in the system downtime

5. Air is precipitated due to a partial vacuum developed at restricted points in a hydraulic system. The spontaneous combustion of the air/oil vapor mixture, as the temperature of the air increases and the nearby oil turns into vapor when the air is subsequently subjected to high pressure, is called:
    a) cavitation
    b) diesel effect
    c) micro-combustion
    d) diffusion

Review Questions

1) Explain the process of cavitation in hydraulic systems.
2) Describe the typical causes of cavitation in hydraulic systems.
3) What are the adverse effects of cavitation on hydraulic pump performance?
4) What are the ways to distinguish cavitation in hydraulic systems?
5) What are the ways to monitor cavitation in hydraulic systems?
6) What is meant by the diesel effect?

Objective-type Questions - Answer key: *1(c), 2(b), 3(a), 4(d), 5(b)*

Chapter 9 | Hydraulic Pressure Intensifiers (Pressure Boosters)

Hydraulic pressure intensifiers are devices that obtain very high pressures from low pressure power sources. They are also called pressure boosters. As shown in Figure 9.1, a pressure intensifier operates on the principle of 'ratio-of-piston-areas.' The high pressure at the output side is regulated by controlling the low pressure at the input side, and the two are directly proportional. However, the pressure intensifier cannot amplify power. It is a constant-power device that ideally delivers the same power at the hydraulic output as the applied input power. However, it may be noted that the output power is slightly less than the input power to the extent of the internal losses.

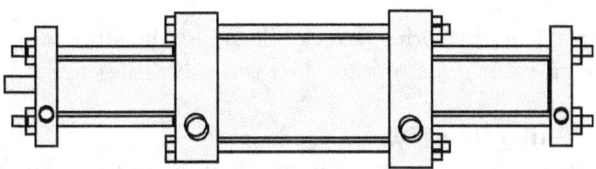

Figure 9.1 | A pressure intensifier

There are three basic types of pressure intensifiers. They are: (1) oil-to-oil intensifier, (2) air-to-oil intensifier, and (3) air-to-air intensifier. The oil-to-oil pressure intensifiers are classified according to their methods of operation as: (1) single-stroke (one-shot) type and (2) reciprocating type. These two types of pressure intensifiers are explained in the following sections.

Single-stroke or One-shot Pressure Intensifier

Figure 9.2 shows the single-stroke (or one-shot) pressure intensifier. It mainly consists of a larger cylinder A and a smaller cylinder B. Let the larger cylinder A has a piston area of A_1 and the smaller cylinder B has a piston area of A_2 ($A_1 > A_2$). The ratio of the larger piston area to the smaller piston area (A_1/A_2) is the pressure intensification ratio. The pistons are connected by a common rod. Check valves separate the suction and discharge lines of the intensifier.

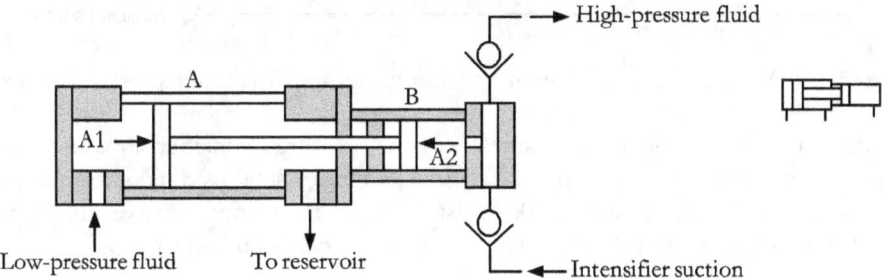

Figure 9.2 | A cross-sectional view of a single-stroke hydraulic intensifier

A low-pressure supply fluid is drawn into the fluid compression chamber of cylinder B through the suction line when the piston-and-piston rod assembly retracts. When a high-flow, low-pressure fluid is applied to the large piston of cylinder A, it develops a force that is mechanically transferred to the smaller piston of cylinder B, pushing it. The smaller piston generates a higher-pressure fluid with each stroke, and a low-flow, high-pressure hydraulic fluid is forced out from the smaller cylinder B. The output pressure increases by a factor equal to the pressure intensification ratio.

That is,

$$\text{Output pressure, } P_{out} = \text{Inlet pressure, } P_{in} \times \frac{\text{Area of the larger piston, } A_1}{\text{Area of the smaller piston, } A_2}$$

$$\text{Output pressure, } P_{out} = \text{Inlet pressure, } P_{in} \times \left[\frac{\text{The diameter of the larger piston, } D_1}{\text{The diameter of the smaller piston, } D_2}\right]^2$$

$$= \text{Inlet pressure} \times \text{Intensification ratio}$$

The larger piston can be retracted by spring, air, or oil. In a single-stroke intensifier, the high-pressure flow is intermittent. A reciprocating type of intensifier can solve this problem.

Reciprocating Type Pressure Intensifier
A reciprocating-type intensifier can enhance the intermittent output flow of the single-shot intensifier. Two back-to-back single-stroke intensifiers feeding the same system with staggered strokes can obtain a nearly continuous flow. Figure 9.3 shows the schematic of the reciprocating-type pressure intensifier. It consists of a centrally mounted low-pressure piston on a double-ended piston rod, with high-pressure chambers at either end. The large piston on the low-pressure side can be made to reciprocate with a suitable control scheme.

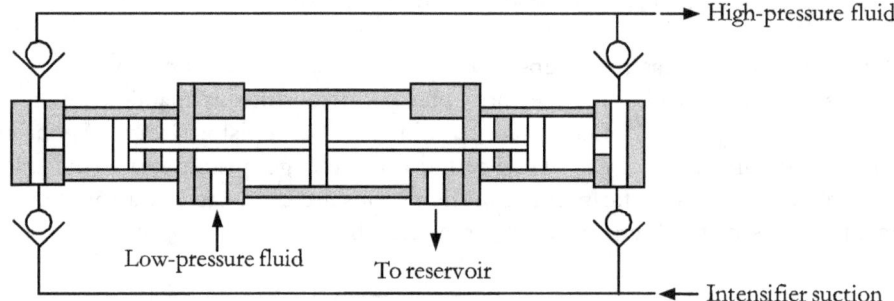

Figure 9.3 | A cross-sectional view of a reciprocating-type hydraulic pressure intensifier

The intensification ratio is an essential parameter for the pressure intensifier. Pressure intensifiers can typically provide intensification ratios up to 20:1. For example, a 3000 psi fluid stream at the input of a pressure intensifier can be intensified to 60000 psi. Pressure intensifiers are available in various sizes and have a maximum flow rate of 20 gpm. They are made of cast iron and steel.

Air-oil Hydraulic Pressure Intensifiers
Air-oil hydraulic pressure intensifiers are primarily used where small amounts of high-pressure oil are needed but where only compressed air is available. They are available with a transmission ratio of up to 51:1. Low-pressure air enters the pressure intensifier's input section and acts against a large-area piston to develop a high force. This force acts upon the hydraulic output fluid section of the pressure intensifier through a small piston to develop high pressure.

Applications of Pressure Intensifiers

Hydraulic pressure intensifiers provide an economical means of producing high pressures for applications with minimal flow.

- They are used to power cylinders, jacks, crimpers, nut splitters, and similar hydraulic-actuated devices.
- They generate high pressures required for work-holding systems and many applications for bending, punching, piercing, forming, crimping, stamping, shearing, or clamping work-pieces.
- Pressure intensifiers can also be used for many power-intensive hydraulic applications, including clamping fixtures, presses, drilling units, die-casting machines, metal forming machines, and molding equipment.
- They can also operate demolition tools, including stone crushing equipment, construction machines, and forklifts.

They can also conduct rupture tests on hydraulic hoses up to a pressure of 72000 psi.

Pressure intensifiers are usually used only in single-cylinder applications. However, an intensifier can drive more than one cylinder if the cylinders work in unison. Alternatively, two or more cylinders can be sequenced using additional intensifiers.

Objective-type Questions

1. What is **not** true about a hydraulic pressure intensifier?
    a) It can amplify the power
    b) It operates on the ratio of piston areas
    c) It can produce very high pressures with a low-pressure power source
    d) It is a combination of two cylinders with differently-sized pistons connected with a common rod

2. The pressure intensification in a hydraulic pressure booster is directly proportional to:
    a) the ratio of the area of the smaller piston to the area of the larger piston
    b) the ratio of the area of the larger piston to the area of the smaller piston
    c) the ratio of the diameter of the larger piston to the diameter of the smaller piston
    d) the ratio of the diameter of the smaller piston to the diameter of the larger piston

3. Which is not a type of pressure intensifier:
    a) oil-to-oil intensifier
    b) oil-to-air intensifier
    c) air-to-oil intensifier
    d) air-to-air intensifier

Review Questions

1) What is the principle of operation of a pressure booster for a hydraulic system?
2) What are the different types of hydraulic pressure intensifiers?
3) Explain the operation of a single-stroke hydraulic pressure booster.
4) Explain the operation of a reciprocating-type hydraulic pressure intensifier.
5) What are the applications of hydraulic pressure intensifiers?

Objective-type Questions - Answer key: 1(a), 2(b), 3(b)

Chapter 10 | Pressure Regulators

A pump in a hydraulic system merely creates a flow through the system. The pressure build-up is caused by the opposition to the flow, probably due to the application of a load. Now, if the load increases excessively, the system can over-pressurize. Overpressure can affect weak points in the system, which, in turn, can lead to mechanical failure of the individual components. Overpressure also tends to increase the load on lubricated surfaces in the system, leading to surface abrasion and even catastrophic system failure. Indeed, the pump can develop pressure to the point of its destruction. Therefore, the system's working pressure must remain within safe limits to protect operating personnel from injury and system components from damage. This function can be achieved using a pressure relief valve (PRV).

A PRV used in a hydraulic system provides an alternative path for the system fluid to flow back into the system reservoir when no flow can be directed to the system's working parts. The PRV modulates the system flow to keep the system's working pressure at the preset level.

Proper sizing, selection, installation, and maintenance of PRVs are critical for achieving maximum protection for hydraulic systems. They must also be designed with materials compatible with a wide range of fluid media. A PRV in a hydraulic system is usually set to 220 psi above the calculated pressure to meet the system's heaviest load. PRVs may be designed for a system with a continuous duty cycle.

Types of Pressure Relief Valves (PRVs)

Pressure relief valves must be designed to operate smoothly and reliably. Depending on the type of construction, there are two basic types of PRVs. They are: (1) Direct-acting PRVs and (2) Pilot-operated PRVs. Pressure relief valves are also available with proportional solenoids (instead of springs) and integrated electronic controls to provide the necessary force and control. Cartridge-type PRVs can be employed in hydraulic systems for easy installation in specialized manifold blocks.

Direct-Acting PRV

A direct-acting PRV consists of a body with an inlet port (P) and an outlet (tank) port (T). It also consists of a poppet that remains pressed against the valve seat by a heavy-duty spring. The valve is usually closed with the spring bias. In most of these valves, an adjusting screw is provided to vary the spring tension externally and thus set the system pressure.

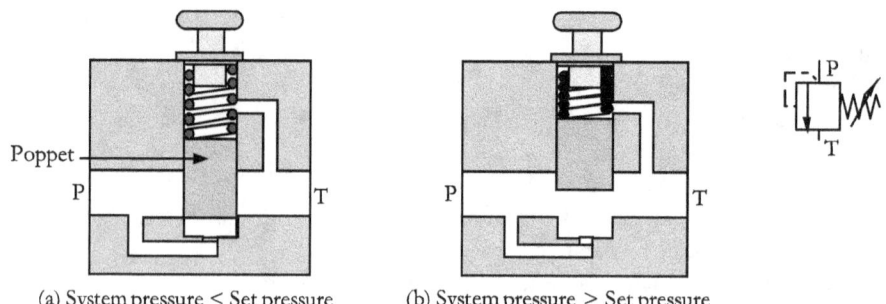

(a) System pressure < Set pressure   (b) System pressure > Set pressure

Figure 10.1 | The positions of a direct-acting PRV for different system pressure levels

Two critical positions of a simplified direct-acting PRV used in a hydraulic system are shown in Figure 10.1. The system pressure acts directly on the spring-biased poppet within the PRV. When the system pressure is below the PRV's set pressure, the flow through the valve is blocked, as shown schematically in Figure 10.1(a). The poppet is lifted off its seat when the system pressure exceeds the preset pressure, as shown in Figure 10.1(b). This action opens an auxiliary passage, diverting the fluid away from the pressurized section of the system to the reservoir. When the excess pressure is relieved, the auxiliary passage closes again. In this way, the system pressure remains at the value set by the PRV.

It may be noted that the poppet cannot be lifted if the fluid leaks to the chamber at the spring side of the valve. Therefore, a drain path is usually provided to relieve any fluid leakage.

The operating characteristics of a direct-acting PRV are affected by the extent of its spring's non-linearity and any backpressure on its discharge side.

A partial Hydraulic Circuit with a PRV

Figure 10.2 shows two positions of a partial hydraulic circuit with a PRV and other system components, such as a pump and a reservoir. The PRV is the first component downstream of the pump and is mounted between the pump output and the reservoir. The maximum system pressure can be set by adjusting the PRV's spring tension. When the system pressure is below the PRV's set pressure (P_s), the PRV remains closed, and the flow is directed to the working parts of the hydraulic system, as shown in Figure 10.2(a). As the system pressure (P) exceeds the set pressure (P_s), the PRV cracks open, diverting the excess flow to the reservoir, as shown in Figure 10.2(b). This way, the PRV protects the circuit from overpressurization.

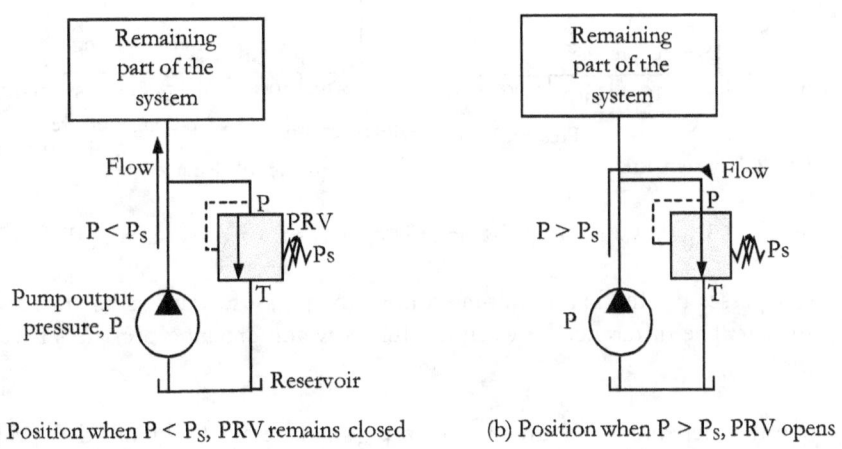

(a) Position when $P < P_s$, PRV remains closed   (b) Position when $P > P_s$, PRV opens

Figure 10.2 | Two critical positions of a hydraulic circuit with a PRV

Characteristic Behavior of Direct-acting PRVs

Let us first understand the ideal pressure-flow characteristic of a hypothetical PRV with the help of Figure 10.3(a). In ideal conditions, we assume no pressure loss across the PRV when fluid flows through it. The poppet of the ideal PRV fully opens at the set pressure, allowing full flow through the valve. However, this characteristic is difficult to achieve in practical hydraulic systems.

A typical characteristic curve for a real direct-acting PRV is shown in Figure 10.3(b). The value of increasing inlet pressure at which the PRV first begins to divert the flow (refer to point A) is called the 'cracking pressure.' At this point, the valve poppet lifts sufficiently so that the flow through the valve becomes continuous. This flow can be treated as a leakage flow. As the flow increases, the poppet is progressively lifted from its seat, increasing spring compression.

The valve's initial opening characteristic, as shown in section B, depends on the poppet's cone angle.

Section C shows that the PRV's behavior depends on the spring rate (lb/in) and the poppet's design. The curve's gradient tends to be steeper for smaller spring rates.

As more flow passes through the PRV, the curve corresponds to section D. The inlet pressure when the PRV passes its rated maximum flow is called the 'full-flow pressure,' which denotes the PRV's rating.

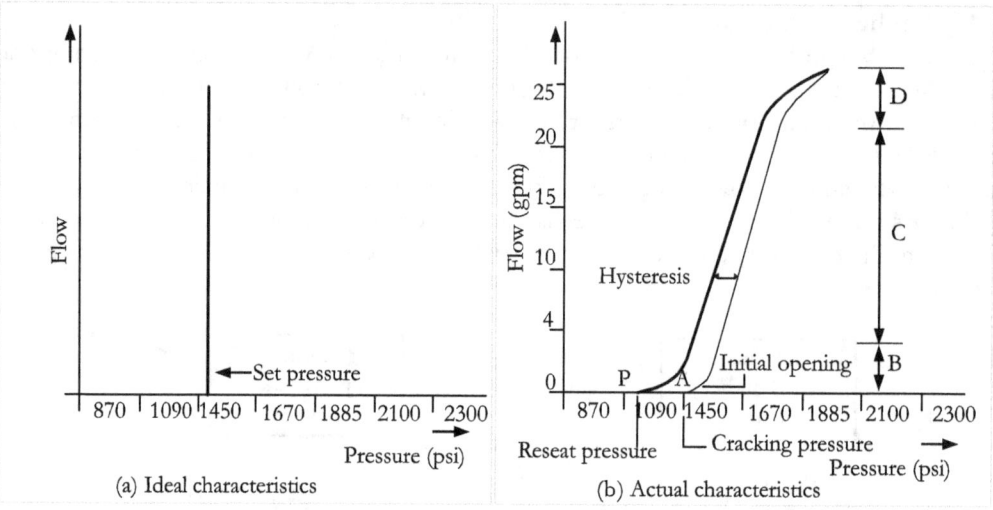

Figure 10.3 | Pressure-flow characteristic curves for a direct-acting PRV

When the PRV bypasses the fluid at its full-rated flow, the pressure exceeds its cracking pressure by a considerable amount. The difference between the full-flow and cracking pressures is called the PRV's pressure override.

The pressure override depends primarily on the spring's stiffness. The higher the spring stiffness (or spring rate), the higher the pressure override of the PRV. A large pressure override in a PRV results in considerable wasted power.

As system pressure decreases, the valve tends to close, reducing flow through it.

However, the path of the closing curve is different from the path of its opening curve, as shown in Figure 10.3(b). The pressure when the valve is fully closed is called its 'closing pressure,' P_c. The closing pressure is lower than the cracking pressure.

The pressure difference when the valve opens and when it closes for a given flow rate is called the valve's hysteresis. The reseat and repeatability characteristics of the valve depend on its hysteresis.

Advantages and Disadvantages of Directing-acting PRVs

Table 10.1 gives the advantages and disadvantages of direct-acting pressure relief valves.

Table 10.1 | Advantages and disadvantages of directing-acting PRVs

Advantages	Disadvantages
- Simplest and low cost	- Excessive seat leakage
- Rugged design	- Sensitive to the effects of backpressure
- High-temperature capability	- Unsuitable for high flows
- Quick actuation	- Lower precision

Pilot-operated PRV

Pilot-operated PRVs are designed to overcome the shortcomings of the direct-acting PRVs. A pilot-operated PRV is also known as a compound-relief valve. Further, it is a two-stage PRV. Figure 10.4 shows the cross-sectional view of a pilot-operated PRV. It consists of a body with an inlet port (P) and an outlet (tank) port (T). It also includes a spring-loaded main spool with a control chamber, and a spring-loaded pilot spool with a knob. The pressure can be set by controlling the spring tension acting on the pilot spool using a knob. The main spool closes the valve in the normal position. There is also a narrow passage from the inlet side of the main valve to the control chamber. By deliberate design, the active area of the main spool at the inlet side is smaller than that on the opposite side.

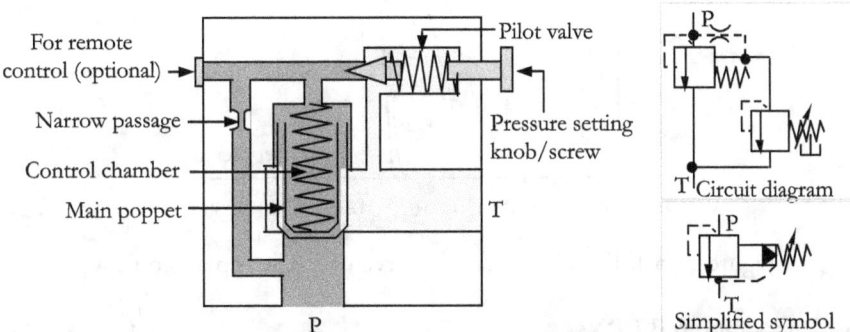

Figure 10.4 | A cross-sectional view of a pilot-operated pressure relief valve

Assume that the pilot-operated PRV is installed in a hydraulic system with a pump, reservoir, and load. When the system pressure is below the PRV setting, the pump draws fluid into the system, increasing the pressure. This pressure also acts on the main spool of the PRV, and the fluid travels to the PRV's control chamber at a controlled rate. As a result, the pressure builds up in the control chamber. The system pressure now acts on both sides of the main spool, causing it to be pressure-balanced. Any increase in the system pressure works on both sides of the main spool, and the spring-biased main spool remains tightly closed.

At the same time, the system pressure also acts on the pilot spool of the PRV. When the system pressure exceeds the PRV's preset value, the pilot valve opens, allowing pressurized fluid from the control chamber to return to the reservoir. The pressure in the control chamber then decreases, causing the main spool to move off its seat. This allows the flow from the pump to return to the reservoir through the main opening in the PRV, thereby releasing the excess pressure in the system. The main spool assumes its normally closed position whenever the system pressure drops sufficiently.

In contrast to a direct-acting PRV, a pilot-operated PRV allows flow through the valve over a narrow pressure range. Therefore, pilot-operated PRVs have a much lower pressure override than direct-acting PRVs with comparable ratings. This advantage allows pilot-operated PRVs to use a reduced spring rate.

Characteristic Behavior of Pilot-operated PRVs
Figure 10.5 shows a typical pressure-flow curve of a pilot-operated PRV in bold lines. The pressure-flow curve for a direct-acting PRV (shown with dotted lines) is superimposed for comparison. The curve of the pilot-operated PRV exhibits a nearly sharp pressure cut-off limit compared to the direct-acting type PRV.

A numerical example further reinforces this idea. A pilot-operated valve may crack open at 1784 psi and fully open at 1885 psi, with a pressure override of 102 psi. In contrast, a direct-acting valve may crack open at 1450 psi and fully open at 1885 psi, with a pressure override of 435 psi.

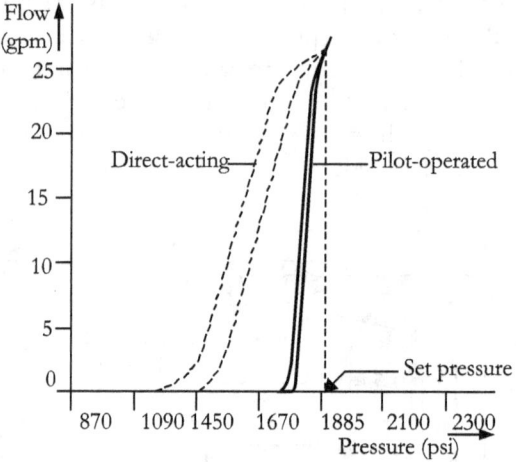

Figure 10.5 | The characteristic curve of a pilot-operated PRV

Advantages of Pilot-operated PRVs
A pilot-operated PRV can be set to fully open over a narrow pressure range, unlike a direct-acting PRV. The pilot-operated PRV can pass large flows with minimal pressure buildup. It provides higher precision than the direct-acting PRV. It can respond quickly to flow changes caused by rapidly closing valves in the associated circuit, preventing excess pressure build-up. Further, it is balanced against the effects of backpressure, as explained in the latter section.

The Disadvantage of Pilot-operated PRVs
A pilot-operated PRV is an appropriate primary pressure control device in a hydraulic system. However, due to its two-stage design, it is unsuitable for safety applications that require a high operating speed. If the pressure increases rapidly, the system with a pilot-operated PRV experiences a longer pressure spike than a system with a direct-acting PRV of comparable rating.

Further, it may be noted that a pilot-operated PRV is more expensive than a direct-operated PRV of comparable ratings. However, the advantages of pilot-operated PRVs are overwhelming, and they are the most widely used PRVs in industrial hydraulic circuits.

Backpressure on PRVs

In a PRV, the force of the pressure acting on the inlet side of its poppet is counterbalanced by the sum of opposing forces exerted by the spring and the pressure at the PRV outlet. Changes in backpressure on the valves directly affect the PRV's characteristics. The backpressure in the PRV can cause many problems in the associated circuit. It tends to modify the PRV's set value, reduce valve capacity, and cause instability and chattering.

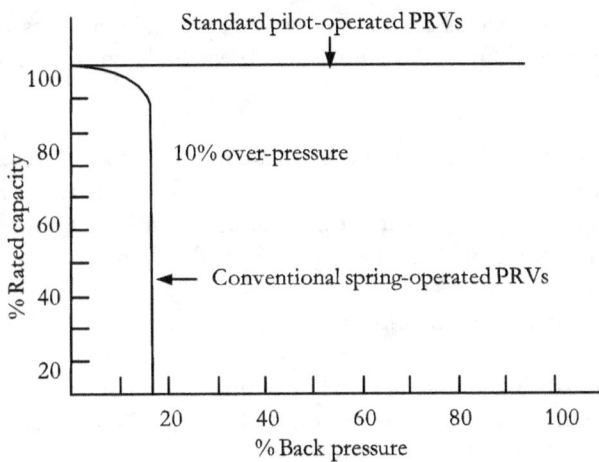

Figure 10.6 | Characteristic curves showing the effect of built-up backpressure on PRVs

Two types of backpressures manifest in a PRV. They are: (1) Superimposed backpressure and (2) Built-up backpressure.

The superimposed backpressure is the pressure existing on the discharge side of the PRV when it is closed. Typically, this condition arises when multiple pressure sources discharge into the common header system.

The built-up backpressure of the PRV is the discharge pressure of the PRV when it is open and fluid is flowing through it. Using Figure 10.6, we now examine the effects of backpressures on conventional spring-operated PRVs and standard pilot-operated PRVs.

A direct-acting PRV is directly affected by the superimposed backpressure. For example, 73 psi of superimposed backpressure on a direct-acting PRV with a set pressure of 1305 psi results in an effective set pressure of 1378 psi. If the 73 psi superimposed backpressure is always constant, the PRV might be intentionally set low by that amount to achieve the desired set pressure. However, constant superimposed backpressure is quite uncommon in a direct-acting PRV.

A pilot-operated PRV is inherently balanced against the effects of the superimposed backpressure on the set pressure of the PRV. In terms of minimizing the effects of built-up pressure, the pilot-operated PRV is superior to all other types of PRVs. The ability of the direct-acting and pilot-operated PRVs to handle the effect of built-up backpressure is compared graphically in the Figure. The standard pilot-operated PRV provides a constant output irrespective of any backpressure acting on the valve poppet. However, the conventional spring-operated PRV provides a drastically reduced output as the backpressure on the valve poppet increases.

Sizing of Pressure Relief Valve

Sizing a pressure relief valve for a specific application is a crucial process that ensures the safety and efficiency of a hydraulic system. It is vital to understand the system flow rate requirements, the maximum system normal and shock pressures, the back pressures in the tank lines, the expected maximum system operating temperature, fluid properties such as type, viscosity, and cleanliness requirements, seal compatibility issues, and the materials used in valve construction.

The following sections provide details on flow rate capacity, maximum pressure development, system operating temperature specifications, seal materials and their fluid compatibility, and valve construction materials.

Flow Rate Capacity: Determine the maximum flow rate in the hydraulic system to identify the highest flow rate the pressure relief valve must handle under all operating conditions. Also, choose the porting configuration. DO (NG or CETOP) is a standardized designation for mounting patterns, and each DO (NG or CETOP) size corresponds with a range of nominal flow rates.

Consult the manufacturer's data sheet for flow rate specifications.

Typical DO (NG, CETOP) Sizes and Flow Rates: While exact specifications vary, these examples illustrate the typical flow rate capacities for common DO (NG, CETOP) sizes:

Size	Flow rate (lpm)	Flow rate (gpm)
DO 2 (NG 4, CETOP 2)	20	5
DO 3 (NG 6, CETOP 3)	40	10
DO 3 (NG 10, CETOP 5)	80	20
DO 3 (NG 16, CETOP 7)	100	30
DO 3 (NG 25, CETOP 8)	240	60

Maximum System Pressure: Calculate the maximum system pressure and shock pressure requirements. The pressure relief valve must be capable of limiting system pressure and managing shock pressure within the calculated limits. High back pressure may necessitate a different valve type, such as a pilot-operated PRV.

Materials of Construction: A wide range of materials is available to suit various fluids and operating environments. Common materials include carbon steel, stainless steel, and brass, among others. Stainless steels are often selected for corrosive environments.

Seal Materials: The compatibility of seal materials with fluids and operating temperature ranges is also important. Buna-N is a typical seal material. Manufacturers also offer optional seal materials, including Fluorocarbon, EPDM, and Silicone.

Fluid Compatibility: The materials chosen for the pressure relief valve must be compatible with the fluid.

Temperature Rating: The materials used in a PRV must function effectively throughout the expected temperature range.

Comparison of PRVs

Table 10.2 compares the differentiating characteristics of direct-acting and pilot-operated PRVs.

Table 10.2 | Comparison of pressure relief valves

Direct-acting PRVs	Pilot-operated PRVs
- Single-stage construction	- Two-stage construction
- Employs a large control spring	- Uses a small control spring
- Large pressure override	- Small pressure override
- Suitable for low-pressure systems	- Suitable for high-pressure systems
- Used for smaller flows	- Used for heavier flows
- Lower precision	- Higher precision
- The right amount of seat and reseat tightness	- Excellent seat and reseat tightness
- Sensitive to the effects of backpressures	- Insensitive to the effects of backpressures

Terminology - PRVs

Some essential factors in the operation of hydraulic pressure regulators include set pressure, cracking pressure, full-flow pressure, pressure override, closing pressure, overpressure, blowdown, and backpressure. These useful terms are defined in the following sections:

Set Pressure: It is the inlet pressure at which a PRV is set to open under the specified service conditions.
Cracking (Opening) Pressure: It is the value of increasing the inlet pressure of a PRV at which there is a measurable lift of its poppet and continuous discharge of the fluid through it.
Full-flow Pressure: The pressure at the inlet of a PRV, when it is passing its rated maximum flow, is called the full-flow pressure.

Pressure Override: The difference between a PRV's full-flow pressure and its cracking pressure is called the pressure override, or the pressure build-up above its set point.
Closing Pressure: It is also known as 'reseat' pressure. The closing pressure of a PRV is the inlet pressure at which the poppet re-establishes firm contact with the seat.
Overpressure: It is the pressure above the set pressure of a PRV, expressed in pressure units or as a percentage.

Backpressure: It is the pressure at the outlet of a PRV due to the pressure in the discharge system. It may be subdivided into superimposed and built-up backpressures.
Superimposed Backpressure: This is the backpressure acting on the outlet of a closed PRV. It is the effect of the pressure in the discharge system coming from multiple sources.
Built-Up Backpressure: It is the increase in pressure in the PRV's discharge header when fluid flows through it.

Objective-type Questions

1. Mark the <u>incorrect</u> statement.
    a) A direct-acting PRV is directly affected by the superimposed backpressure
    b) Direct-acting PRVs are suitable for low-pressure systems, and pilot-operated PRVs are suitable for high-pressure systems
    c) A pilot-operated PRV allows the flow through it over a narrow range of pressures
    d) A pilot-operated PRV is suitable for applications where the speed of operation is essential

2. Mark the valve used to limit the pressure over a narrow range of pressures.
   a) Direct-acting PRV
   b) Pilot-operated PRV
   c) Pressure-reducing valve
   d) Counterbalance valve

3. Mark the <u>correct</u> statement.
   a) Direct-acting type PRVs are used for heavier flows
   b) A direct-acting PRV employs a small control spring
   c) Pilot-operated PRVs are insensitive to backpressures
   d) The pilot-operated PRVs have lower precision as compared to that of the direct-acting types

4. The difference between the full-flow pressure and the cracking pressure of a PRV is called its:
   a) Closing pressure
   b) Pressure override
   c) Overpressure
   d) Backpressure

Review Questions
1) Why is it essential to have a relief valve in a hydraulic system with a positive-displacement pump?
2) What are the consequences of over-pressurization in hydraulic systems?
3) Describe the operation of a basic PRV used in a hydraulic system.
4) What is the function of a pressure-relief valve used in a hydraulic circuit?
5) Where is the appropriate place to mount a pressure relief valve in a hydraulic system?
6) How does a PRV provide safety to a hydraulic circuit?
7) What are the consequences of malfunctioning a pressure relief valve in a hydraulic system?
8) Explain the operation of a direct-acting type pressure relief valve.
9) Give the graphical symbol for the pressure relief valve.
10) Explain the following terms that pertain to pressure relief valves: (a) cracking pressure, (b) full flow pressure, and (c) pressure override.
11) How could pressure override in PRVs be disadvantageous?
12) State an advantage and a disadvantage of directing-acting PRVs.
13) Explain the pressure-flow curve of the directing-acting type PRVs.
14) What is a pilot-operated PRV? Explain its operation.
15) How do the balanced-piston pressure relief valves reduce the pressure override?
16) Draw the pressure-flow curves of the direct-acting and the pilot-operated types of PRVs and describe their implications.
17) Differentiate the direct-acting and the pilot-operated pressure relief valves.
18) What are the advantages of the pilot-operated pressure relief valves?
19) What is the effect of backpressure on the direct-acting type PRVs?
20) What are the disadvantages of the pilot-operated pressure relief valves?

Objective-type questions - answer key: *1-d, 2-b, 3-c, 4-b*

Chapter 11 | Hydraulic Cylinders

A hydraulic actuator is a positive-displacement device used in a hydraulic system to drive the attached load and perform useful work. Its primary function is to convert hydraulic power into mechanical power. The resulting output motion can be either linear or rotary. Accordingly, there are two basic types of hydraulic actuators. They are: (1) Linear actuators and (2) Rotary actuators.

The linear actuators convert hydraulic energy into straight-line mechanical energy, and the rotary actuators convert hydraulic energy into rotary mechanical energy. An example of a linear actuator is a cylinder, and a rotary actuator is a hydraulic motor. This chapter presents the details of the linear actuators.

Linear Actuators
A linear hydraulic actuator converts hydraulic power into a controllable linear force, motion, or both. Technically and economically, hydraulic and pneumatic cylinders are the optimum form of linear actuators. However, a cylinder in the high-pressure hydraulic system can deliver much higher force than a comparable-sized cylinder in the low-pressure pneumatic system.

Basic Cylinder Working
Figure 11.1 shows the cross-sectional view of a typical hydraulic cylinder. It consists of a barrel, piston-and-piston-rod assembly, end caps, and necessary seals and ports. The end caps are securely fastened to the barrel. The piston and piston rod assembly with tight sealing forms two fluid chambers (the piston chamber and the piston rod chamber) and can move within the barrel with two bearing surfaces. The cylinder is provided with two ports for the entry or exit of the system fluid. They are: (1) piston-side (cap-end) port X and (2) rod-side (head-end) port Y.

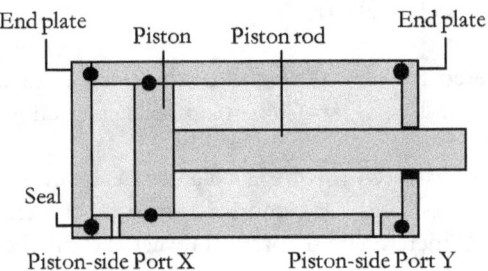

Figure 11.1 | A primary hydraulic cylinder

If the fluid is pumped into the piston chamber through port X and the fluid in the piston rod chamber is discharged through port Y, the piston-and-rod assembly extends with a definite force (thrust). If the system fluid is pumped through port Y and the fluid in the piston chamber is discharged through port X, the piston-and-rod assembly retracts with a definite force (pull). This type of cylinder is called a 'double-acting' cylinder, as its extension and retraction are hydraulically controlled. In this way, reciprocating linear motions can be obtained quite easily using hydraulic cylinders.

If the motion of the cylinder is obtained in only one direction hydraulically, then such a cylinder is called a 'single-acting' cylinder. A spring, gravity, or any other external force can obtain motion in the opposite direction.

Terms and Definitions - Hydraulic Cylinders
Some essential parameters concerned with hydraulic cylinder operation and applications are bore diameter, piston rod diameter, force (thrust and pull), stroke length, speed, and piston rod buckling.

Maximum Operating Pressure (P)
It is the pressure that overcomes all resistances in the system, including useful work and losses. Alternatively, it is the maximum working pressure that the cylinder can sustain without adverse consequences.

Bore Diameter (D)
It refers to the diameter of the cylinder's bore. It can be used to calculate the cylinder's bore area. It is also equal to the piston diameter in a close-fitting hydraulic cylinder.

Piston Rod Diameter (d)
It refers to the diameter of the piston rod of the cylinder.

Stroke Length
It is the distance through which the cylinder's piston and piston rod assembly move through the cylinder.

Maximum Stroke Length
It is the maximum linear movement that a cylinder can produce. For standard double-acting cylinders, stroke lengths can reach 80 inches; for special designs, they can reach 240 inches.

Cylinder Thrust/Pull (F)
The theoretical thrust (F) during the forward stroke or pull (F) during the return stroke of the cylinder can be determined by multiplying the effective area (A) of the piston by the working pressure (P) to which it is subjected, according to Pascal's law.

The active area (A_{ext}), considered for the calculation of the cylinder thrust, is the full area (A_p) of the cylinder piston (bore) and is given by ($\pi.D^2/4$). The parameter 'D' denotes the piston diameter.

Further, the active area (A_{ret}), considered for the calculation of the cylinder pull, is the area (A_p) of the cylinder bore minus the piston rod area (A_r), and it is given by the expression [$\pi.(D^2 - d^2)/4$]. The parameter d is the piston rod diameter. The theoretical thrust and the theoretical pull are given by:

    Thrust, F (Newton)   = P (Pascal) x A_{ext} (m²)
    Pull, F (Newton)   = P (Pascal) x A_{ret} (m²)
Where,
    A_p is the piston area
    A_r is the piston rod area
    A_{ext} is the active area during extension: ($A_{ext} = A_p$)
    A_{ret} is the active area during retraction: ($A_{ret} = A_p - A_r$)

Table A9.1 of Appendix 9 gives the theoretical forces of hydraulic cylinders in the English system of units. These figures do not account for seal or packing friction in these cylinders, which is estimated to affect the thrust by about 10%.

Example 11.1 | A high-pressure double-acting hydraulic press cylinder with an effective piston area of 11 in² for push stroke and a piston-rod area of 3.41 in², operating at 10153 psi, does produce what theoretical forces for the push stroke and pull stroke? [1 ft² = 144 in²]

Solution

Effective piston area, push stroke, A_{push} = 11 in²
Piston Rod area, A_{rod} = 3.41 in²
Pressure, P =10153 psi

Effective piston area, pull stroke, $A_{pull} = A_{push} - A_{rod}$
= (11 − 3.41) = 7.59 in²

Thrust, F_{push} = P × A_{push}
= (10153) × (11) = 111683 lb

Pull, F_{pull} = P × A_{pull}
= (10153) × (7.59) = 77061 lb

Input Power, Cylinder

The hydraulic input power (P_{input}) supplied to the cylinder is given below:

P_{input} (hp) = P (psi) × Q_A (gpm)/1714

Output Power, Cylinder

The mechanical output power (P_{output}) of the hydraulic cylinder is given by the product of the force (F) developed by the cylinder and the resulting velocity (v). The cylinder output power is given below:

P_{output} (hp) = Force (lb) × Velocity (ft/s) /550

The input power supplied must be higher than the required output power to account for losses due to friction and leakage.

Speed, Cylinder

Figure 11.2(a) and (b) show two working positions of a cylinder with the piston in position 1 and position 2, respectively, for determining the cylinder speed during its forward stroke. Assume that the piston-rod assembly of the cylinder moves with a velocity of 'v' when pushed by the system fluid with a flow rate of 'Q.' Further, assume that the cylinder piston of area 'A' has moved a distance 'S' in time 't' to attain the velocity v. Figure 11.2(b) also shows the positions '1' superimposed. Mathematically,

$$v = S/t \quad \text{or} \quad t = S/v$$

We can easily relate the theoretical flow rate (Q) of the system fluid to the speed (v) at which the piston-rod moves if we consider the cylinder volume (V) that must be filled with the fluid and the distance (S) through which the cylinder piston must travel at the specified speed. The volume (V) of the cylinder is simply the length of the stroke (S) multiplied by the piston area (A). The following section gives the flow rate (Q) to achieve the required speed (v).

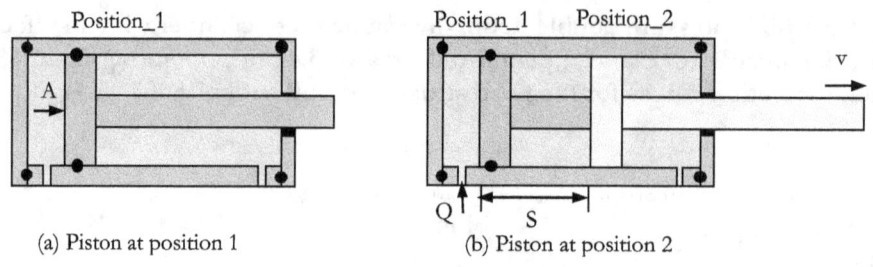

(a) Piston at position 1 (b) Piston at position 2

Figure 11.2 | Two positions of the piston of a hydraulic cylinder

$$Q \text{ (ft}^3/\text{s)} = A \text{ (ft}^2) \times v \text{ (ft/s)}$$

The equation mentioned above shows that a given cylinder's speed (v) depends on the system fluid's flow rate (Q).

The equation also shows that a small-diameter cylinder moves faster than a large-diameter one, while the flow rate remains the same. This equation also shows that a double-acting cylinder has a higher speed during retraction than during extension, provided the system flow rate remains constant. This speed difference is mainly due to the different active areas exposed to the system fluid.

It may be noted that a cylinder with a leak-proof piston and piston-rod seals provides consistent performance at very slow speeds against a wide range of load resistances.

Example 11.2 | **A double-acting hydraulic clamping cylinder must move out with a velocity of 1.64 ft/s on the extension stroke. Calculate the flow rate required by the cylinder operating at 3000 psi and producing a thrust of 11240 lb.**

Solution

Thrust, F = 11240 lb
Pressure, P = 3000 psi
Speed, v = 1.64 ft/s = 19.68 in/s

Area, A_{ext} = F/P
= 11240/3000 = 3.75 in²

Flow rate, Q = A_{ext} × v
= 3.75 × 19.68 = 73.8 in³/s

Types of Hydraulic Loads

Hydraulic actuators are designed to drive a load. There are three types of loads associated with hydraulic systems. They are: (1) Resistive (Positive) load, (2) Overrunning (Negative load), and (3) Inertial load.

A resistive load on a hydraulic actuator tends to oppose its motion. The cutting or shearing operation acting against the cylinder's motion is an example of a resistive load. An overrunning load moves and acts in the same direction as the associated actuator. An example of an overrunning load is a descending heavy weight on an actuator, as in a hydraulic press or winch system. An inertial load tends to resist the acceleration or deceleration of an actuator.

Summary of Relations for Hydraulic Cylinders

Figure 11.3 summarizes the important relations of hydraulic cylinders to aid the reader's understanding and facilitate correlation.

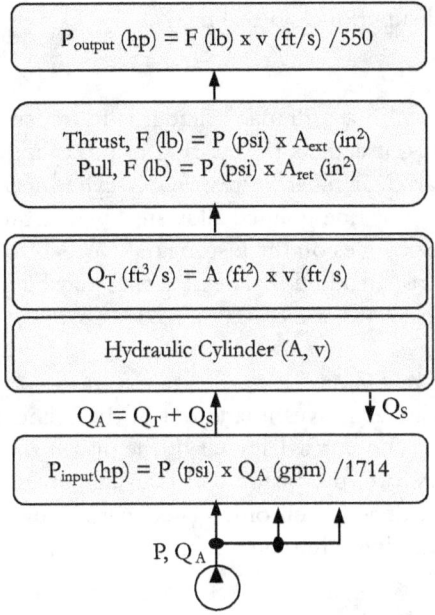

Figure 11.3 | Summary of relations of hydraulic cylinders

Principal Parts of Hydraulic Cylinders

Hydraulic cylinders are designed to operate at high pressures and withstand heavy loads under demanding conditions. Therefore, they must be constructed with high-strength materials and advanced features to provide ruggedness, top-quality performance, and a long service life.

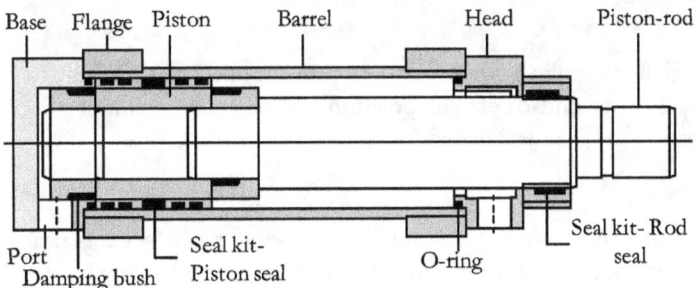

Figure 11.4 | A cross-sectional view of a hydraulic cylinder

A hydraulic cylinder combines various parts, including a barrel, a piston, a piston rod, end caps, cushion seals, wear bands, a piston rod seal/wiper, bearings, rod boots, and a stop tube. It may incorporate advanced features, such as a magnetic piston and end-of-stroke or mid-stroke sensors, to reliably sense piston position. Air bleeds can also be provided to exhaust the air accumulated in the cylinder. Figure 11.4 shows the cross-sectional view of a double-acting hydraulic cylinder.

Barrel

The barrel is made from a high-strength, seamless, drawn tube, precision-machined to a perfect finish. The internal surface of the barrel must be very smooth to control wear and leakage in the cylinder. A top-quality manufacturing process ensures the straightness and roundness of cylinders, which, in turn, guarantee smooth operation and superior fluid sealing.

Piston

The primary function of the piston in a hydraulic cylinder is to transmit the force to the load attached to its piston rod. Apart from this, it acts as the bearing in the cylinder barrel. The piston must be a perfect fit inside the cylinder barrel. It must be reasonably cylindrical and finely finished to ensure a smooth output motion. It is made of fine-grained alloy steel with a bronze coating. It can be threaded or welded to the piston rod. The grooves on the piston are provided to contain the packing or seals. As required in heavy-duty applications, an additional bearing ring may be fitted to the piston to improve its wear resistance.

Piston Rod

The piston rod of a hydraulic cylinder moves in and out of the cylinder barrel and comes into contact with its surrounding atmosphere. The outer diameter of the piston rod must have a smooth, hard, and corrosion-resistant surface. Therefore, the piston rod is usually made of induction-hardened steel or stainless steel. It is also surface-hardened or chrome-plated, with an ultra-fine finish, to ensure resistance to wear and corrosion. Good slide rings are required in the piston rod for excellent wear resistance and proper sealing.

End Caps

The end caps are attached to the ends of the cylinder barrel to enclose the pressure chamber. They are cast from iron or aluminum or made from high-quality steel. Further, they may be designed in square or round shapes to match the barrel shape. They can be fixed by tie-rods, or threaded or welded to the barrel. They also incorporate threaded entries for ports. They must withstand any bending stress and shock loads to which they may be subjected. End-of-travel shocks in the cylinder can be absorbed by the cushion valves built into its end caps.

Cushion

A cushion is a device in a hydraulic cylinder, incorporated at one or both ends, to minimize shock loads as the piston approaches its end-of-stroke position. A hydraulic cylinder should be provided with cushions when the piston velocity is expected to be over 0.1 m/s.

Seals

A significant amount of leakage from a hydraulic component can pose serious problems, including energy waste, reduced efficiency, environmental contamination, and safety hazards. It can be controlled by using appropriate sealing devices. A seal is an elastomeric part used in a hydraulic component to close the imperfections in their mating surfaces. Installing proper seals in the cylinder helps maintain the system's pressure, prevents fluid loss, reduces friction, and keeps contamination out of the system. However, it should be noted that no sealing device can be 100% reliable. Sometimes, a seal for a hydraulic cylinder may be deliberately designed to allow controlled internal leakage to lubricate its moving parts. A small amount of internal leakage up to 3 in^3/min across the piston of a hydraulic cylinder is considered normal. A good seal must be compatible with a wide range of fluids and withstand the harsh industrial environment.

Piston Wear Bands
Wear rings/bands in a hydraulic cylinder limit the wear on its piston and piston rod seals and provide superior protection against side loads. They are usually made from glass-reinforced nylon, which has excellent wear resistance and non-scoring properties.

Piston Rod Seal/Wiper
The specially designed piston rod seal serves as both a pressure seal and a contaminant wiper. The pressure seal prevents pressurized hydraulic fluid in a hydraulic cylinder from leaking out, while allowing the piston rod to move freely back and forth through the piston rod gland. The wiper/scraper part of the seal prevents external contamination, such as dust and dirt, from entering the cylinder through the piston rod gland. The wiper is usually made of metal-clad polyurethane. Polyurethane is highly abrasion-resistant.

Piston Rod Bearing
The piston rod bearing in a hydraulic cylinder guides the cylinder piston rod as it passes back and forth through the rod gland. It also supports the weight of the piston rod, the attached load, and any side loads. Piston rod-end bearings are made of brass or bronze to withstand side loads on the piston rod and ensure proper lubrication.

Air Bleeds
Hydraulic cylinders are considered self-bleeding when cycled full stroke. However, air bleeds can be provided at one or both ends of the cylinder body to vent air.

Piston Rod Boots
When a cylinder is used in a highly contaminated area, its exposed piston rod should be covered with a boot. The piston rod boot is a collapsible cover made from neoprene-coated fabric that is impervious to oil, grease, and water. This arrangement protects the piston rod, seals, and bearing.

Body Styles of Hydraulic Cylinders
Hydraulic cylinders are manufactured in various body styles according to how their outer parts are assembled. Based on their assembly types, there are four basic types of cylinders. They are: (1) Tie-rod cylinders, (2) Mill cylinders, (3) Threaded-end cylinders, and (4) Welded cylinders. Any of these body styles can be selected based on the application and the surrounding environment.

Tie-rod Hydraulic Cylinders
This type of cylinder is used most often in industrial hydraulic applications. In a tie-rod cylinder, four or more high-strength (typically with a minimum yield of 100000 psi) threaded steel tie-rods run the length of the cylinder. These rods actively hold the barrel and the two end caps together. The size and strength of the tie-rods and their fasteners determine the cylinder's strength. The high-tensile tie-rods in a hydraulic cylinder commonly bear a large portion of the applied load and absorb internal energy stresses. A large-bore, high-pressure tie-rod cylinder may have as many as 24 tie rods connecting the end caps to the barrel. Figure 11.5 shows the end views of a tie-rod cylinder.

The tie-rod cylinders offer rugged design, low maintenance, and a long service life. They can be completely disassembled for service and repair. Therefore, heavy-duty cylinders used in the automotive, mobile, and machine tool industries are of the tie-rod design. These sturdy, dependable cylinders can also be used in various light- and medium-duty hydraulic applications.

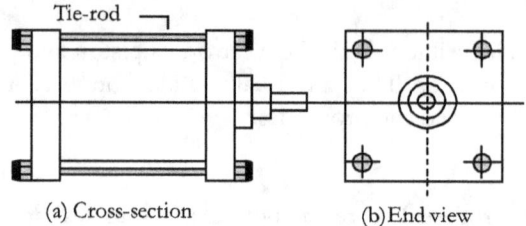

Figure 11.5 | The front and end views of a tie-rod cylinder

Mill-Type Hydraulic Cylinders
They are heavy-duty hydraulic cylinders designed for harsh environments and demanding service conditions, as in steel mills, mines, and furnaces. A mill-type hydraulic cylinder typically features a thick-walled tubular housing that provides exceptional strength and heat resistance. It may also include optional features such as cushions, a stop tube, and oversized ports. A hydraulic cylinder of this type has its end caps fastened to the mating flanges with bolts or cap screws.

Mill-type hydraulic cylinders are built to withstand buckling under compressive loads and can be used in corrosive and non-corrosive environments. Apart from applications in steel mills, furnaces, and mines, they are also well-suited for harsh-environment applications, such as offshore, marine, and metal-forming industries.

Welded Hydraulic Cylinders
In a welded hydraulic cylinder, the heavy-duty barrel is welded along the entire outside diameter to its end caps. The ports are also welded to the barrel, and the mountings are welded to the cylinder ends.

Figure 11.6 shows the sketch of the welded hydraulic cylinder. Welded hydraulic cylinders provide an incredibly strong bond and high strength. They are also very compact. Moreover, they have smooth exterior surfaces, free of tie-rods and fasteners. These constructional features mean fewer places for dirt and debris to accumulate. The round body of a welded cylinder is free of obstructions, making it easy to clean.

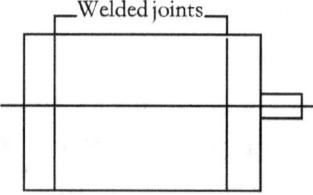

Figure 11.6 | A welded hydraulic cylinder

Welded hydraulic cylinders have many advantages over tie-rod cylinders. They are compact and usually designed for high strength. They are also space-efficient in their overall length. Because of these advantages, they dominate the mobile hydraulic equipment market and the heavy equipment industry.

However, they have the disadvantage that maintenance technicians cannot disassemble them for servicing.

Threaded-end Hydraulic Cylinders

In this type of cylinder construction, the end caps are threaded to fit the internal thread of the tubular housing. Such cylinders are used in applications where space is an important consideration. The food-processing and food-packaging industries extensively use this type of cylinder body style.

Piston Rod Size

The maximum thrust force that a cylinder can practically provide is limited by its piston rod diameter, overall length, and mounting type. In cylinders with longer piston rods, the piston rod must handle the thrust forces generated by the application. The cylinder must also be supported adequately. Note that a head-end mounting provides greater column strength than the cap-end mounting due to the smaller distance between the mounting points in the head-end mounting than that in the cap-end mounting. The piston rod size can be selected from the size charts using its free buckling length and the load imposed on the cylinder.

Side Loads in Hydraulic Cylinders

A long-stroke hydraulic cylinder often experiences significant side-load forces while in operation, especially when fully extended. The side load is the force component that acts laterally across the axis of the cylinder's bearings, piston rod seals, or piston.

Many reasons can be attributed to the development of side loads on the cylinder. An off-center load on the cylinder, as shown in Figure 11.7, causes side loads on its bearings, seals, and piston. Similarly, the cylinder experiences side loads if it is misaligned or improperly mounted. Even the weight of the cylinder or its piston rod exerts side loads on the rod and piston bearings. Standard hydraulic cylinders are not designed for handling heavy eccentric loads, so they are vulnerable to side loads.

The side loads on a hydraulic cylinder are liable to cause stress on its bearings, seals, piston, mounting flange, and mounting bolts. They ultimately shorten the cylinder's service life. Therefore, reducing the side loads imposed on the cylinder is essential. The proper mounting arrangement of the cylinder can reduce it. The cylinder's axis must be correctly aligned with the attached load's axis. A technician must take great care when mounting the cylinder so that the load does not exert side load forces on the piston rod. Adequate bearing support on the piston and in the piston rod gland area also helps reduce the side loads on the cylinder.

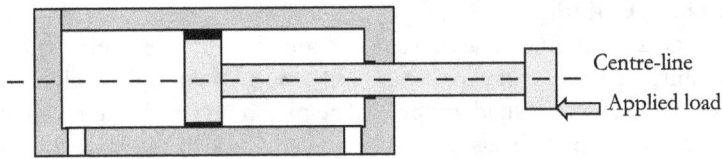

Figure 11.7 | A hydraulic cylinder with an off-center load

Cylinder Drift

Under ideal conditions, a hydraulic cylinder should hold its position when stopped. However, it tends to drift from its intended position at times. The cause of the drifting is often in the associated control valve rather than the cylinder itself. If the valve spool is fully blocked in its neutral position, the cylinder should remain in place. However, if the valve spool leaks in its neutral position, possibly due to a damaged seal, both the working ports of the valve may get connected to the system reservoir, and consequently, the cylinder may drift from its intended position.

Piston Rod Buckling

Hydraulic cylinders capable of giving very long strokes are essential for some applications. However, if a compressive axial load is to be applied to the piston rod of such a long-stroke cylinder, it must be within the safety limit to prevent the piston rod from buckling. Piston rod buckling can occur when the piston rod bends under the load. In other words, buckling can occur in the cylinder if the piston rod is not sized to match its stroke and the connected load. For example, the buckling of the piston rod is likely to occur if it is longer and thinner.

Classification of Hydraulic Actuators

Hydraulic actuators are available in a wide range of types, sizes, and models to meet various application requirements. Table 11.1 gives a broad classification of hydraulic actuators.

Table 11.1 Classification of hydraulic actuators

Main types	Sub-types	Examples
Linear actuators	Single-acting	Spring-returned
		Gravity returned
	Double-acting	Non-cushioned type
		Cushioned type
	Variants	Plunger/Ram cylinder
		Position sensing cylinder
		Double-rod-end cylinders
	Special assemblies	Tandem cylinder
		Telescopic cylinder
Rotary actuators	Semi-rotary type	Vane type
		Helical-spline Type
		Rack-&-Pinion Type
	Rotary type (Hydraulic motors)	External gear motors
		Internal gear motors
		Screw motors
		Vane motors
		Piston motors

Single-acting Hydraulic Cylinders

Figure 11.8 shows the cross-sectional view of a single-acting hydraulic cylinder. It consists of a barrel, a piston-and-rod assembly, a spring, end caps, seals, and a port. Next, a fluid chamber is formed in the cylinder by the barrel, piston, and cap-end endplate. The piston-and-rod assembly is a tight fit inside the barrel and is biased by the spring. Seals primarily prevent fluid leakage in the system. The port is integrated into its cap-end, which admits or relieves the system fluid.

Applying pressure through the port moves the piston-and-rod assembly in one direction to provide the working stroke. The piston-and-rod assembly moves in the opposite direction, either by force due to a spring, gravity, or an external force. In the cylinder with a spring-assisted retraction, the spring is designed not to carry any load but to retract the piston and piston rod assembly with sufficient speed. In the alternative design, the spring can be mounted on the piston side of the cylinder, with the port at the head end. The single-acting cylinder can work only in one direction of its motion, hence the name 'single-acting cylinder.'

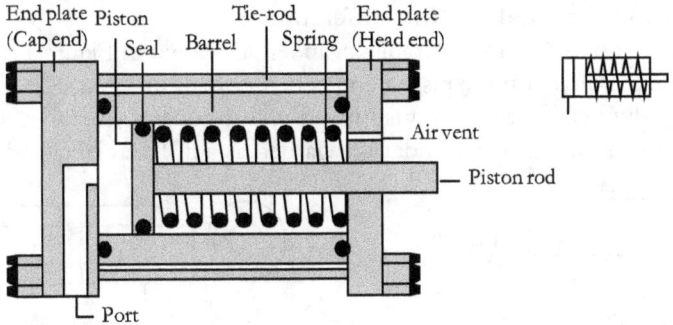

Figure 11.8 | A single-acting hydraulic cylinder

A single-acting cylinder converts system pressure and flow into mechanical force and motion. Its ease of operation makes it particularly suitable for applications such as clamping, pressing, cutting, holding, ejecting, feeding, and lifting.

Double-acting Hydraulic Cylinders

Figure 11.9 gives a cross-sectional view of a double-acting hydraulic cylinder. It consists of a barrel, a piston-and-rod assembly, end caps, seals, and two ports. The double-acting cylinder has fluid ports at both ends: the piston-side port and the piston rod-side port.

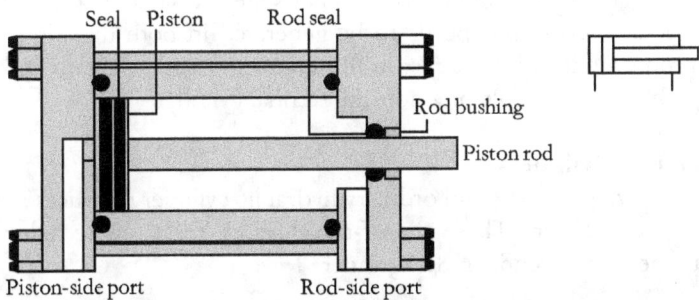

Figure 11.9 | A double-acting cylinder

Applying pressure through the piston-side port extends the cylinder, provided that the pressure on the piston-rod side is relieved. In the same way, applying pressure to the piston-rod side port retracts the cylinder, provided that the pressure on the piston side is relieved. Therefore, the cylinder can convert the pressure and flow into force and motion. A double-acting cylinder can work in both directions of its motion, hence the name 'double-acting cylinder.'

Ideally, a conventional double-acting hydraulic cylinder can be designed with an unlimited stroke length. However, the maximum stroke length is practically limited to about 6.5 ft due to the extended cylinder's possible bending and buckling with a very long piston rod.

Double-acting hydraulic cylinders are the most common type of cylinder in modern industries. They are used in applications involving linear mechanical motions. They offer high pushing and pulling forces. They perform well in applications, especially where precise low-speed control is required. They are designed for heavy-duty lifting and positioning in mobile applications, industrial production, and assembly jobs.

Hydraulic Cylinders - Differential vs. Non-differential

In a typical single-ended, double-acting hydraulic cylinder, as shown in Figure 11.10(a), the piston area (A1) exposed to the fluid contact on the piston end is higher than the area (A2) on the piston rod end. The active area of the fluid contact on the piston rod end is equal to the bore area minus the cross-sectional area of the piston rod. Such a cylinder is regarded as a differential cylinder.

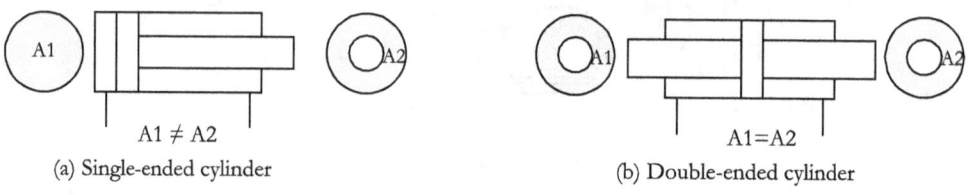

(a) Single-ended cylinder (b) Double-ended cylinder

Figure 11.10 | The basic types of hydraulic cylinders

A differential cylinder produces a greater force and lower speed when extending than when retracting, provided the pressures applied at both ends of the cylinder remain constant. A differential cylinder's area ratio (A1/A2) depends on its construction, application, and stroke length. For standard applications, the area ratio is typically about 6:5. For heavy-duty applications with large-size piston rods, an area ratio of about 2:1½ may be selected.

The fluid contact areas on the piston and piston rod ends are the same in a non-differential hydraulic cylinder. This allows equal forces and speeds to be generated in both directions of piston travel. A double-ended piston rod cylinder with the piston rod extending through both ends of the cylinder, as shown in Figure 11.10(b), is an example of a non-differential cylinder.

Cushioning in Hydraulic Cylinders

The load-attached fast-moving piston of an ordinary hydraulic cylinder produces shock loads or impact forces when it strikes its end covers. These end-of-travel shock loads can be reduced by incorporating cushioning devices at one or both ends of the cylinder.

The cushioning devices work by decelerating the piston-and-rod assembly as it approaches its end-of-stroke position, thereby preventing excessive mechanical stresses. The cushions may be either fixed or adjustable. Hydraulic cylinders can be designed with simple or adjustable cushions to operate at higher speeds initially and lower speeds at the end-of-stroke position. However, non-cushioned cylinders are appropriate for full-stroke operation at low speeds.

Hydraulic Cushion Cylinder

Figure 11.11 shows a cross-sectional view of a double-acting hydraulic cylinder with adjustable cushioning devices. Each cushioning device comprises a throttle valve integrated into the cylinder's end caps and cushion sleeves attached to its piston. When the piston starts moving forward, the entire system flow exits unrestricted through the large orifice in the fluid chamber at the head end of the cylinder. As the cushion sleeve enters its chamber, the fluid's normal exit path is blocked. This blockage forces the flow through the throttle valve, restricting it and progressively slowing piston movement. The piston continues to move to its end-of-stroke position at a controlled speed without damaging the cylinder. The cushioning can be adjusted by using the adjusting screw. Typically, the cap-end cushion functions the same way as the head-end cushion. The cushions are usually designed to operate over the final ¾ inches of the piston stroke.

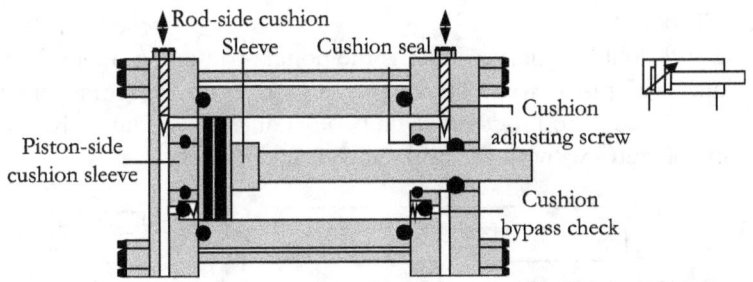

Figure 11.11 | An adjustable cushion hydraulic cylinder

A bypass check valve can be incorporated into the cushioning device to allow an unrestricted entry of the system fluid into the cylinder. There is always some pressure intensification on the piston rod end, but this is usually not a significant problem. With the development of enormous forces and high accelerations in a load-coupled cylinder piston, additional measures, such as an external shock absorber, must be taken to assist in load deceleration.

Ram (Plunger) Cylinders
The ram (plunger) cylinder is a variant of the standard single-acting hydraulic cylinder. Figure 11.12 shows a cross-sectional view of the ram cylinder.

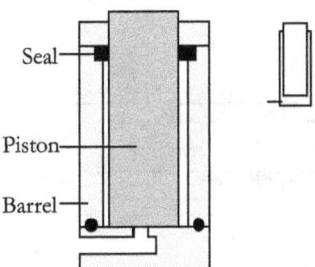

Figure 11.12 | A Plunger/Ram cylinder

It is designed with a robust and thick piston rod (ram) whose diameter is almost equal to that of the cylinder piston. Most ram cylinders do not use return springs. Instead, they use gravity or the loads attached to them to retract the piston rods. The piston and the piston rod in a ram cylinder must be precisely cylindrical and finely finished for smooth output motion. A surface-hardened or chrome-plated piston rod is often used with an ultra-fine surface finish to ensure the long service life of its seal elements.

Ram cylinders are usually upright and most often used for short-stroke applications. They are primarily used for push operations rather than pull operations. A ram cylinder with a hollow rod can be used for both pushing and pulling operations.

Piston rod bending in a basic long-stroke horizontal cylinder unit, or buckling in the basic cylinder under a high vertical load, can be prevented by replacing the primary cylinder with a ram cylinder. Ram cylinders are commonly used in high-pressure applications, such as pressing and stamping machines and lifts in automobile service stations.

Double-rod-end Cylinders

A variant of the standard double-acting cylinder is the double-rod-end cylinder. Figure 11.13 shows a double-rod-end cylinder with piston rods extending from both ends. This arrangement provides better rod alignment, as the attached piston rods move on two bearings. The cylinder has equal areas on both sides of the piston and operates without the differential cylinder effect.

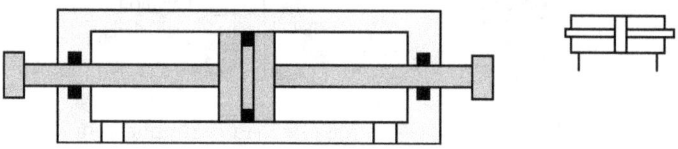

Figure 11.13 | A double-rod-end hydraulic cylinder

A double-rod-end cylinder can perform two functions at either end in staggered cycles. It is also used when the piston speed must be the same in both directions of its motion. Further, it is useful in a robotic mechanism in which the piston rods are clamped at both ends of the cylinder, with the body moving instead.

Tandem Cylinder

In a tandem-type cylinder unit, two (or more) cylinders are assembled in-line, with the first cylinder's piston rod attached to the second cylinder's piston, and so on. Figure 11.14 shows the cross-sectional view of the two-stage tandem cylinder.

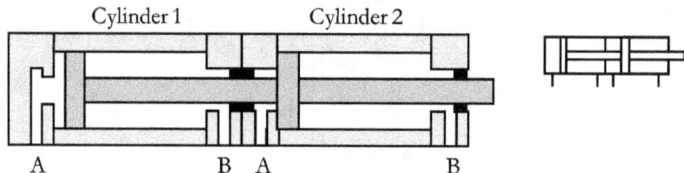

Figure 11.14 | A tandem cylinder

When the pressurized fluid is applied to the piston-side ports (marked as 'A'), the cylinder extends with a force twice as much as the force produced by a single-stage cylinder of the same bore size while extending. In the same way, when the pressurized fluid is applied to the piston-rod-side ports (marked as B), the cylinder retracts with a force that is twice as much as the force produced by a single-stage cylinder of the same piston and piston rod sizes while retracting. The tandem cylinder is suitable for an application where a high output force must be developed within a narrow radial space, but there is a substantial axial length.

Telescopic Cylinders

A multi-stage telescopic cylinder has multiple cylinder bodies nested within one another. That is, the piston rod of the first stage is used as the piston barrel of the second stage, and the second piston rod is used inside the barrel of the second stage. Similarly, there can be up to six stages in the cylinder. Therefore, the total stroke length of the telescopic cylinder can be up to six times the stroke length of the basic cylinder. Remember that the output force is highest in the first stage and decreases in subsequent stages. The telescopic cylinders are constructed of single-acting and double-acting varieties.

Single-acting Telescopic Cylinder

Figure 11.15(a) shows the single-acting telescopic cylinder. It is a multi-stage cylinder with two to six concentric tubular envelopes. Most telescopic cylinders are single-acting, meaning the fluid pressure always acts in one direction—on the extended stroke, to be more precise.

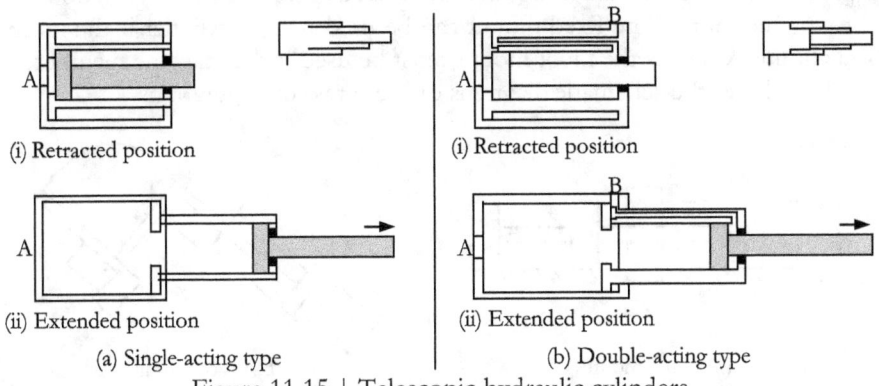

Figure 11.15 | Telescopic hydraulic cylinders

Double-acting Telescopic Cylinder

Figure 11.15(b) shows the double-acting telescopic cylinder. In this telescopic-cylinder type, the system pressure alternately extends and retracts the cylinder. Double-acting telescopic cylinders are highly complicated and must be specially designed and manufactured with a high degree of precision. Therefore, they are much more expensive than regular hydraulic cylinders.

Telescopic cylinders are ideal for applications that require long-stroke cylinders in a space-constrained environment. They are widely used in hydraulic equipment for the agriculture, construction, and heavy engineering industries. They are commonly used in mobile hydraulic systems for tilting truck dump bodies and forklifts, lifting hydraulic cranes, and material handling.

Installation of Hydraulic Cylinders

The two crucial aspects of the design and maintenance of hydraulic systems are mounting a cylinder body to its base and coupling the piston rod to the machine's working part—any mismatch in either results in stresses on the cylinder, reducing its service life. When a force is transmitted to the machine, bearing stresses arise at the cylinder barrel and piston rod, resulting in high edge pressures on the bearing bushes and the piston rod guide bearings. Moreover, increased and uneven stresses may develop in the piston and piston rod seals due to the high lateral side loads acting across their axes.

A hydraulic cylinder should be fixed so that the side loads on the piston rod bearing are minimal. Good engineering practices can reduce them to an acceptable limit. Slide or roller guides can be used to carry the load wherever possible.

The mounting dimensions and methods for mounting hydraulic cylinders must meet the requirements of the relevant ISO/NFPA standards. Cylinder mounting methods should be studied to determine the best approach for optimal installation.

Mounting Methods of Hydraulic Cylinders

A hydraulic cylinder can be mounted to a machine using a permanent or movable mounting.

For example, the body and piston rod of a hydraulic cylinder can be rigidly fastened to the machine, as in the fixed type mounting, or they can be allowed to swivel as part of the linkage in one or more planes, as in the pivot type mounting. The fixed mount can be used for the cylinder if the machine member moves in a straight line, whereas the pivot mount must be used if the machine member must move in an arc. Figure 11.16 shows the schematic diagrams of these two configurations.

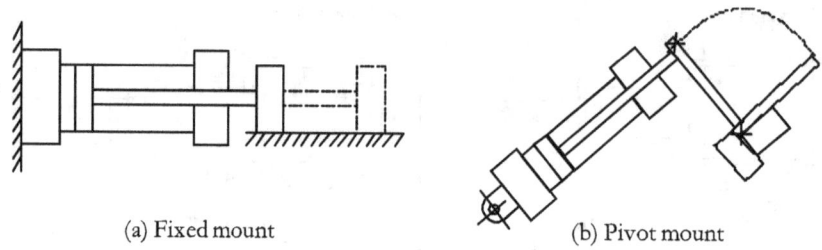

(a) Fixed mount   (b) Pivot mount

Figure 11.16 | Two important mounting arrangements for cylinders

The selection of a mounting method for a hydraulic cylinder depends primarily on the specific application. Factors such as the cylinder stroke length, piston rod diameter, method of connection to the load, and the presence of shock pressures must be considered when selecting the cylinder mounting style.

Mounting Styles of Hydraulic Cylinders

Hydraulic cylinder mounts are available in various configurations. Typical examples of fixed mountings include tie-rod, flange, foot, or bolt mounts. The pivot mounts can be head, cap, center-trunnion, swivel-flange, or rear-cap. Figure 11.17(a) depicts several critical mounting methods for the cylinder body.

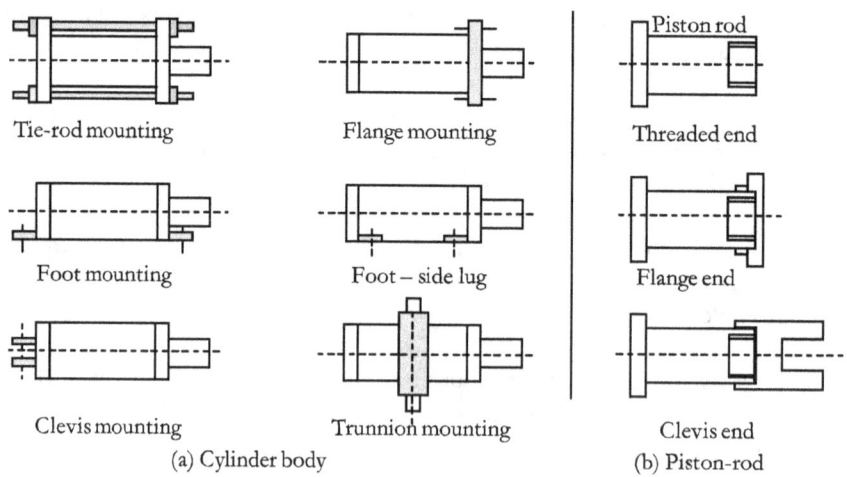

Figure 11.17 | Mounting arrangements of cylinders

Tie-rod Mount
In this type, the extended tie-rods of a hydraulic cylinder are used to mount the cylinder. The tie-rods can be extended on either the piston or the rod side. The free end of the cylinder should be supported to prevent its misalignment or sag. The best application of the tie-rods extended on the piston side is with a thrust load (i.e., with the piston rod in compression), and that of the tie-rods extended on the piston rod side is with a tension load.

Flange Mount
This type of mount provides a rigid mounting with no scope for misalignment. The flange mount is best for installing a cylinder when a load force acts along its centerline. A cap-end flange in the cylinder is best used in a thrust load application, while piston rod-end flange mounts are best used in tension load applications.

Foot or Lug Mount
The foot-mounted or lug-mounted hydraulic cylinder provides relatively rigid mounting. This type of cylinder mounting can tolerate some amount of misalignment.

Pin-and-Trunnion Mounts
A pin-and-trunnion-mounted hydraulic cylinder requires provisions for pivoting at both ends. It is designed to carry shear loads. If the load path is curved or misalignment is a problem, a pivoted centerline mounting should be used. These mounts compensate for non-linear travel in only one plane.

Piston-rod Mounts
Various coupling accessories are used for hydraulic cylinder piston rods, including pinholes, piston rod-end threads, rod clevises, eye brackets, knuckles, and pivot pins. Figure 11.17(b) shows some coupling arrangements for piston rods.

Threads
The essential threads used are the SI system Thread, British Standard Pipe Thread, National Pipe Thread, and Unified Fine Thread.

Advantages of Hydraulic Cylinders
Some key advantages of hydraulic cylinders include their high power density, ruggedness, wide range of stroke options, and cost-effectiveness.

Power Density: A hydraulic cylinder can generate significant force despite its small size, making it suitable for modern machinery with limited space.

Ruggedness: Hydraulic cylinders are designed to withstand extremely tough overload conditions with rugged construction.

Range of Strokes: Small industrial hydraulic cylinders, such as clamping cylinders, may have a maximum stroke of ¼ inch. Large hydraulic cylinders in the heavy equipment industry can have power strokes up to 6.5 ft. Telescopic cylinders can obtain very long stroke lengths.

Economy: Cylinders can be easily manufactured to the precise dimensions specified by end users.

Applications Notes, Hydraulic Cylinders

Manufacturers offer a range of standard and heavy-duty cylinders suitable for various applications and harsh environments. The heavy-duty hydraulic cylinders are designed for high-pressure, high-flow, high-force applications and corrosive environments.

They are used in various engineering applications for pulling and pushing operations, especially when precise low-speed control is required.

The self-explanatory Figure 11.18 gives some typical examples of hydraulically driven mechanisms.

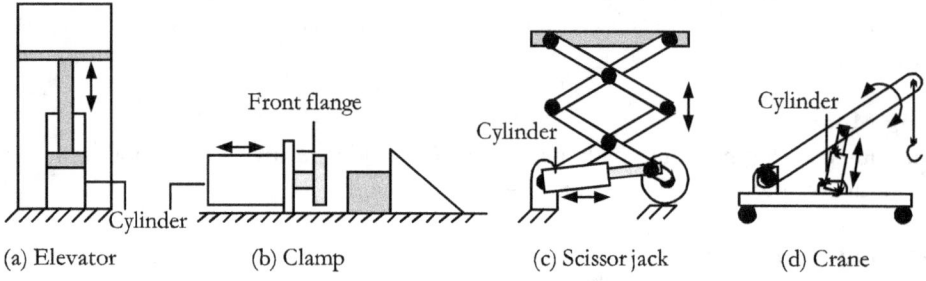

Figure 11.18 | Typical examples of hydraulically-driven mechanisms

The following list highlights some of the application areas for hydraulic cylinders:
- Mining,
- Offshore,
- Steel mills,
- Aerospace,
- Machine tools,
- Defense sector,
- Automotive plants,
- Construction sector,
- Forge-pressing machines.
- Chemical and food industries, and
- Industrial production and assembly jobs,

- Hydraulic cylinders are also used for clamping, pressing, cutting, holding, feeding, lifting, and material handling operations.

- They are also used for heavy-duty digging, hoisting, tilting, and positioning operations in mobile equipment.

Hydraulic Cylinder Standards

Hydraulic cylinders conforming to the ISO and NFPA standards are available in the market. The ISO hydraulic cylinders are manufactured in accordance with standards 6020-1, 6020-2, 6020-3, and 6022.

ISO (or NFPA) hydraulic cylinders have standard installation dimensions and are interchangeable. Table A9.2 in Appendix 9 lists some ISO standards for hydraulic cylinders.

Objective-type Questions

1. What determines the speed of a cylinder used in a hydraulic system?
   a) Fluid flow rate
   b) System pressure
   c) Size of the applied load
   d) Stroke length

2. Wear bands in hydraulic cylinders eliminate:
   a) impact forces
   b) cylinder scoring
   c) leakage
   d) pressure drop

3. Which type of hydraulic cylinder is used to multiply force in a limited lateral space?
   a) Duplex cylinder
   b) Telescopic cylinder
   c) Tandem cylinder
   d) Ram cylinder

4. Which type of hydraulic cylinder is most suitable for harsh service conditions?
   a) Tie-rod cylinders
   b) Mill cylinders
   c) Threaded-end cylinders
   d) Welded cylinders

5. Which component converts the compressed air energy into mechanical energy in linear movement in one direction only?
   a) Hydraulic pump
   b) Hydraulic motor
   c) Single-acting cylinder
   d) Double-acting cylinder

Review Questions

1) Explain the function of a hydraulic actuator by taking a double-acting cylinder as an example.
2) Name two basic types of hydraulic actuators and differentiate them.
3) Describe the relationship between a hydraulic cylinder's fluid flow rate, piston area, and piston velocity.
4) Describe the constructional and design features of standard hydraulic cylinders briefly.
5) Explain the operational and other features of hydraulic cylinders.
6) Name the most general types of hydraulic cylinders.
7) What are the key advantages of the welded style hydraulic cylinders?
8) Explain the purpose of the piston rod boot in a hydraulic cylinder.
9) What are the different seal materials used in hydraulic actuators?
10) Name four different types of mounting arrangements for hydraulic cylinders.
11) Give a brief account of the commonly used hydraulic actuators.
12) What is the difference between the single-acting and the double-acting hydraulic cylinder?
13) What is a double-rod cylinder? When is it used?

14) What is the cylinder cushioning? Explain with a diagram.
15) Why are end cushions used in hydraulic cylinders?
16) Explain the working of a telescopic hydraulic cylinder with a neat diagram.
17) What is a telescopic hydraulic cylinder? When is it used?
18) Briefly explain the working of a tandem hydraulic cylinder
19) List some advantages of hydraulic cylinders.
20) List some applications of hydraulic cylinders.

Numerical Problems

1) A 5.12-inch bore hydraulic cylinder should lift a load of 161862 lb. What operating pressure is required to lift the load? [Ans: 7861 psi]
2) A 1.77-inch bore hydraulic cylinder in a construction site should lift a load. The gauge shows an operating pressure of 4931 psi. What is the weight of the load? [Ans: 12130 lb]
3) What is the speed of a hydraulic cylinder with a piston area of 20.46 in² supplied with fluid at the rate of 101.7 in3/s? [Ans: 5 in/s]
4) A single-acting hydraulic cylinder has an active area of 0.02185 ft² and a stroke of 0.328 ft. How much fluid does the cylinder require per cycle? [Ans: 0.007165 ft³]
5) At what speed will a hydraulic cylinder with an effective area of 4.65 in² move when powered by a pump with a no-load flow rate of 6.1 in³/s? [Ans: 1.31 in/s]
6) Calculate the forward speed of a double-acting hydraulic cylinder with a bore diameter of 3.281 ft and a flow rate of 6102 in³/min. [Ans: 0.084 in/s]
7) A double-acting hydraulic cylinder with a single piston-rod, used for lifting and lowering heavy loads in a marine application, must produce a thrust of 17985 lb and move out with a velocity of 0.0098 ft/s on the out-stroke. The operating pressure is 1450 psi. Calculate the cylinder's bore diameter and the required flow rate. [Ans: 3.97 in, 1.46 in³/s]
8) A single-acting hydraulic cylinder has a piston of 2.95 in diameter and is supplied with fluid at 1160 psi and a flow rate of 0.562 cfm. Calculate the thrust, velocity, and power. [Ans: 7923 lb, 2.37 in/s, 2.845 hp]
9) A double-acting hydraulic cylinder is used to reciprocate in an application. The relief valve setting is 1015 psi. The piston area is 24.955 in², and the rod area is 6.99 in². If the pump flow is 85.43 in³/s, find the speed and load-carrying capacity of the cylinder for its: (i) extension stroke and (ii) retraction stroke. [Ans: 3.42 in/s, 12.22 in/s, 25329 lb, 7095 lb]
10) A hydraulic cylinder has a bore dia of 3.1496 in and a piston-rod diameter of 1.5748 in. If the cylinder receives a flow of 101.7 in³/s at 1740 psi pressure, find: a) extension as well as retraction speeds, and b) extension as well as retraction load-carrying capacities of the cylinder. [Ans: 13.06 in/s, 26.07 in/s, 13555 lb, 6786 lb]
11) A 3-in-diameter hydraulic cylinder has a 1.5-in-diameter piston-rod. What is the flow rate leaving through the piston-rod port of the extending cylinder when the flow rate that enters its piston port is 8 gpm? [Ans: 32 gpm]
12) A 3¼-inch-diameter hydraulic double-acting cylinder has a 1¾ diameter rod. If the cylinder receives flow at 30 gpm/115.5 in³/s and 1740 psi, find the: (1) thrust and pull forces produced by the cylinder and (2) extension and retraction speeds of the cylinder, and compare. [Ans: 14442 lb, 1.16 ft/s, 1.63 ft/s]

Objective-type questions - answer key: *1-a, 2-b, 3-c, 4-b, 5-c*

Chapter 12 | Hydraulic Motors

Hydraulic rotary actuators are the muscles behind the rotary motions in industrial and mobile hydraulic systems. They are positive-displacement devices that convert hydraulic energy into rotary mechanical energy when supplied with a pressurized fluid. A rotary actuator converts the system pressure and flow to a controllable rotary force (torque), rotary motion, or both. Here, the fluid pressure is converted to torque, and the flow rate to rotary speed. The rotary actuator's torque and motion can be used to obtain the rotary operation in industrial machinery.

Figure 12.1 | A graphic representation of a hydraulic motor*
Courtesy: Penton Business Media, Inc., U.S.A.
*Note: "The graphic is the copyrighted property of and is reprinted with the permission of Penton Business Media, Inc."

Hydraulic rotary actuators can be classified into two types. They are: (1) Semi-rotary actuators and (2) Hydraulic motors. A semi-rotary actuator (or oscillating motor) can produce only limited rotation and twist objects along a partial arc. On the other hand, a hydraulic motor can produce continuous rotation and impart continuous rotary motion to the connected load. Figure 12.1 shows the cut-section view of a hydraulic motor.

Basic Motor Operation

A hydraulic motor mainly consists of a set of moving elements, such as gears, vanes, or pistons, connected to the motor's output shaft and enclosed in a single housing. The shaft rotates when the pressurized system fluid is applied to the motor's rotating parts. In this way, the motor can convert the applied pressure into rotary mechanical force, thereby driving the load attached to it. The fluid returns to the system reservoir after passing through the motor.

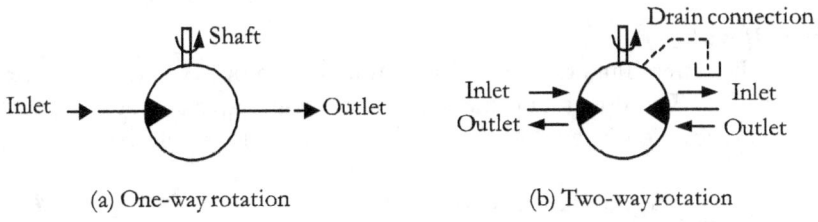

Figure 12.2 | Symbolic diagrams of hydraulic motors

It may be noted that leakage fluid can accumulate in the pockets of the shaft seal and bearing. When the fluid is exposed to high pressure, it tends to dislodge the seals. Therefore, a drain connection is provided in the motor to return the leakage fluid to the reservoir. A small amount of internal leakage is deliberately allowed in some motors to lubricate and cool internal parts.

Figures 12.2(a) and 12.2(b) show the symbolic diagrams of a unidirectional and bidirectional hydraulic motor, respectively. The symbols are necessarily circles with triangles inside. The triangle points inward to show the fluid sinking into the motor. The unidirectional hydraulic motor provides rotation in only one direction, whereas the bidirectional hydraulic motor can rotate in both clockwise and anticlockwise directions. The direction of rotation of the motor's shaft can easily be reversed by changing the direction of the fluid flow through the motor ports.

Difference Between Hydraulic Pumps and Motors

Hydraulic motors are very similar to hydraulic pumps in design and construction. However, they have many differences.

The main difference between a pump and a hydraulic motor is that, in a pump, the moving parts, connected to a hydraulic system, push the system fluid to create flow and pressure. In contrast, in a motor, the pressurized fluid pushes its moving elements to produce rotary mechanical motion and force.

Another difference is that the pump is always coupled to its prime mover, whereas the motor is always coupled to a load.

Hydraulic pumps for general industrial applications are usually built for a specific rotation direction, whereas most hydraulic motors must reverse that direction.

Terms and Definitions - Hydraulic Motor

Some critical factors relevant to the operation and applications of every hydraulic motor are its operating pressure, displacement, flow rate, input power, output power, torque output, and efficiency.

Operating Pressure (P)

It is the pressure in a hydraulic system that overcomes all resistances, including useful work and losses. The rated pressure of a hydraulic motor is the maximum pressure the manufacturer recommends for the motor.

Motor Displacement (V_D)

It refers to the volume of the system fluid required to turn the motor's output shaft through one revolution. Some units of motor displacement are m^3/rev, lit/rev, cc/rev, or in^3/rev.

Theoretical Flow Rate (Q_T)

It is the quantity of the system fluid that must flow through a motor per unit of time, provided there is no leakage in the system. The flow rate is measured in cubic-inch per minute (in^3/min) or gpm. The equation for the theoretical flow rate (Q_T) of the hydraulic motor is as follows:

$$Q_T(gpm) = \frac{V_D(in^3/rev) \times N(rpm)}{231}$$

Slippage in Hydraulic Motors

It is the internal leakage of the system fluid that passes through unintended paths within the motor without performing any useful work. As slippage in the hydraulic motor increases, more of the available flow intended for useful work is lost, resulting in a loss of power.

Speed, Hydraulic Motor

It is directly related to the theoretical flow rate of a motor and inversely related to the motor's displacement. It can be expressed by the following equation:

$$\text{Speed, N (rpm)} = \frac{\text{Theoretical flow to the motor, } Q_T \text{ (gpm)} \times 231}{\text{Motor displacement, } V_D \text{ (in}^3/\text{rev)}}$$

Therefore, it can be observed that, for a given flow rate, increasing the motor displacement decreases the motor speed and vice versa.

Maximum motor speed is the speed of a hydraulic motor at a particular inlet pressure that it can sustain for a limited period without damage to the motor.

Minimum motor speed is the slowest, continuous, rotational speed obtainable from the output shaft of a hydraulic motor.

Theoretical Torque (T_T), Hydraulic Motor

The theoretical torque of a hydraulic motor is a function of the motor's displacement and the differential pressure across the motor. The following section presents the equation for the motor's theoretical torque. The theoretical figures represent the torque at the motor shaft, assuming no mechanical losses.

$$\text{Theoretical Torque, } T_T \text{(in.lb)} = \frac{V_D \text{(in}^3/\text{rev)} \times \Delta P \text{(psi)}}{2\Pi}$$

The **breakaway (Starting) torque** of a hydraulic motor is the rotary force required to turn a stationary load connected to the motor. More torque is required to turn the stationary load than to keep it moving. This is because the motor must initially overcome the load's inertia. Therefore, the motor needs a breakaway (starting) torque large enough to turn the load.

- **Running torque** of a hydraulic motor refers to the torque required to run a load connected to the motor. Remember, the hydraulic motor's running torque changes whenever the associated system pressure varies.

- **Stalling torque** of a running hydraulic motor is the torque needed to stop the motor from reaching a standstill.

Torque ripple of a hydraulic motor is the difference between the minimum torque and maximum torque delivered by the motor at a given pressure during its one cycle of rotation.

Actual Torque (T_A), Hydraulic Motor

It is the torque a motor develops to drive the attached load. It equals the theoretical torque minus the frictional torque losses in the motor.

Input Power (P_{in}), Hydraulic Motor

The equation for the input power of the hydraulic motor is as follows:

$$\text{Input Horse Power (hp)} = \frac{P(\text{psi}) \times Q_A(\text{gpm})}{1714}$$

Output Power (P_{out}), Hydraulic Motor

The equation for the output power of a hydraulic motor is as follows:

$$\text{Output Horse Power (hp)} = \frac{T_A(\text{in.lb}) \times N(\text{rpm})}{63025}$$

Motor Efficiency

The efficiency of a hydraulic motor is the ratio of its output power to its input power. An ideal hydraulic motor has no leakage and frictional losses and is 100% efficient. In practice, however, leakages and frictional losses occur in the motor. Accordingly, two basic types of efficiencies are identified for the motor. They are: (1) Volumetric efficiency and (2) Mechanical efficiency. Overall efficiency can be derived from these two types of efficiencies. Generally, mechanical friction and flow losses in motors and pumps amount to 10 to 25% of the input power. All power losses are converted to heat in the fluid.

Volumetric Efficiency (η_v) of the hydraulic motor is the ratio of the theoretical flow rate responsible for developing the actual motor speed to the total flow rate consumed by the motor, including the leakage in the motor. Remember, the motor theoretically consumes more flow than it should due to leakage. The equation for the volumetric efficiency of the motor is as follows:

$$\text{Volumetric efficiency, } (\eta_v) = \frac{\text{Theoretical flow rate } (Q_T)}{\text{Actual flow rate } (Q_A)}$$

Mechanical Efficiency (η_m) of the hydraulic motor is the ratio of the actual torque delivered by the motor to the theoretical torque of the motor. The hydraulic motor produces less torque than it should, due to frictional losses within the motor. The equation for the mechanical efficiency of the hydraulic motor is as follows:

$$\text{Mechanical efficiency, } (\eta_m) = \frac{\text{Actual torque, } (T_A)}{\text{Theoretical torque } (T_T)}$$

Overall Efficiency (η_o) of the hydraulic motor is the ratio of the 'brake' power delivered by the motor to the hydraulic power delivered to the motor. It is also the product of its volumetric efficiency and its mechanical efficiency and is expressed mathematically as:

$$\text{Overall efficiency, } (\eta_o) = \frac{\text{Brake power delivered by motor}}{\text{Hydraulic power delivered to the motor}}$$

$$= \eta_v \times \eta_m$$

Summary of Relations for Hydraulic Motors

Figure 12.3 summarizes the important relationships of hydraulic motors to help the reader easily understand and correlate them.

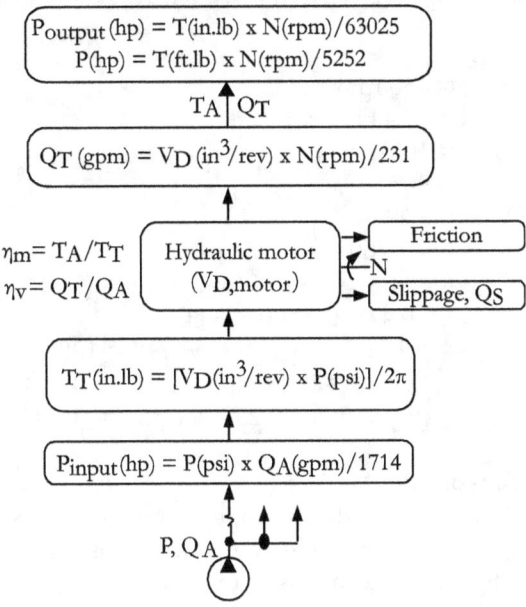

Figure 12.3 | Summary of relations for a hydraulic motor

Example 12.1 | A skid steer broom used to clean a construction site has a hydraulic motor with a displacement of 3.2 in³/rev and operates at 3000 psi. What is the theoretical torque the motor is capable of producing?

Solution

Volumetric displacement, V_D = 3.2 in³/rev
Pressure, P = 3000 psi
Theoretical Torque, T_T = V_D (in³/rev) x P (psi) / (2π) in.lb
= 3.2 x 3000 / (2π) = 1527.89 in.lb

Example 12.2 | A hydraulic gear motor consumes 11.7 gpm while running at 500 rpm. Assume the motor's volumetric efficiency to be 90%. What is the volumetric displacement of the motor?

Solution

Actual flow rate, Q_A = 11.7 gpm
Motor speed, N = 500 rpm
Motor speed, n = 8.33 rps
Volumetric efficiency, η_v = 0.9

Theoretical flow rate, Q_T = Q_A x η_v = 11.7 x 0.9 = 10.53 gpm
Volumetric displacement, V_D = Q_T (gpm) x 231 /N (rpm) = 10.53 x 231 / 500 = 4.86 in³/rev

Example 12.3 | A hydraulic motor rotates at a speed of 450 rpm with a nominal displacement of 0.244 in³/rev. The pressure differential across the hydraulic motor is 1088 psi. The overall efficiency is 80%, and the volumetric efficiency is 90%. Calculate the following: (1) Theoretical flow rate, (2) Actual flow rate, (3) Power input, (4) Shaft power, and (5) Shaft torque.

Solution

Speed, N = 450 rpm
Displacement, V_D = 0.244 in³/rev
ΔP = 1087.785 psi
η_v = 90%
η_o = 80%
Speed, n = 450/60 rps = 7.5 rps
Theoretical flow rate, Q_T = V_D (in³/rev) x N (rpm)/231 = (0.244 x 450)/231 = 0.475 gpm
Actual flow rate, $Q_A = Q_T / \eta_v$ = 0.475/0.9 = 0.528 gpm
Power input P_{in} = ΔP (psi) x Q_A (gpm) /1714 = 1088 x 0.478/1714 = 0.3 hp
Shaft power, P_{out} = P_{in} x η_o = 0.3 x 0.8 = 0.243 hp
Shaft torque, T = (P_{out} (hp) x 63025)/N (rpm) = (0.243x63025)/450 = 34 in.lb

Constructional Features of Hydraulic Motors

The general constructional features of hydraulic motors are similar to those of hydraulic pumps. However, hydraulic motors require some design modifications due to their unique application requirements. For example, a hydraulic motor has to overcome high starting torque at low speeds and the effects of side loading.

The nature of applications usually dictates the materials used in the construction of hydraulic motors. These include cast iron, ductile iron, bronze, cast steel, and stainless steel. Unique features of hydraulic motors may include rotary seals, drain connections, integrated flushing valves, and integrated brake valves.

Rotary Seals: A rotary seal is a critical element of a hydraulic motor that prevents leakage of system fluid through any clearances between the motor's mating parts.

Drain Connections: Every hydraulic piston motor has case drains with ports to drain the leakage fluid to the associated system reservoir. A case drain line must be of adequate size and connected directly from the drain port to the reservoir, without any restrictions. The line must be piped to prevent siphoning and terminate below the minimum fluid level in the reservoir. The normal pressure of the leakage fluid may not exceed 10 psi, and pressure surges at the case drain connection may not exceed 25 psi.

Side Loads on Hydraulic Motors

The performance of hydraulic motors is affected by thrust and side loads. A thrust load occurs in a hydraulic motor when a compressive load acts along the longitudinal axis of the motor shaft. A side load occurs in the motor when it is coupled to the load through a pulley or gear system, or when its shaft bears the weight of the attached load. A crucial consideration during motor operation is to keep thrust and sideload to a minimum, as these loads affect the service life of the bearings and seals. This adverse condition can lead to eventual shaft breakage.

Mounting of Hydraulic Motors

Side loads on a hydraulic motor can cause excessive wear of its parts. Therefore, the motor should be mounted to avoid side loads on its output shaft. When side loads on the motor are unavoidable, it is necessary to support the motor's output shaft and the attached load with auxiliary bearings. The coupling flange must be correctly aligned to the motor shaft. A flexible shaft coupling can be used whenever possible to avoid side loads caused by shaft misalignment.

Classification of Rotary Actuators

Hydraulic rotary actuators can be classified as semi-rotary actuators or motors based on the extent of rotation. They are also classified by several relevant parameters, including the type of internal moving elements, the nature of displacement, and the torque/speed requirements of hydraulic rotary actuators. The following sections elaborate on these classifications.

Based on the type of their internal moving elements, hydraulic motors are classified as: (1) Gear motors, (2) Vane motors, and (3) Piston motors.

Based on the nature of displacement, hydraulic motors are divided into fixed-displacement and variable-displacement types.

The fixed-displacement motor displaces a fixed amount of system fluid with each revolution. Its displacement cannot be varied except by changing the flow rate of the system fluid. A fixed-displacement motor typically produces a constant torque. Gear, vane, and piston motors can be designed for the fixed-displacement operation.

The variable-displacement motor is equipped with an adjustment mechanism that changes its fluid displacement per revolution. This adjustment mechanism allows the motor speed to be adjusted from zero to its maximum, keeping the pump's delivery constant. The motor's torque can also be varied by varying its displacement. With the motor's input flow and operating pressure held constant, varying its displacement can adjust the torque-to-speed ratio to suit the load requirements. Only vane and piston motors can be designed for variable-displacement operation.

Based on their torque-speed characteristics, hydraulic motors can be classified into two basic types: high-speed, low-torque (HSLT) and low-speed, high-torque (LSHT).

LSHT motors can transmit high torque despite their relatively small envelopes. The growing need for hydraulic motors capable of driving high-inertia loads at low speeds led to the expansion of LSHT hydraulic motor technology. A typical low-speed hydraulic motor can operate at 0.1 to 1000 rpm and provide a starting torque of 75% to 90% of its maximum torque. On the other hand, a typical high-speed hydraulic motor can operate at 1000 to 5000 rpm, but it typically lacks high starting torque.

Semi-rotary Hydraulic Actuators

A semi-rotary hydraulic actuator is a device that rotates its shaft using pressurized fluid through a fixed arc, usually less than 360°. The actuator's output torque varies directly with the fluid pressure. Semi-rotary hydraulic actuators can perform various operations, including turning, tilting, transferring, indexing, mixing, clamping, feeding, and agitating.

Vane Type Semi-rotary Actuator

Figure 12.4 shows the schematic diagram of a vane type semi-rotary actuator. It consists of one or more vanes enclosed in a cylindrical chamber, with a block integral to its casing for separating the high-pressure side from the low-pressure side, and with necessary inlet and outlet ports. The vanes are attached to the actuator shaft.

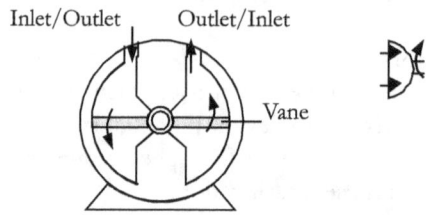

Figure 12.4 | A vane type semi-rotary actuator

When the fluid enters at one port, the vane is pushed in that direction, and the driveshaft turns in the same direction. When the fluid enters the second port, the vane is pushed in the opposite direction, and consequently, the driveshaft turns in the opposite direction.

Rack-and-Pinion Type Semi-rotary Actuator

Figure 12.5 shows the schematic diagram of the rack-and-pinion type semi-rotary hydraulic actuator. It comprises a rack-and-pinion gear assembly and a cylinder part enclosed in a shared housing. Further, the pinion gear is attached to the actuator's driveshaft.

The system fluid entering at one port of the actuator and exiting at the other port pushes the rack across the pinion gear. This action causes the shaft to rotate through a given arc in one direction. Reversing the flow reverses the shaft's direction of rotation. The torque produced in the actuator is a function of the piston area (A), the input pressure (P), and the radius of the pinion gear (R).

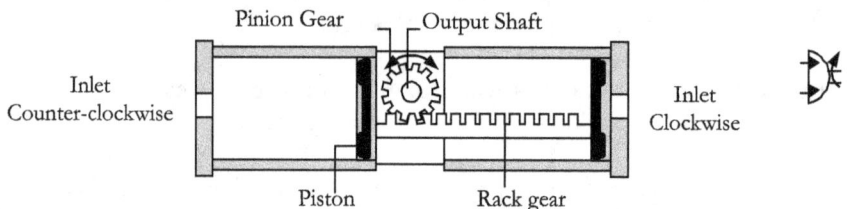

Figure 12.5 | A rack & pinion type semi-rotary actuator

The rack-and-pinion semi-rotary actuator, available in single- and dual-cylinder designs, may be equipped with cushions and stroke limiters for smooth, adjustable stops. Its rotary motion can be used for turning, tilting, indexing, mixing, and clamping.

Hydraulic Motors

A hydraulic motor is a positive-displacement device that continuously turns its rotating elements and shaft using the pressurized fluid in the associated system. The motor shaft can be directly connected to the load or through clutches and gears. The following sections provide details on the gear, vane, and piston motor.

Gear Motors

Gear motors are positive-displacement devices that produce torque by allowing fluid pressure to act on their gears. By design, they are always fixed-displacement motors. The constructional features of gear motors are similar to those of gear pumps, but their operation is the opposite. That is, every gear pump pushes the system fluid to create pressure, whereas the pressurized fluid pushes every gear motor to develop torque. Two types of gear motors have extensive applications in various machines. They are: (1) external-gear motors and (2) internal-gear motors.

Gear motors have many advantages and some disadvantages. They are generally the least expensive type of hydraulic motor. They can tolerate contamination in their fluid media better than other types of hydraulic motors. However, their efficiency tends to decrease at lower speeds.

External-gear Motor

Figure 12.6 shows the schematic diagram of an external gear motor. It consists of a pair of matched gears enclosed in a housing, an inlet port, an outlet port, and a driveshaft. One gear unit is connected to the motor's output shaft, and the other is an idler. System fluid enters the housing through the inlet port. The fluid follows the periphery of the housing and forces the gears to rotate. Finally, it exits through the outlet port at a low pressure. The motor's rotary force is available at its output shaft for useful work. The tight-fitting gears in the housing help control the motor's internal fluid leakage and increase the motor's volumetric efficiency.

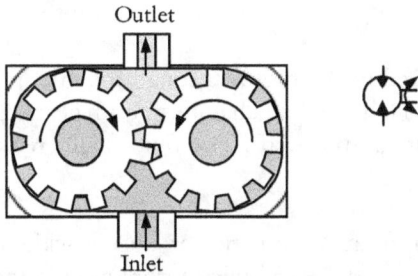

Figure 12.6 | An external gear motor

External gear motors are compact and less expensive than piston and vane motors. They work best in high-speed operations and are suited for bi-directional operation. The slippage losses in them are reasonably uniform for speeds over 500 rpm. Moreover, they can deliver relatively constant torques. However, they are the noisiest and the least efficient of the three types of hydraulic motors.

Gear motors are usually built for high-speed, low-torque (HSLT) outputs. They are suitable for applications with light starting loads but high running loads. They are used as drives in many types of agricultural, construction, and mining equipment.

Gerotor/Geroler (LSHT) Motors

Gerotor and Geroler motors are quite similar but have some minor differences. These positive-displacement motors are internal-gear motors, classified as low-speed, high-torque (LSHT) hydraulic motors. They are commonly used in mobile hydraulics and are especially useful for generating torque for traction-intensive construction, agriculture, and forestry applications.

Gerotor Motor

As shown in Figure 12.7(a), this motor arrangement can be considered a gear-within-a-gear type motor. It consists of two sets of interlocking gears—one on the inner rotor and another on an oblong rotor ring —with inlet and outlet ports and an output shaft. A typical gerotor motor has a 7-tooth outer ring and a 6-tooth rotor, thus forming six fluid chambers. Remember, the inner rotor has one tooth fewer than the rotor ring. At any point, three consecutive fluid chambers at one end are pressurized, and the three chambers at the other end are connected to the system reservoir. The inner rotor is mounted on the motor's driveshaft and is eccentric to the rotor ring. Each tooth of the rotor gear is always in contact with the internal surface of the rotor ring. Fluid chambers are formed between the gear teeth and the housing.

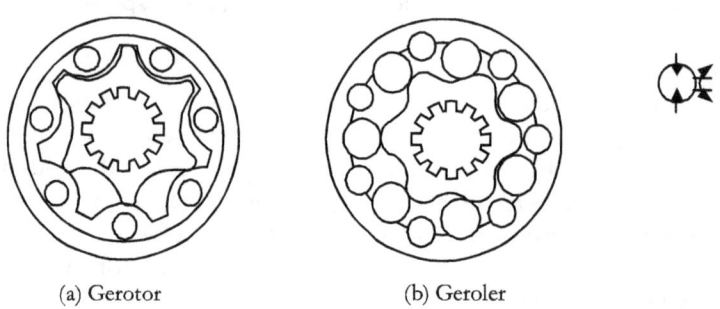

(a) Gerotor   (b) Geroler

Figure 12.7 | Internal gear motors

The high-pressure fluid is delivered to the motor through the inlet port, then flows through the fluid chambers. As the fluid passes through the fluid chambers, both gears rotate. The motor's drive coupling transmits the rotor's motion to its output shaft. The fluid finally returns to the reservoir through the outlet port.

Gerotor motors can deliver significant output power over a wide speed range. They are the most common low-speed, high-torque (LSHT) hydraulic motors used in industrial and mobile applications. They are used in agricultural and forestry equipment, construction machines, food processing machines, machine tools, lawnmowers, road rollers, excavators, and winches. Compact Gerotor motors are the natural choice for many applications, including plastic injection molding, CNC tool changer drives, conveyor drives, clamping, and drilling.

Geroler Motor

A variant of the gerotor motor is the Geroler motor. Figure 12.7(b) shows the schematic diagram of the motor. Note that the operation of the Geroler motor is similar to that of the gerotor motor. However, the motor has its teeth fitted with rollers to reduce friction. This type of motor typically provides higher torques at lower speeds. The typical features of Geroler motors include their compact yet heavy-duty construction, smooth running even at low speeds, and reduced pressure spikes.

Geroler motors are well-suited for applications requiring quick start and stop cycles and rapid speed reversals. An application requiring a speed below 100 rpm should consider a Geroler motor. Typical applications of geroler motors include agriculture and forestry equipment, construction machines, material handling, lifting gear and winches, machine tools, and many other areas of industrial and mobile hydraulics.

Vane Motor

A vane-type hydraulic motor is essentially a positive displacement device. Figure 12.8 shows the schematic diagram of the vane motor. The vane motor consists of a slotted rotor with close-fitting vanes. The rotor is mounted on the motor's driveshaft. It can move within the motor's elliptical cam ring. Springs or centrifugal forces push the vanes against the cam ring. The motor also has two ports: one serves as the inlet port and the other as the outlet port, depending on the required direction of motion. While the fluid pushes half the vanes in sequence, the other half helps the motor discharge the spent fluid through its outlet port. The motor develops the output torque by allowing fluid pressure to act on its vanes.

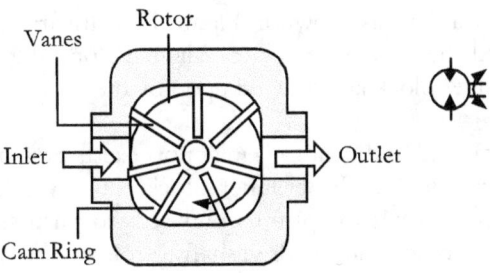

Figure 12.8 | A vane motor

Vane motors operate well at speeds ranging from 20 rpm to 6000 rpm. Moreover, they are suited for bidirectional operation. They can operate at significantly lower noise levels than other hydraulic motors. The cost of vane motors is higher than that of gear motors but less than that of piston motors of comparable power ratings.

However, vane motors are more prone to fluid contamination. They are susceptible to internal leakages, especially at low speeds. They are also less efficient than piston motors. The service life of the vane motors is shorter than that of the piston motors but longer than that of the gear motors.

Vane motors are widely used in industrial hydraulic applications. They are extensively used on machine tools, plastic molding machines, winches, and hydrostatic transmissions, and are rarely used in mobile applications other than high- and low-speed drilling.

Piston Motors

A piston motor is a positive-displacement device. It consists of a cylinder block with many pistons, a cam/swashplate, and a driveshaft.

Next, piston motors can be classified according to various parameters. According to the arrangement of the cylinder blocks relative to their driveshafts, they are classified as axial-piston or radial-piston motors. In the axial piston motor, the cylinder block is arranged axially, whereas in the radial piston motor, the cylinders are arranged radially. Further, piston motors can be classified as fixed-displacement or variable-displacement motors based on their displacement.

Piston motors deliver the highest torque, speed, and power for medium- and heavy-duty applications. They are generally the most efficient and versatile hydraulic motors, but also the most expensive.

Axial Piston Motors

An axial piston motor uses an axially-mounted piston block to generate mechanical power. Next, axial piston motors can be classified into inline and bent-axis types. They can also be either fixed-displacement or variable-displacement. The fixed-displacement axial piston motor has a stationary cam plate, whereas the variable-displacement unit has mechanical means to vary the angle of its cam plate. The motor's torque and speed depend on the cam plate angle.

In-line Axial Piston Motor

In-line piston motors are the most commonly used rotary actuators in hydraulic systems. An in-line axial piston motor consists of a cylinder block with many pistons, an angled cam/swashplate, inlet and outlet ports, and a driveshaft. The cylinders, as shown in Figure 12.9, are arranged in a circle, parallel to one another. The motor shaft and the cylinder block are aligned along the same axis. The swashplate is located at one end of the cylinder block and is acted upon by the pistons.

Applying high-pressure system fluid to the motor's inlet port exerts pressure on the ends of the cylinder pistons, which then reciprocate within the cylinder block. The cylinders are filled with fluid in a particular sequence. This action moves the pistons outwards to push sequentially against the angled swashplate, causing the cylinder block and the driveshaft to rotate. The fluid is then swept back to the system reservoir at low pressure on the piston's return stroke. The torque produced by the in-line axial piston motor is related to the swash-plate angle and the area of the pistons.

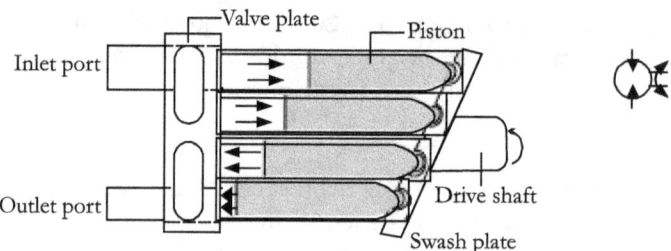

Figure 12.9 | An in-line axial piston motor

The inline axial piston motors are available in fixed-displacement and variable-displacement variants.

The swashplate angle is set in the fixed-displacement type.

In contrast, the variable-displacement type can be varied in various ways, ranging from a simple lever to sophisticated servo controls. Increasing the swashplate angle improves the motor's torque capacity but reduces the shaft speed, and vice versa. Most manufacturers recommend a minimum swashplate angle of 15° to 17° for best results. A maximum angle of 40° to 45° gives good torque output and extended motor life.

Axial-piston motors are well-known for their high volumetric efficiency at both high and low speeds. They are the most efficient hydraulic motors.

They work best in low-speed high-torque (LSHT) applications. In-line axial piston motors find applications in agricultural and construction equipment.

Bent-axis Axial Piston Motor

Figure 12.10 gives the cross-sectional view of a bent-axis piston motor. It consists of a cylinder block with pistons, a cam/swashplate, inlet and outlet ports, and a driveshaft. Unlike the case of the in-line axial piston motor design, the axis of the cylinder block and the axis of the driveshaft in the bent-axis axial piston motor design are arranged at an angle. Torque is developed in the motor in response to the system pressure acting on its reciprocating pistons. Because the shaft axis and the cylinder block axis are set at an angle, the force acting on the joint is resolved into axial and tangential components. The bearings take up the axial load while the tangential component develops the torque at the motor's shaft. The bent-axis axial piston motor has the same operating characteristics as that of the in-line axial piston motor.

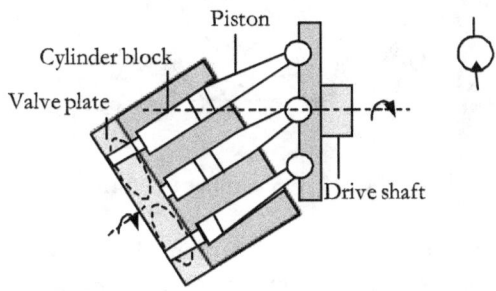

Figure 12.10 | A bent-axis axial piston motor

Bent-axis piston motors are available in fixed-displacement and variable-displacement types. Variable-displacement types can be controlled mechanically or via pressure compensation. The angle of the cylinder block with the driveshaft of a bent-axis piston motor determines its torque and speed ranges. The greater the angle, the higher the torque and the lower the motor speed.

Bent-axis piston motors are rugged and capable of handling higher operating pressures. Since the piston shoes do not slide in a bent-axis piston motor, they tend to generate less friction and higher torque for a given power rating. However, bent-axis piston motors are hefty, particularly the variable-displacement type.

They find many applications in earthmoving machines, construction equipment, forestry equipment, marine equipment, offshore equipment, industrial conveying systems, heavy-duty winches, and high-power crushers.

Radial Piston Motors

Radial piston motors have various designs within the primary radial configuration. Figure 12.11 shows the cross-sectional view of one type of radial piston motor for a hydraulic system. It consists of a block of radially arranged cylinders and pistons arranged within the housing, a thrust ring, a pintle, and a driveshaft. The cylinder block with five to seven radial bores is attached to the driveshaft. Further, each piston reciprocates in its respective radial bore. The outer piston ends bear against the thrust ring.

The pressurized fluid flowing through the pintle at the motor's center pushes the pistons outward. As a result, the pistons are pushed against the thrust ring, and the reaction forces developed in the motor rotate the barrel.

Motor displacement can be varied by shifting the cylinder block laterally. When the centerlines of the cylinder block and the housing coincide, there is no fluid flow and, therefore, no barrel rotation. The motor's rotational movement can be reversed by moving the cylinder block from its current side to the opposite side, past the housing center.

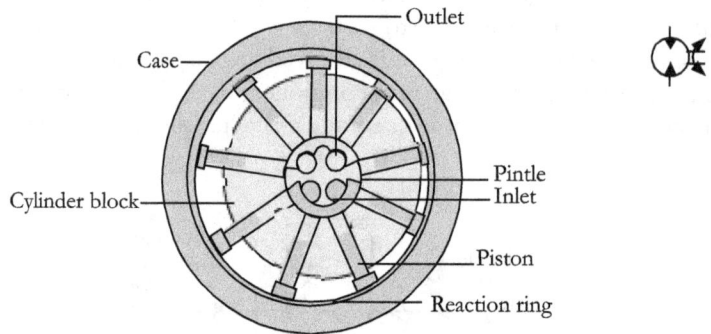

Figure 12.11 | A radial piston motor

Radial piston motors are robust, compact, and very efficient. They have excellent low-speed capabilities and long service lives and are well-suited to starting under loads. However, they have only limited high-speed capabilities and are costly.

LSHT radial piston motors drive compact machines such as skid steers and mini excavators. They are also used as wheel motors and in other suitable applications, such as forklifts.

Performance Characteristics of Hydraulic Motors
The performance of a hydraulic motor is influenced by both the associated system and motor parameters. The performance can be analyzed using different characteristics, such as the torque vs. speed, pressure vs. volumetric efficiency, and flow vs. speed curves. The following sections give these characteristics.

Torque-Speed Characteristic
Figure 12.12(a) shows the typical torque-speed characteristics of a hydraulic piston motor. Motor speed is shown on the x-axis, and torque on the y-axis. The horizontal curves correspond to the characteristics at different pressures. Sloping vertical curves represent the fluid flow rate through the motor.

Pressure-Volumetric Efficiency Curves
Figure 12.12(b) gives the pressure–volumetric efficiency curves for the gear, vane, and piston motors for a direct comparison. For different types of motors in hydraulic systems, fluid leakage through them increases, and consequently, their efficiencies decrease in varying degrees as the corresponding system pressure increases.

The volumetric efficiency of the gear motor decreases linearly with increasing pressure. The volumetric efficiency of the vane motor is typically higher than that of the gear motor. Typically, the vane motor has the highest volumetric efficiency for $p < 1000$ psi. At higher pressures, the piston motor has higher volumetric and overall efficiencies.

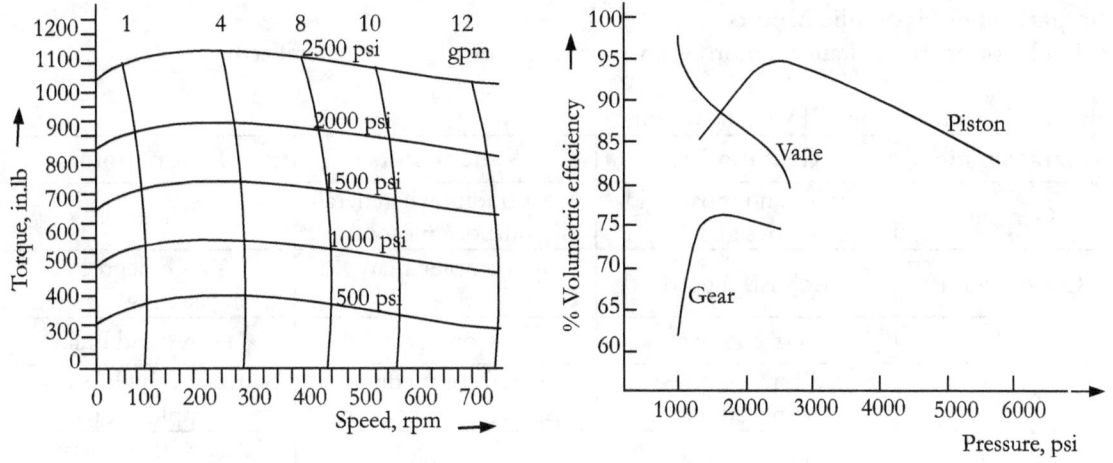

Figure 12.12 | Typical characteristics of hydraulic motors

Torque and Flow Curves Against Speed

Figure 12.13 shows the typical flow/torque/power characteristics of a hydraulic motor plotted against its speed.

The lower line shows the flow vs. speed characteristic, which shows a linear relationship between flow rate and speed. It may be noted that the slope of the line is the motor's displacement. The governing equation for the characteristic is $Q \propto V_D \times N$.

The upper line is the torque vs. speed characteristic. As shown, the motor torque tends to drop off at higher speeds.

The figure also shows the motor's output power curve as a function of speed. The governing equation for this characteristic is Power $\propto T \times N$. For a given torque, the power increases with the speed.

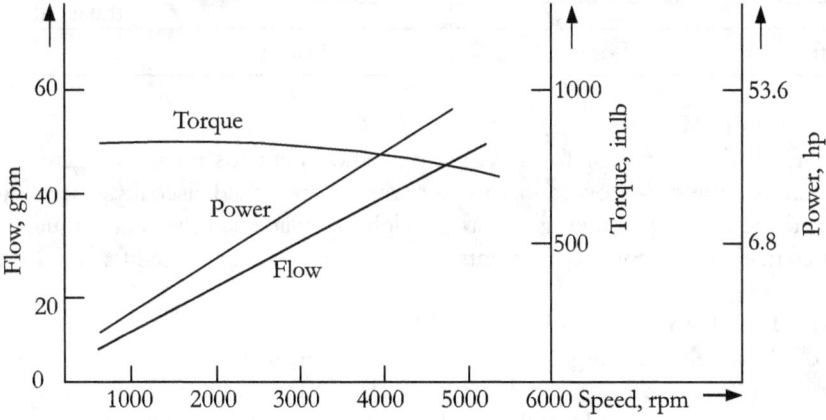

Figure 12.13 | Typical torque-speed, power-speed, and flow-speed curves of a hydraulic motor

Comparison of Hydraulic Motors

Table 12.1 compares hydraulic motors against typical parameters in a generalized way.

Table 12.1 | Comparison of hydraulic motors

Characteristics	Gear motors	Vane motors	Piston motors
General	Simple and most rigid design	Most traditional general-purpose motors	Versatile
Construction	Robust, rugged	More complex than gear types	A high degree of precision is required
Size	Very compact	Compact	Heavy and bulky
Pressure	Suited for low-pressure applications	Suited for medium-pressure applications	Suited for high-pressure applications
Speed	500 - 3000 rpm	100 – 4000 rpm	Axial: 10–4500 rpm Radial: 0.1–2000 rpm
Low-speed operation	Not well-suited	Inefficient at low speeds	Radial piston motors are an excellent choice
High-speed operation	Best suited	Right choice	Axial piston motors are an excellent choice
Duty	Light	Light/Medium	Heavy
Efficiency	Least efficient, especially at lower speed ranges	Moderately efficient, not so efficient at low speeds	Most efficient
Cost	Least expensive	Moderate	Most expensive
Slippage	Uniform slippage losses	A higher percentage of internal leakage	Provide the best sealing
Noise	Most noisy	Least noisy	Noisy
Contamination tolerance	More tolerance to fluid contamination	Less tolerance to fluid contamination	Sensitive to contamination, especially the axial piston motors
Service life	Long	Shorter	Long

Selection of Hydraulic Motors

The selection of a hydraulic motor for a given application depends on many factors. These factors include the system parameters — operating pressure range, flow, fluid viscosity, operating temperature, and noise — and the motor parameters — size, weight, displacement, speed, torque, volumetric and mechanical efficiencies, mounting requirements, cost, and estimated service life.

Advantages and Disadvantages of Hydraulic Motors

Hydraulic motors have relatively large power-to-weight ratios, which helps develop compact systems. They run smoothly at low speeds and are simple and reliable. Each can provide infinitely variable speed control and stall capability even under full load. The motors can also quickly reverse rotation.

The main issues they face while in service are seal failures, excessive leakage, and associated noise.

Applications of Hydraulic Motors
Hydraulic motors are well-suited for various applications due to their advantages. They are widely used for operations such as opening and closing, drilling, mixing, agitating, dumping, feeding, material handling, pushing, pulling, and lowering. They are used for heavy-duty applications, including steel mills, machine tools, agriculture, construction, crane drives, winches, material handling, defense, aerospace, marine, and mining. They are also used in military vehicles, excavators, and drilling rigs. Hydraulic motors in fluid-drive transmissions are becoming increasingly popular.

Objective-type Questions
1. Which of the following statements is <u>incorrect</u>?
    a) Gear motors work best in high-speed, low-torque applications.
    b) Gear motors have higher volumetric efficiency than other types of motors.
    c) Hydraulic vane motors operate at a lower noise level than other motors.
    d) Piston motors are the most efficient but most expensive motors.

2. Which of the following hydraulic motors is more tolerant of contamination?
    a) Gear motor
    b) Vane motor
    c) Axial piston motor
    d) Radial piston motor

3. Which of the following hydraulic motors is an excellent choice for high-speed operation?
    a) Gear motor
    b) Vane motor
    c) Axial piston motor
    d) Radial piston motor

4. Identify the component:

Figure 12.14

   a) Semi-rotary pneumatic actuator
   b) Semi-rotary hydraulic actuator
   c) Hydraulic motor
   d) Pneumatic motor

5. Identify the component:

Figure 12.15

   a) Semi-rotary pneumatic actuator
   b) Semi-rotary hydraulic actuator
   c) Hydraulic motor
   d) Pneumatic motor

6. Which power transmission system provides stepless control of speed, torque, and power?
   a) Mechanical
   b) Electrical
   c) Hydrostatic
   d) Pneumatic

Review Questions
1) Explain the working principles of a hydraulic motor.
2) Draw the symbols for the hydraulic semi-rotary actuators and motors.
3) Write two differences between the hydraulic motors and the hydraulic pumps.
4) Define the term 'operating pressure' of a hydraulic motor.
5) What determines the selection of a hydraulic motor operating speed?
6) Define hydraulic motor torque.
7) Define the following terms of hydraulic motors: (a) starting torque, (b) running torque, and (c) stalling torque.
8) What factors determine the torque output of a hydraulic motor?
9) Write the formula for calculating the torque of a hydraulic motor.
10) What determines the operating speed of a hydraulic motor?
11) Define a hydraulic motor's fluid power and shaft power.
12) Write the formula for calculating the input horsepower.
13) Write the formula to calculate the mechanical horsepower output of a hydraulic motor.
14) Define the nominal displacement of hydraulic motors.
15) Explain the relationship between a hydraulic motor's flow rate and shaft speed.
16) What is an ideal hydraulic motor?
17) Define the following terms of hydraulic motors: (1) Motor displacement, (2) Volumetric efficiency, and (3) Mechanical efficiency.
18) Define the terms concerning hydraulic motors: (1) Volumetric efficiency and (2) overall efficiency.
19) Explain the interaction of the flow rate and the pressure during the operation of a hydraulic
20) Determine hydraulic motors' power, torque, and flow rates.
21) What are the functions of hydraulic rotary seals?
22) Give a brief note on the case drain connection in hydraulic motors.
23) How are hydraulic motors classified?
24) What are the different ways of constructing hydraulic motors?
25) How can the output speed of a variable displacement motor be changed?
26) Name two types of positive displacement hydraulic motors.
27) Describe the general constructional features of one type of semi-rotary hydraulic actuator.
28) Explain how a vane-type semi-rotary hydraulic actuator works with a simple sketch.
29) Explain the working of a rack-and-pinion type semi-rotary hydraulic actuator with a sketch.
30) Explain the operation of a gear motor.
31) Describe the general constructional features of gear motors.
32) Give two advantages and disadvantages of gear motors.
33) What are the applications of gear motors?
34) Explain the operation of a gerotor motor.
35) Describe the general constructional features of gerotor motors.
36) What are the advantages and disadvantages of gerotor motors?
37) What are the applications of gerotor motors?
38) Explain the construction features of a Geroler motor.

39) What are the typical characteristics of Geroler motors?
40) Describe the operation of a hydraulic vane motor.
41) Describe the general constructional features of hydraulic vane motors.
42) Give the advantages and disadvantages of hydraulic vane motors.
43) What are the applications of hydraulic vane motors?
44) Explain the difference between the vane motor and the vane pump used in a hydraulic system.
45) Explain how an in-line axial piston motor works.
46) Describe the general constructional features of in-line axial piston motors.
47) What are the advantages and disadvantages of in-line axial piston motors?
48) How is the fluid displacement adjusted in a variable displacement axial-piston motor?
49) Explain the operation of a bent-axis axial piston motor.
50) Describe the general constructional features of bent-axis axial piston motors.
51) What are the advantages and disadvantages of bent-axis axial piston motors?
52) What are the applications of bent-axis axial piston motors?
53) Explain the operation of a radial piston motor.
54) Describe the general constructional features of radial piston motors.
55) What are the advantages of radial piston motors?
56) What are the applications of radial piston motors?
57) Draw the torque-speed characteristics of a hydraulic motor and explain.
58) Draw the pressure-volumetric efficiency curves of the gear, vane, and piston motors.
59) Write two factors that affect the rating and selection of a hydraulic motor for an application.
60) Briefly explain the areas of hydraulic motor applications.
61) Compare various performance factors of gear, vane, and piston hydraulic motors.

Numerical Problems

1) A hydraulic motor theoretically consumes 48 in^3/min while running at a speed of 2000 rpm. What is the volumetric displacement of the motor? [Ans: 0.024 in^3/rev]
2) A hydraulic piston motor operates with a pressure drop of 4000 psi across its ports. The measured flow rate to the motor is 52.8 gpm. What is the input hydraulic power? [Ans: 123 hp]
3) A hydraulic piston motor must produce an output power of 104.56 hp with a pressure differential of 2900 psi. The overall efficiency is 91%. What is the flow rate of fluid required? [Ans: 67.9 gpm]
4) What is the output power of a hydraulic vane motor if the measured torque produced by the motor is 531 in.lb while running at 1000 rpm? [Ans: 8.43 hp]
5) Calculate the output power of a hydraulic gear motor with an actual flow rate of 79.3 gpm and a pressure differential of 1160 psi. The overall efficiency is 65%. [Ans: 34.88 hp]
6) How fast will a 4 in^3/rev motor turn with 10 gpm steady input flow? What is the motor's volumetric efficiency if the actual speed is 500 rpm? [Ans: 6.5%]
7) A hydraulic motor operates with a pressure drop of 2000 psi across its ports. The measured flow rate to the motor is 10 gpm. What is the input hydraulic power? What is the output power if the measured torque is 1080 in.lb at 536 rpm? Also, calculate the overall efficiency. [Ans: P_{in}= 11.67 hp, P_{out} =9.19 hp, η_o= 78%]
8) The torque required for turning a sugar mill drive using a hydraulic motor is 1814 in.lb. The drive unit is powered by a power pack delivering 55.5 gpm at 2990 psi. Assume the mechanical efficiency of the motor is 88%. What is the required displacement and speed of the hydraulic motor? [Ans: V_D=4.46 in^3/rev, N=2870 rpm]

Objective-type questions - answer key: 1-b, 2-a, 3-c, 4bc, 5-c, 6-c

Chapter 13 | Directional Control Valves and Control Circuits

A hydraulic system primarily transfers power from a pump to one or more actuators through a fluid medium for useful work. The hydraulic circuit is essentially the path the fluid follows from the power source to the actuators and then back to the tank. The actuator's direction, speed, or both will be controlled during the work process. At times, the system is controlled based on pressure conditions in certain parts of the circuit. Any simple or complex control requirements can be realized using an appropriate combination of valves.

What is a Hydraulic Valve?

As shown in Figure 13.1, a hydraulic valve consists of a body with an internal moving element, such as a poppet or spool, actuating mechanisms, and many ports. Hydraulic valves are control devices that can direct or restrict the fluid flow. Another class of valves can also provide controls based on specified pressure conditions in some parts of the associated circuits. Accordingly, hydraulic valves can be classified as directional control valves, flow control valves, and pressure control valves.

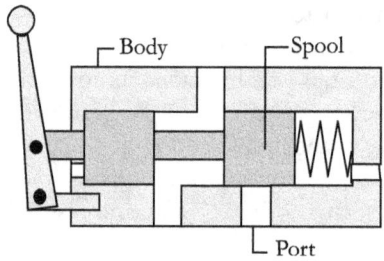

Figure 13.1 | A typical hydraulic valve

Valves can generally be discrete (on/off) or infinitely variable. In a discrete-valve type, the moving element assumes one of its discrete positions in response to control signals. In the infinitely variable type, the valve spool can be placed at any desired position with adjustable electronic signals. Electronically controlled valves are categorized as either (1) Proportional valves or (2) Servo valves. This book/chapter discusses discrete-valve types.

Functional Classification of Discrete Hydraulic Valves

Figure 13.2 classifies discrete hydraulic valves. A directional control (DC) valve (or way valve) controls the path taken by the fluid. This valve allows the flow in a particular direction for a given switching position. The DC valve can also control the direction of motion of an actuator connected to the system.

A non-return valve (NRV) allows pressurized fluid in a hydraulic system to flow in only one direction and blocks it in the opposite direction.

A flow control valve restricts the flow rate of pressurized fluid in a hydraulic circuit.

A pressure control valve limits or reduces the fluid's pressure or generates a control signal when a set pressure in a system part is reached, initiating a subsequent action. The pressure control valves include pressure-reducing, unloading, sequence, counterbalance, and brake valves.

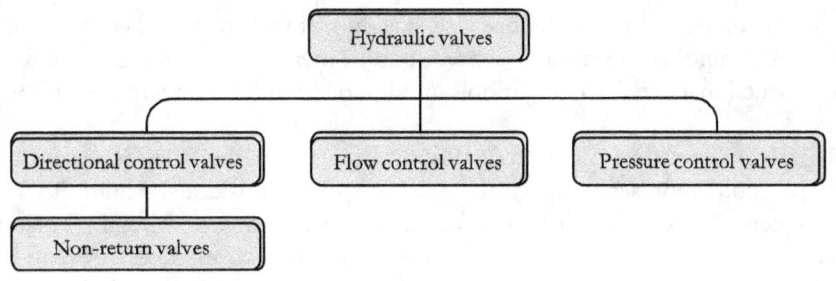

Figure 13.2 | Classification of hydraulic valves

Types of Directional Control Valves
There are two basic types of valves. They are: (1) Poppet valves and (2) Spool/Slide valves.

Poppet Valves
In a poppet valve, a spring-biased disc, ball, or cone, along with the valve seat, creates a leak-free enclosure and controls flow. The poppet valve quickly opens a relatively large orifice in short travel to permit the full flow of the fluid. It has the intrinsic characteristic of fast response and is used to generate and convey control signals.

Spool/Slide Valves
In a spool/slide valve, a close-fitting spool moves axially within the body to control the direction of the fluid flow. The spool valves exhibit excellent shifting characteristics. They produce reduced shock pressures during their spool transitions. They are primarily used as main valves to supply power signals to the actuators in the system.

Graphical (Symbolic) Representation of Directional Control Valves

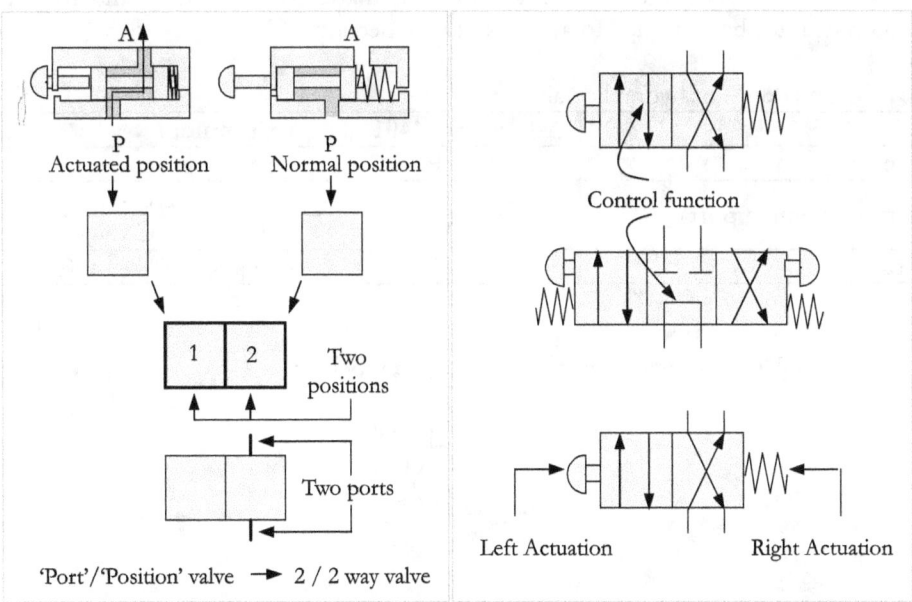

Figure 13.3 | Graphic representation of directional control valves

Symbols represent hydraulic components because sketches of their complex control functions are too difficult to draw. A component's symbolic representation merely specifies the part's function without indicating its constructional details. The symbols are described in the standard ISO 1219.

A hydraulic DC valve is specified as a 'port/position valve,' where the 'port' represents the number of ports and the 'position' represents the number of switching positions. Figure 13.3 illustrates the port/position concept. Thus, a 3/2-DC valve has three ports and two switching positions. The lines inside the valve represent the valve's function. The valve's actuation method is shown on the left and right sides.

Symbols of Basic DC Valves

Symbols aid in the functional identification of the components in the circuit diagrams of fluid power systems. Figure 13.4 shows symbols of the essential directional control valves.

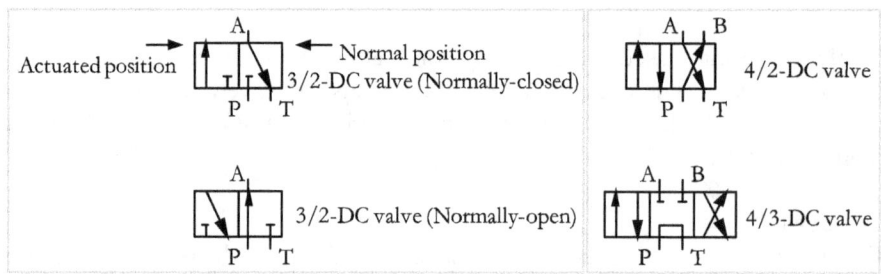

Figure 13.4 | Symbols of essential directional control valves

Port Markings

As per current practice, ports of hydraulic valves are designated using a letter system in accordance with ISO 4401. Table 13.1 presents the designations for port markings as per the standard. All DC valve inputs and outputs must be identified to avoid faulty connections.

Table 13.1 | Port markings of hydraulic valves

Port	As per ISO 4401	Comment
Pressure port	P	Supply port
Working port (Service port)	A	In a 3/2-DC valve
	A, B	In a 4/2-DC valve
Tank port	T	In a 3/2 or 4/2-DC valve

Example

An example of a 4/2-DC valve with port markings is given below:

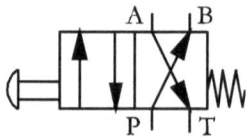

Figure 13.4(a)

Methods of Valve Actuation

An essential feature of directional control valves is their actuation method. These valves can be actuated manually, mechanically, hydraulically, or electrically. Figure 13.5 shows the symbols of various methods of hydraulic valve actuation.

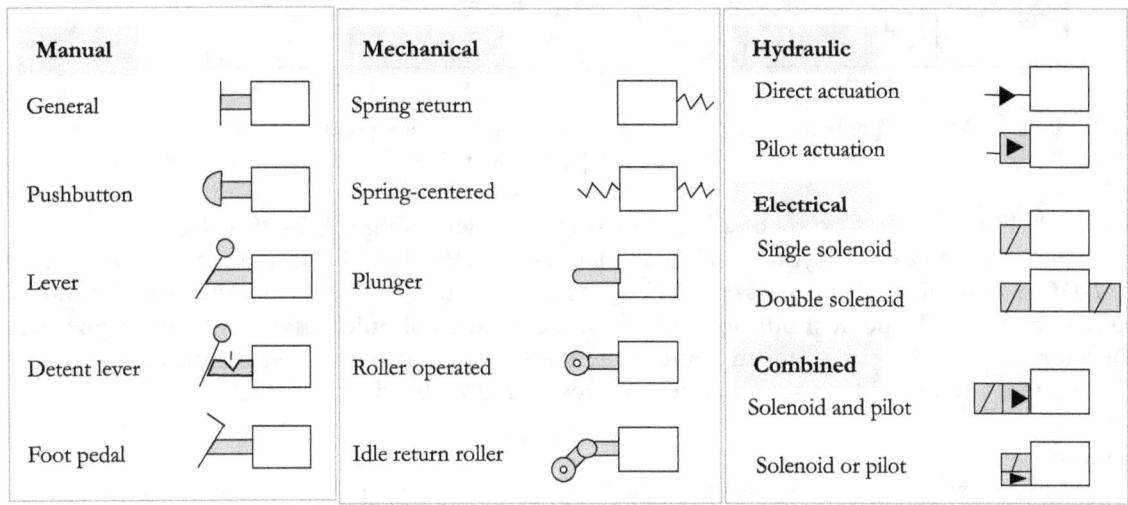

Figure 13.5 | Methods of valve actuation

2/2-DC Valve

Figure 13.6 shows the cross-sectional views of a 2/2-directional (DC) valve in the normal and actuated positions. In the normal position of the valve, the pressure port P and the working port A are blocked. In the actuated position of the valve, the working port A is open to the pressure port P. When the actuating force is removed, the compressed spring brings the spool back to its normal position. The 2/2–way valve can block or open the flow passage in a hydraulic system. Leakage fluid, if any, can be removed through a drain line.

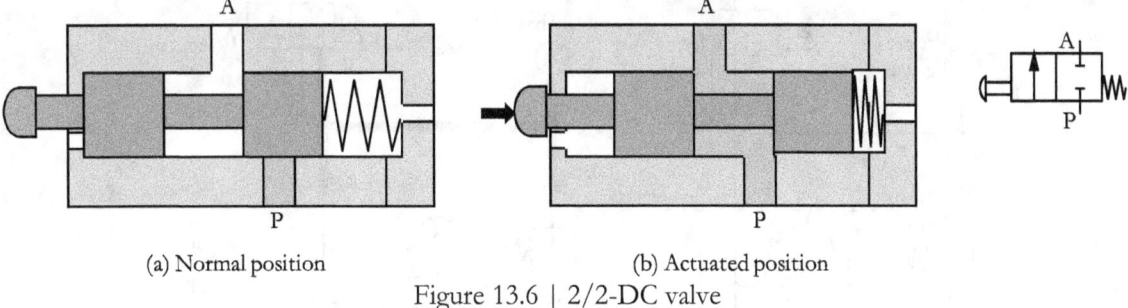

(a) Normal position (b) Actuated position

Figure 13.6 | 2/2-DC valve

3/2-DC Valve

Figure 13.7 shows the cross-sectional views of a 3/2-directional (DC) valve in the normal and actuated positions. In the normal position of the valve, the working port A is closed to the pressure port P and open to the tank port T. The pressure port is blocked in the normal position of the valve. In the actuated position of the valve, the working port A is open to the pressure port P and closed to the tank port T. The 3/2-way valves can be used to control single-acting hydraulic cylinders and other valves.

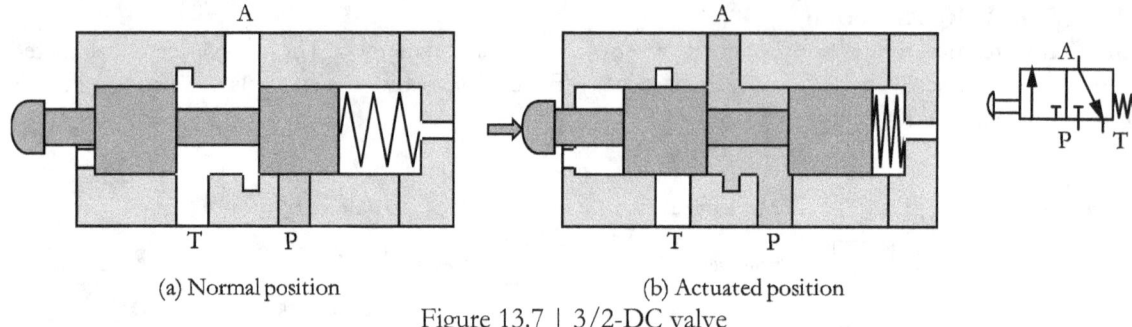

(a) Normal position   (b) Actuated position

Figure 13.7 | 3/2-DC valve

Example 13.1 | Direct Control of a Single-acting Cylinder Using a 3/2-DC valve
A single-acting hydraulic cylinder of small piston diameter should clamp a component when a 3/2-DC pushbutton valve is pressed. As long as the pushbutton is pressed, the cylinder should remain clamped. If the pushbutton is released, the cylinder should retract to its home position. Develop a hydraulic circuit to implement the control task using a fixed-displacement pump. The system shall be able to set a maximum pressure of 1450 psi.

Solution
Figure 13.8 shows two positions of the hydraulic circuit in the normal and actuated positions of the valve for implementing the control task given in Example 13.1. The power supply unit consists of a hydraulic pump driven by an electric motor, a reservoir, and an integral pressure relief valve. The pump delivers the pressurized fluid to the circuit with constant displacement. As shown, a separate pressure relief valve (PRV) can be used to set the maximum/operating pressure (1450 psi) in the system.

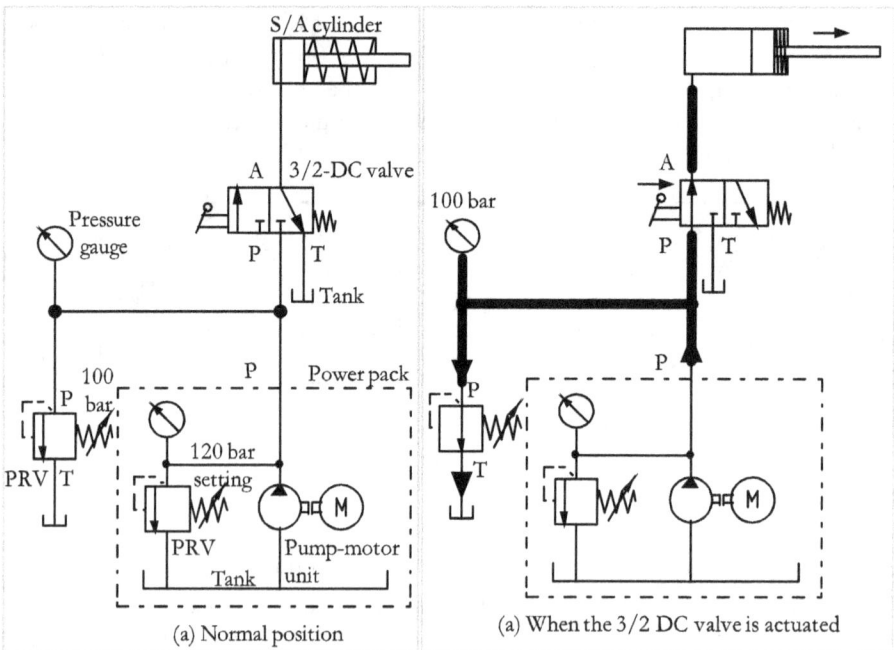

Figure 13.8 | Two positions of the hydraulic circuit for the direct control of a single-acting cylinder (Example 13.1)

The single-acting cylinder is controlled by a manually-actuated 3/2-DC valve, as shown in Figure 13.8.

In the actuated position, as shown in Figure 13.8(b), the valve allows the flow from the pump to the cylinder. The cylinder then extends with minimum flow resistance and pressure development. When the cylinder reaches its fully extended position, the flow encounters greater resistance and develops higher pressure. The pump flow is bypassed through the pressure relief valve (PRV) when the PRV reaches its setting.

In its normal position, the 3/2-DC valve blocks flow from the pump to the cylinder, as shown in Figure 13.8(a). The cylinder then retracts to its home position. When the pressure reaches the PRV's setting, the pump flow is bypassed through the PRV.

The maximum pressure-limiting action of the pressure relief valve protects the system against over-pressurization. However, during the period when the relief valve is bypassing, all of the energy delivered by the pump is converted into heat. It may be noted that the arrow alongside the valve indicates the direction of the applied force (F).

Remember, it is customary to place the individual tank symbols close to the relevant component rather than have confusing lines running from most components to the reservoir throughout the circuit.

Typical Structure of Hydraulic Circuits
It may be noted that when drawing a conventional circuit for a hydraulic system, the power source is typically shown at the bottom of the circuit, the actuators at the top, and the control valves in between (See Figure 13.8). The energy is shown to flow from the bottom part to the top part of the circuit through the control valves.

Knowing the symbols of components and types of lines is essential for drawing a hydraulic circuit.

4/2-DC Valve
Figure 13.9 shows the cross-sectional views of a 4/2-directional (DC) valve in the normal and actuated positions. In the normal position of the valve, paths from port P to working port B and from working port A to port T are open. When the valve is actuated, paths from port P to working port A and from working port B to tank port T are open. This valve can serve as the final control element to drive a double-acting hydraulic cylinder or a bi-directional hydraulic motor.

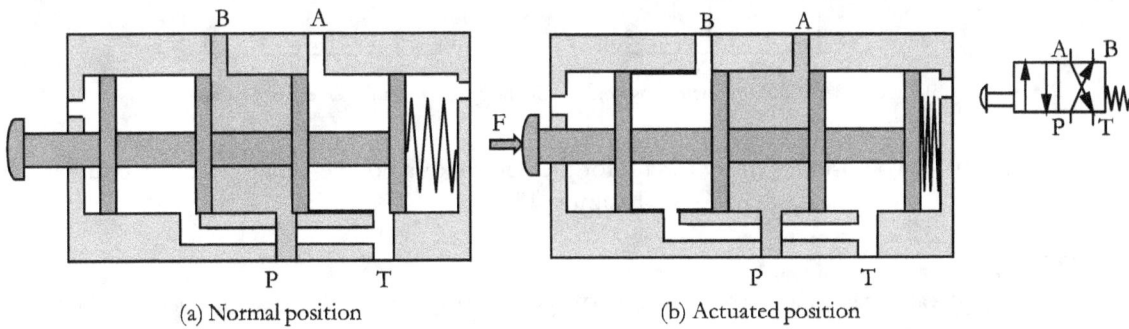

(a) Normal position (b) Actuated position

Figure 13.9 | 4/2-DC valve

Example 13.2 | Direct Control of a Double-acting Cylinder Using a 4/2-DC valve

A double-acting hydraulic cylinder should extend and clamp a workpiece when a pushbutton valve is actuated. The cylinder should remain clamped as long as the valve is actuated. If the pushbutton is released, the cylinder is to retract. Develop a hydraulic circuit to implement the control task. A fixed-displacement hydraulic pump is used as the power source. The system shall be able to set a maximum pressure of **1450 psi**.

Solution

Figure 13.10 shows two positions of the hydraulic circuit in the normal and actuated positions of the valve for the control task in Example 13.2. The double-acting cylinder can be controlled using the manually actuated 4/2-DC valve. The power supply unit comprises a hydraulic pump and an integral pressure relief valve (PRV). The pump delivers pressurized fluid to the circuit with a constant displacement. Next, the integral pressure relief valve limits the power pack's maximum pressure rating. The operating pressure (say 1450 psi) in the system can be set by using a separate pressure relief valve (PRV), as shown.

When the valve is actuated, as shown in Figure 13.10(b), the fluid flow is directed to the piston-side port of the cylinder, which then extends. In the normal position of the DC valve, as shown in Figure 13.10(a), the fluid flow is directed to the piston rod-side port of the cylinder, which retracts to its home position.

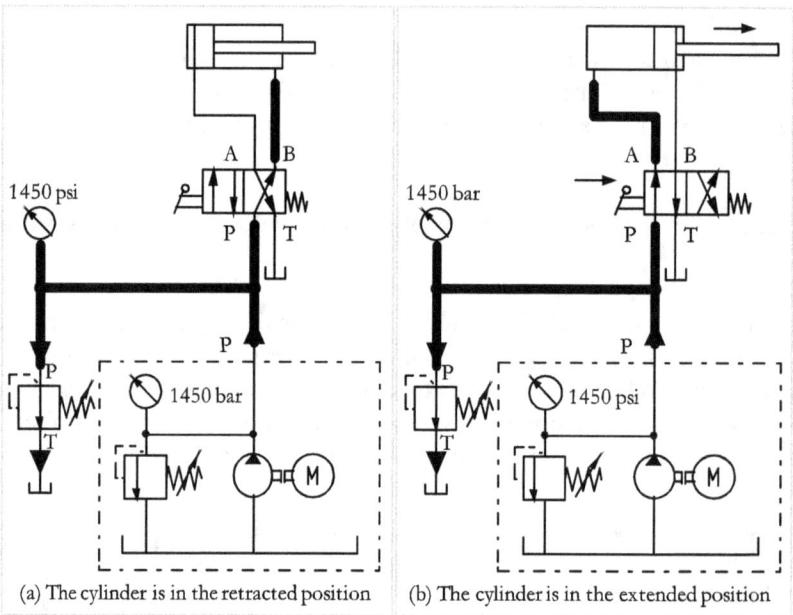

(a) The cylinder is in the retracted position   (b) The cylinder is in the extended position

Figure 13.10 | Two positions of the circuit for the control of a double-acting hydraulic cylinder (Example 13.2)

4/3-DC Valves

A 4/3-DC (way) valve has four ports and three switching positions. In the normal position, the valve remains centered by springs mounted at either end of its spool shaft. The center position is the valve's position when the associated hydraulic component is not required to move.

Center Positions of 4/3-DC Valves

Typically, a DC valve in a hydraulic system with a pump and an actuator connects the pump to the actuator and provides the direction control of the actuator. In addition to direction control, the application also requires additional controls.

Usually, there are two requirements concerning the pump. Either the pump flow must be blocked to the system, or it may be required to unload the pump flow back to the system reservoir, necessarily at atmospheric pressure.

Likewise, there are a few requirements regarding actuators. One requirement is that a moving cylinder must be abruptly stopped and hydraulically locked so that an external force cannot drive it. Another requirement is that a hydraulic motor carrying a load should be stopped smoothly, without causing jerkiness.

In addition to the two usual valve switching positions for the forward and return strokes of the actuator, an additional spring-centered switching position can be incorporated to meet the requirements described above. These additional requirements have led to the development of numerous designs of 4/3-DC valves.

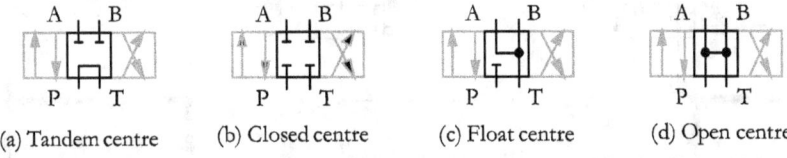

(a) Tandem centre   (b) Closed centre   (c) Float centre   (d) Open centre

Figure 13.11 | Various center positions of 4/3-DC valves

Figure 13.11 shows many designs of 4/3 DC valves. The most important center positions are tandem, closed, open, and float. It may be remembered that all these variants have identical flow path configurations at the other two switching positions.

Tandem-center Position

Directional Control valves with the tandem-center (recirculation-center) position are the most commonly used 4/3-DC valves for constructing low-power hydraulic equipment with constant delivery pumps, mainly due to their energy-saving feature. In the tandem-center design, as shown in Figure 13.12, the pressure port P is connected to the tank port T, and the working ports A and B are blocked when the valve is in its neutral position. The use of the tandem-center valve provides energy savings for the system.

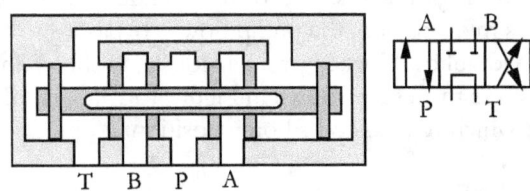

Figure 13.12 | A hydraulic valve with a tandem-center position

Example 13.3 | Control of a Double-acting Cylinder Using a 4/3-DC Tandem-center Valve

A 4-way, 3-position directional control valve is used to advance and retract a double-acting cylinder used in a hydraulic system. The cylinder should clamp a workpiece when it is fully extended. A fixed-displacement pump is used for the system. Additionally, the circuit should be designed to lock the cylinder hydraulically for a long duration and manually unload the pump to save energy, especially when the cylinder starts holding the job at the set pressure. Develop a hydraulic circuit to implement the control task. The system shall be able to set a maximum pressure of 1450 psi.

Solution

The double-acting hydraulic cylinder can be controlled using the manually actuated 4/3-DC valve with the tandem-center position. The power supply unit includes a pump that delivers pressurized fluid to the circuit with constant displacement. A pressure relief valve (PRV) can set the system operating pressure (Example, 1450 psi).

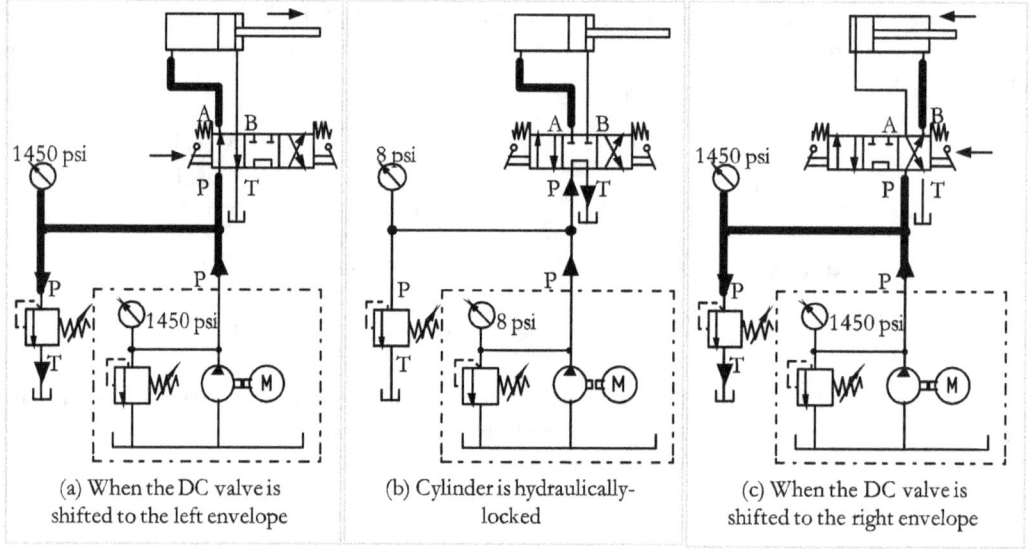

Figure 13.13 | Three positions of the hydraulic circuit for the direct control of a double-acting cylinder using a tandem-center, 4/3-DC valve (Example 13.3)

Figure 13.13 shows three positions of the hydraulic circuit with the fixed-displacement pump for the control task in Example 13.3. When the 4/3-DC valve is actuated towards the right, as shown in Figure 13.13(a), the flow is directed to the piston side port of the cylinder, and the cylinder extends. When the 4/3-DC valve is brought to its mid-position, the pump flow is bypassed to the reservoir at atmospheric pressure, and the cylinder is hydraulically locked simultaneously, as shown in Figure 13.13(b). When the 4/3-DC valve is actuated towards the left, as shown in Figure 13.13(c), the flow is directed to the piston rod side port of the cylinder, which retracts to its home position.

It may be noted that the cylinder tends to creep with the valve centered. Further, tandem-center valves are unsuitable for a multiple-actuator system with a single pump, as the pump flow cannot be used independently for other parts of the circuit. However, these valves can control a series-operated multiple-cylinder system when the cylinders must operate one after another.

Closed-center Position

In the closed-center design of a 4/3-DC hydraulic valve, as shown in Figure 13.14, all four ports are blocked from one another in its neutral position. Consider a hydraulic system with a pump and multiple double-acting cylinders. For independent cylinder control, 4/3-DC closed-center valves can be used. As the pressure port in each valve is blocked in its neutral position, the pump flow can be directed to a particular cylinder independently when the associated valve is actuated. At the same time, every actuator connected to the working ports of the corresponding valve can be hydraulically locked when the valve is brought to its center position.

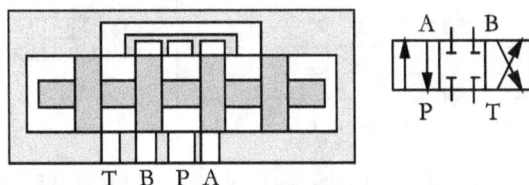

Figure 13.14 | A hydraulic valve with a closed-center position

Example 13.4 | Control of a Double-acting Cylinder Using a 4/3-DC, All-closed-center Valve

A 4-way, 3-position DC valve is used to advance and retract a double-acting cylinder in a hydraulic system with a fixed-displacement pump. When fully extended, the cylinder should clamp a workpiece. The circuit should support the independent control of other cylinders connected to the system. Develop a hydraulic circuit to implement the control task. The system shall be able to set a maximum pressure of 1450 psi.

Solution

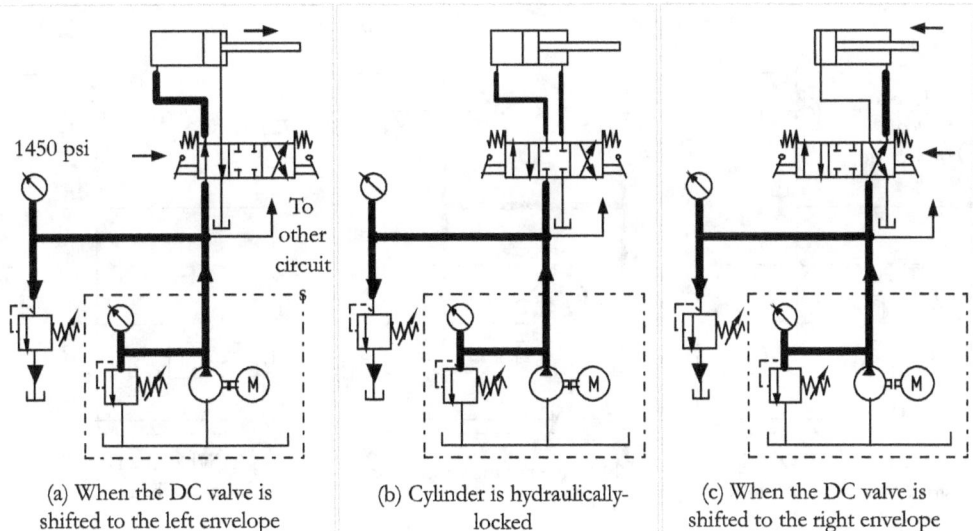

(a) When the DC valve is shifted to the left envelope

(b) Cylinder is hydraulically-locked

(c) When the DC valve is shifted to the right envelope

Figure 13.15 | Three positions of the hydraulic circuit for the control of a double-acting cylinder using a closed-center, 4/3-DC valve (Example 13.4)

Figures 13.15 (a), (b), and (c) show three positions of the hydraulic circuit in the normal and actuated positions of the valve for the control task in Example 13.4. In the center position of the valve, the cylinder is hydraulically locked, and the pump is relieved against the PRV's high-pressure setting.

Float-center Position

Another commonly used configuration of the 4/3-DC valve is the float-center design, as shown in Figure 13.16. In the neutral position of the valve, the working ports A and B are connected to the tank port T, and the pressure port P is blocked. In this position, the valve allows the pressurized fluid on both sides of the associated actuator to drain back into the tank. Therefore, the moving elements in the actuator can float in this position. Further, this valve enables a soft shut-off while stopping a hydraulic motor. Therefore, these types of valves are often used to control hydraulic motors.

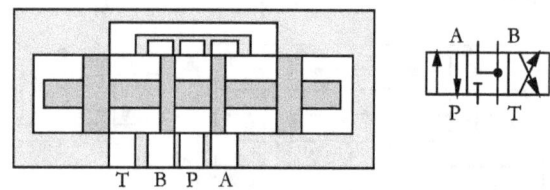

Figure 13.16 | A hydraulic valve with a float-center position

Example 13.5 | Control of a Hydraulic Motor Using a 4/3-DC Float-center Valve

A 4-way, 3-position DC valve should control a hydraulic motor in clockwise and anti-clockwise directions in a hydraulic system with a fixed-displacement pump. The circuit should provide a soft stop for the motor and be designed to support the independent control of multiple actuators connected to the system. Develop a hydraulic circuit to implement the above control requirements. The system shall be able to set a maximum pressure of 1450 psi.

Solution

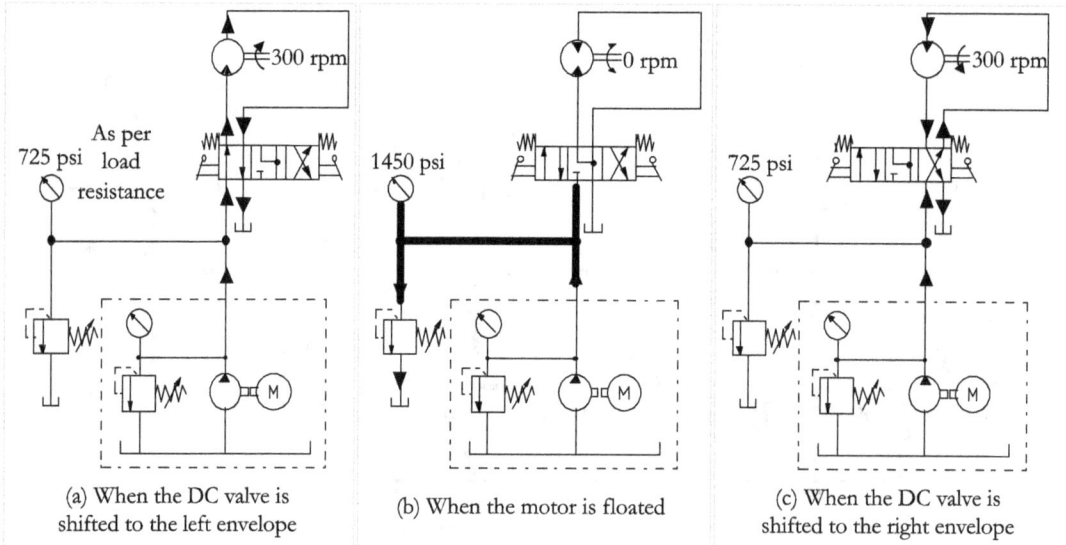

(a) When the DC valve is shifted to the left envelope

(b) When the motor is floated

(c) When the DC valve is shifted to the right envelope

Figure 13.17 | Three positions of the hydraulic circuit for the direct control of a hydraulic motor using a float-center, 4/3-DC valve (Example 13.5)

Figure 13.17 shows three positions of the hydraulic circuit in the normal and actuated positions of the float-center valve for the control task in Example 13.5. In the left-envelope position of the valve, as shown in Figure 13.17(a), the hydraulic motor runs in one direction (say clockwise). In the right-envelope position, as shown in Figure 13.17(c), the motor runs in the opposite direction.

Suppose a closed-center valve is used, and the valve is brought to its center position to stop the motor. In that case, the motor tends to squeeze the trapped oil and develop high pressure in its working lines, alternately, owing to the inertia of the running motor. That is, the motor tends to jerk before coming to a stop. Therefore, relieving the fluid in the working lines to the tank is essential to obtain a soft motor stop. This function is realized by the float-center position of the valve, as shown in Figure 13.17(b).

Open-center Position
In the neutral position of the 4/3-DC valve's open-center design, as shown in Figure 13.18, all four ports are connected. This valve allows the associated pump to unload and the associated cylinder to float, reducing heat build-up in the system.

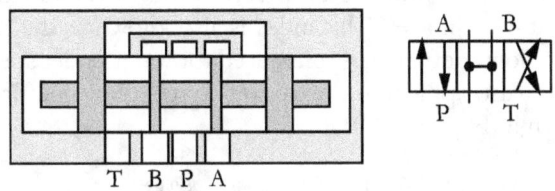

Figure 13.18 | A hydraulic valve with an open-center position

Example 13.6 | Control of a Hydraulic Motor Using a 4/3-DC All-open-center Valve
A 4-way, 3-position DC valve should control an intermittently operated hydraulic motor in clockwise and anti-clockwise directions. The circuit should be capable of providing a soft motor stop. A fixed-displacement pump is used. The circuit should provide an energy-saving feature, as the motor has a long idle time. Develop a hydraulic circuit to implement the above control task. The system shall be able to set a maximum pressure of 1450 psi.

Solution

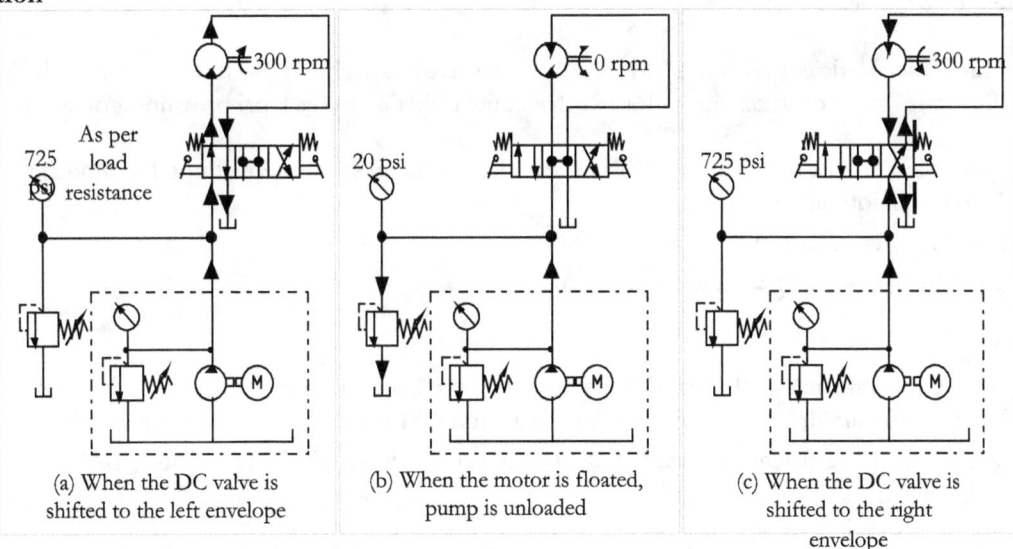

Figure 13.19 | Three positions of the hydraulic circuit for the direct control of a hydraulic motor using an all-open-center, 4/3-DC valve (Example 13.6)

Figure 13.19 shows two positions of the hydraulic circuit in the normal and actuated positions of the valve for the control task in Example 13.6. In the left-envelope position of the valve, the hydraulic motor runs in one direction. In the right-envelope position, the motor runs in the opposite direction. In the center position of the valve, the circuit provides an energy-saving feature and a soft motor stop. However, this circuit is unsuitable for the multiple-actuator circuit.

Flow Rate Coefficient of Control Valves
When a fluid flows through a hydraulic valve, the pressure drop across the valve increases, and consequently, some energy gets wasted. A hydraulic valve needs to be rated by its ability to pass fluid, with pressure drop and energy loss kept within limits. Every valve is assigned a unique capacity index by its manufacturer to measure the pressure drop across the valve and to make a fair comparison amongst similar valves of different makes. This index is also known as the flow-rate coefficient or the valve-sizing coefficient. The flow coefficient is referenced for water under specific operating conditions. However, the water formulas also apply to ordinary liquids. The flow coefficient is measured using the standard test setup, as shown in Figure 13.20.

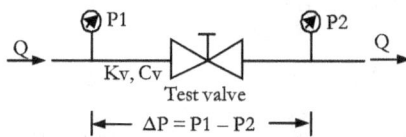

Figure 13.20 | The test setup to calculate the flow coefficient of a valve

The valve's flow coefficient can be expressed in several ways. They are:

- Kv is the flow coefficient term used to measure the valve's capacity in the SI system units. It is defined as the flow rate (m³/h) of water through the valve at a temperature between 5 and 30°C that causes a 1 bar pressure drop across it.

- Cv is the flow coefficient term used to measure the valve's capacity in the English units. It is defined as the flow rate (gpm) of water through the valve at 60°F that causes a 1-psi pressure drop across it.

The general definition of the flow coefficient can also be expressed as equations for modeling liquid flow. These equations are given below:
In the SI system units, $Q = Kv \times \sqrt{(\Delta P/SG)}$
In the English units, $Q = Cv \times \sqrt{(\Delta P/SG)}$

Where,
Q is the flow rate in lpm (In the SI units), gpm (In the English units)
ΔP is the pressure drop across the valve in kPa (In the SI units), psi (In the English units)
Kv is the flow coefficient in 'lpm/$\sqrt{kPa}$' and Cv is the flow coefficient in 'gpm/$\sqrt{psi}$'
SG denotes the specific gravity.

The approximate relationships between Kv and Cv are as follows:
Kv = 0.865 x Cv | Cv = 1.156 x Kv

Example: 13.7 | A hydraulic valve should carry fluid with a specific gravity of 0.9 at the rate of 114 m³/hour, and the permissible pressure drop across the valve is 1 bar. Find the valve's capacity coefficient in metric units.

Solution
Flow rate, $Q = 114 \text{ m}^3/\text{hour}$
Specific gravity = 0.9
Permissible pressure drop, $\Delta P = 1 \text{ bar} = 100 \text{ kPa}$

Valve coefficient, $K_v = Q \text{ (m}^3/\text{hour)} / \sqrt{[\Delta P(\text{kPa})/SG]}$
$= 114/\sqrt{100/0.9} = 10.82 \text{ (m}^3/\text{hour)}/\sqrt{\text{kPa}}$

Example: 13.8 | A hydraulic valve should carry fluid with a specific gravity of 0.9 at the rate of 100 gpm, and the permissible pressure drop across the valve is 1.98 psi. Find the capacity coefficient of the valve in the English units.

Solution
Flow rate, $Q = 100$ gpm
Specific gravity = 0.9
Permissible pressure drop, $\Delta P = 1.98$ psi

Valve coefficient, $C_v = Q \text{ (gpm)} / \sqrt{[\Delta P(\text{psi})/SG]}$
$= 100/\sqrt{1.98/0.9} = 67 \text{ (gpm)}/\sqrt{\text{psi}}$

ΔP vs. Q Characteristics of DC Valves

An essential characteristic of a hydraulic valve is the pressure drop (ΔP) across the valve as a function of the flow rate (Q) through it. Figure 13.21 shows a typical ΔP-Q characteristic of a 2/2-way valve with a bypass check valve for both flow directions (For the flow from A to B and the flow from B to A). This characteristic shows that the pressure drop in each direction increases with increasing flow rate.

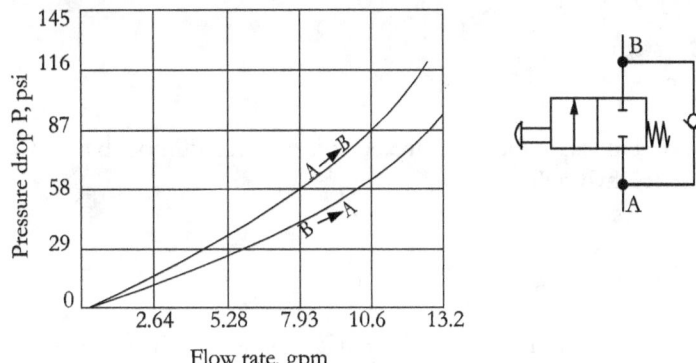

Figure 13.21 | Pressure drop vs. flow rate (ΔP vs. Q) characteristic

Note: A Summary of Controls for Hydraulic Systems - Basic Level is given in Chapter 22

Specifications, Hydraulic DC Valves
Essential performance specifications for a DC valve include its maximum pressure rating, maximum flow rate, number of ports, center-position configuration, actuation method, and mounting style. The most general requirements for hydraulic valves are quick response, minimal pressure drop, leak-free operation, and excellent switching reliability.

Selection, Hydraulic Valves
Proper selection of hydraulic valves is crucial for any given application. Knowing the necessary control functions to select suitable valves for a hydraulic system being developed is essential. Next, the valves' sizes (flow capacities) must be determined to limit the pressure drop across the valves and the heat buildup in the system to reasonable levels. Other factors to consider when selecting hydraulic valves include actuation methods, quality, cost, and mounting styles. All materials used in valve construction must be carefully selected based on the application and the type of fluid.

Objective-type Questions
1. A directional control valve can:
    a) Stop fluid flow
    b) Permit fluid flow
    c) Change the direction of fluid flow
    d) All of the above

2. Mark the class of hydraulic systems with infinitely variable valves and integrated electronic controllers intended mainly for open-loop control.
    a) Conventional Electro-hydraulic system
    b) Proportional valve system
    c) Servo valve system
    d) Cartridge valve system

3. Mark the hydraulic system with infinitely variable valves and integrated electronic controllers intended for closed-loop control.
    a) Proportional valve system
    b) Servo valve system
    c) Cartridge valve system
    d) Electro-hydraulics

4. Mark the hydraulic system class where a valve bank is integrated into a block.
    a) Conventional electro-hydraulic system
    b) Proportional valve system
    c) Servo valve system
    d) Cartridge valve system

5. The direction of motion of a double-acting hydraulic cylinder can be controlled using:
    a) 2/2-way valve
    b) 3/2-way valve
    c) 4/2-way valve
    d) Check valve

6. Indicate the correct statement.
   a) A pilot-operated check valve always permits fluid flow in one direction and blocks the flow in the opposite direction.
   b) A servo valve system is an open-loop system with an electronic controller.
   c) In tandem-center 4/3 DC valves, working ports are blocked in the center position.
   d) A non-return valve controls the fluid flow rate.

7. Mark the center position of a 4/3-DC valve with the pressure port connected to the tank port, and the working ports are blocked from one another.
   a) Tandem-center position
   b) Open-center position
   c) Closed-center position
   d) Float-center position

8. Which of the following is the center position of a 4/3-DC valve, where the pressure port is blocked and the working ports are connected to the tank port?
   a) Tandem-center position
   b) Open-center position
   c) Closed-center position
   d) Float-center position

9. Indicate the _incorrect_ statement.
   a) The spool valves exhibit excellent shifting characteristics and produce reduced shock pressures during the transition.
   b) In a cartridge valve system, a bank of valves can be built into one block and controlled electrically/electronically.
   c) The proportional valve system typically uses the spool feedback for the closed-loop operation.
   d) The pressure control valves restrict the flow rate at which the pressurized fluid is transferred in a circuit.

Review Questions
1) What is the primary function of a hydraulic valve?
2) Briefly explain the conventional hydraulic valve system.
3) Briefly explain the electro-hydraulic systems and their types.
4) What is a hydraulic proportional valve system?
5) What is a hydraulic servo valve system?
6) What is a cartridge valve system used in hydraulic systems?
7) What are the advantages of cartridge valves used in hydraulic systems?
8) Briefly explain the importance of the standard valve bases and manifolds in hydraulic systems.
9) What are the essential control functions of valves in hydraulic systems?
10) How are the discrete hydraulic valves functionally classified?
11) Briefly explain the functions of the following hydraulic valves: (1) Directional Control valves, (2) Flow control valves, and (3) Pressure control valves.
12) Give one application to each of the following hydraulic valves: (1) Directional Control valves, (2) Non-return valves, (3) Flow rate control valves, and (4) Pressure control valves.
13) What is a poppet valve? Mention the advantages and disadvantages of this type of hydraulic valve.
14) What is a spool valve? Mention some pros and cons of this kind of hydraulic valve.

15) What is a two-way hydraulic valve?
16) Draw two representative symbols for hydraulic directional control valves and mention one application of each valve.
17) Explain the different port configurations of hydraulic directional control valves.
18) What are the many ways hydraulic valves can be actuated? Explain
19) Describe the constructional features of hydraulic directional control valves.
20) Identify each of the valves in Figure 13.22 and explain the possible function of each valve.

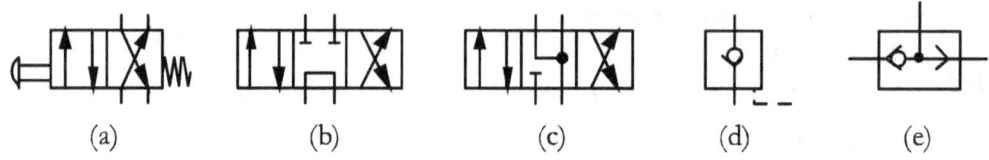

Figure 13.22

21) Sketch and describe the working of a 3/2 directional control hydraulic valve.
22) Draw a typical circuit for manually controlling the forward motion of a single-acting hydraulic cylinder.
23) Explain the working of a four-way, two-position (4/2) hydraulic directional control valve.
24) What is a 4/2-DC valve as used in hydraulic systems? What are the different methods of actuating the valve?
25) Explain the operation of a pilot-operated 4/2-way directional control hydraulic valve with the help of its symbolic representation.
26) Draw a typical hydraulic circuit for controlling the forward motion of a double-acting hydraulic cylinder with manual actuation.
27) Draw the graphical symbol of a four-way, three-position, closed-center, spring-centered, manually operated hydraulic directional control valve.
28) Explain the operation of a four-way, three-position (4/3) tandem-center hydraulic DC valve.
29) Draw the ISO symbol of the 4/3-way directional control hydraulic valve with a tandem center.
30) Draw the symbols of four 4/3-way hydraulic valves with different flow path configurations in their center positions and give their applications.
31) Differentiate between the 4/2-DC and 4/3-DC valves used in hydraulic systems.
32) What is the recirculation center position in the 4/3-way hydraulic valve, and why is it used?
33) What is the essential advantage of a 4/3-way tandem-center hydraulic valve?
34) What is the purpose of blocking the working ports in a hydraulic DC valve in its neutral position?
35) Explain why a closed-center valve is essential in hydraulic systems if it is in a bank of valves supplied by the same power source.
36) What is the advantage of an open-center 4/3-DC hydraulic valve?
37) Differentiate the 4/3-way open-center hydraulic valve and 4/3-DC closed-center hydraulic valve.
38) What is the advantage of using the float-center 4/3-way valve in hydraulic systems?
39) A 4/3-way valve should control a double-acting hydraulic cylinder driving a heavy load. Develop a circuit to hold and lock the load in position for extended periods using pilot-operated check valves.

Objective-type questions - answer key: *1-d, 2-b, 3-b, 4-d, 5-c, 6-c, 7-a, 8-d, 9-d*

Chapter 14 | Non-return Valves and Control Circuits

A non-return valve (NRV) allows pressurized fluid in a hydraulic system to flow in only one direction and blocks it in the opposite direction. A check valve is the basic NRV.

Check Valve

A check valve permits flow in one direction and prevents it entirely in the reverse direction. Therefore, the valve can also be considered a one-way directional control valve. As shown in Figure 14.1, the valve consists of a valve body and a spring-biased ball poppet or cone poppet, apart from inlet and outlet ports ('A' and 'B'). The spring holds the poppet against the valve seat. Check valves can be designed for in-line or manifold mounting with a right-angle design, depending on the relative positions of their ports. The inline check valve is unsuitable for high-flow-rate applications because the pressure drop across it tends to be excessive.

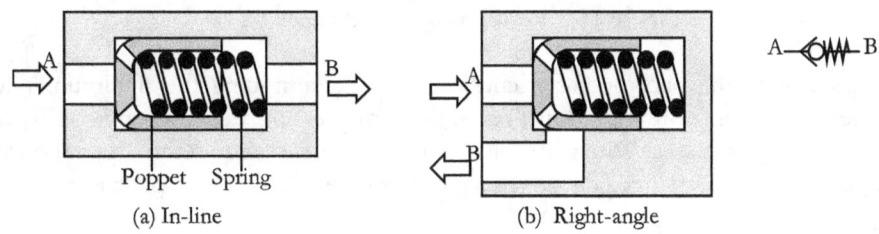

Figure 14.1 | Two designs of check valves

When pressure is applied to port A of the check valve, the poppet is pushed off its seat, allowing the system fluid to flow freely through the valve (port A to port B) with a low pressure drop. The flow through the valve is blocked when the intended flow direction is from port B to port A by poppet reseating. The minimum pressure at which the valve begins to open is called its cracking pressure. The cracking pressure depends on the valve's spring rate. The larger the spring rate, the higher the cracking pressure.

Hydraulic check valves are built ruggedly to operate in hydraulic systems with low-pressure drops. They close smoothly without pounding and resist shock pressures. A normal check valve requires about 5 psi to move its poppet against the spring. However, some check valves have stiff springs with cracking pressures up to 125 psi.

A ball poppet check valve is the least expensive type. However, leakage can occur if the poppet is worn or not guided correctly. A soft-seated check valve can be used to prevent leakage. Ball poppet check valves are unsuitable for systems with pressures above 3000 psi. It is to be noted that check valves that seal satisfactorily at high pressures may leak at lower pressures.

Pilot-operated check valve

A pilot-operated check valve is a variant of the primary check valve. It consists of a valve body, a poppet biased by a spring, ports A and B, and a pilot port X. The poppet has a pilot piston attached to the poppet stem. Pressure applied to port A lifts the poppet from its seat against the spring force, allowing flow from port A to port B with a low pressure drop across the valve

The flow through the valve is blocked when the intended flow direction is from port B to port A by poppet reseating. However, the poppet can be held open by applying an external pilot signal through the pilot port X. Considering the forces generated by the pressures and areas on both sides of the poppet, the pressure at the pilot port displaces the poppet from its seat, allowing fluid to flow from port B to port A.

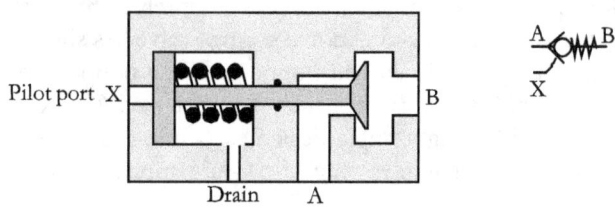

Figure 14.2 | A pilot-operated check valve

The primary issue with the pilot-operated check valve is determining the minimum pilot pressure required to open the valve for reverse flow. The minimum pressure value depends on the ratio of the pilot piston area to the valve area. With a 4:1 pilot ratio, the pressure required to open the valve is about 25% of the opposing load pressure. This valve can appropriately be termed a pilot-to-open check valve.

Check Valve with Pilot-to-Close
A pilot-to-close check valve consists of a valve body with ports A and B, a spring-biased poppet, and a pilot port X [Not shown]. The valve allows flow from port A to port B and blocks flow from port B to port A when no signal is applied to the pilot port X. However, the valve prevents flow in either direction when pilot pressure is applied to the pilot port X.

Valve Combinations with Check Valves
Pilot-operated check valves can be integrated into directional control valves for applications requiring positive load holding. Figure 14.3 shows the design of a combination valve involving a directional control valve and two pilot-operated check valves. In this valve, each pilot section is cross-connected to the opposite line, providing automatic pilot operation in either direction of the fluid flow. It should be noted that the directional control valve must have a float-center spool to vent the pilots, allowing the check valve poppets to seat correctly. Further, the directional control valve and the double-pilot check valves are available as separate standardized subassemblies that can be stacked to form the combination.

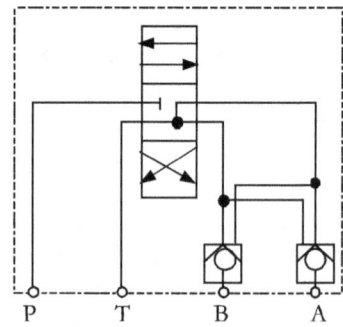

Figure 14.3 | A valve combination with pilot-operated check valves

Typical Applications of Non-Return Valves

There are many low-pressure and high-pressure hydraulic applications where the check valves and their variants can be used. Some typical applications are shown in Figure 14.4. A check valve can be installed in a hydraulic circuit between the pump and the rest of the downstream circuit to prevent pressure surges from the downstream circuit from impacting the pump. Other typical applications of check valves include accumulator charging circuits for energy storage. A check valve is often used as a bypass valve in many hydraulic components, including coolers/filters. A pilot-to-open check valve can be used to control, hold, or lock a descending load.

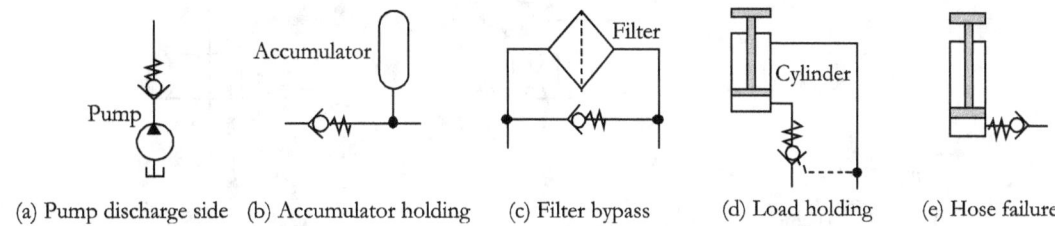

(a) Pump discharge side (b) Accumulator holding (c) Filter bypass (d) Load holding (e) Hose failure

Figure 14.4 | Typical applications of non-return valves

Load Control, Load Holding, and Load Locking

Hydraulic cylinders or motors are widely used to lower elevated, high-inertia loads. An elevated load should move down safely when hydraulic pressure is applied. However, the loaded cylinder tends to descend uncontrollably due to gravity, leading to cavitation. Therefore, necessary controls must be incorporated into these systems to bring the load under control.

Basically, three application requirements for bringing an elevated heavy load are well recognised. It is required to (1) bring the load down to the intended final position in a controlled manner, (2) hold the load in mid-positions, and (3) lock the load positively in a mid-position. Pilot-operated check valves can be used to realise these functions. The following examples provide solutions to these requirements. A simple solution to positively hold or lock a load in position is to use a 4/3-DC closed-centre valve with the load ports blocked in its centre position. However, this solution does not work satisfactorily as the actuator drifts due to the DC valve's spool leakage.

Example 14.1 | A Basic Hydraulic Circuit for Controlling a Descending Load

Develop a basic circuit to move a heavy load attached to the piston rod end of a vertically mounted double-acting hydraulic cylinder downward to the final position in a controlled manner. A pilot-operated check valve and a 4/2-DC valve are used to realise the control function. The system shall be able to set a maximum pressure of 3000 psi.

Solution

Figure 14.5 shows the three positions of the basic hydraulic circuit with a pilot-operated check valve connected to the piston-rod end of the vertically mounted double-acting cylinder, controlling the descent of the heavy load attached to the cylinder. The pilot signal for the check valve is taken from the line connected to the piston side. The manually actuated 4/2-DC valve controls the direction of the cylinder's motion. The system operating pressure of 3000 psi can be set by using a pressure relief valve, as shown. When the 4/2-DC valve is shifted to its right envelope (normal position), as shown in Figure 14.5(b), the check valve allows the fluid to enter the piston-rod end of the cylinder, retracting the piston-rod and raising the load hydraulically.

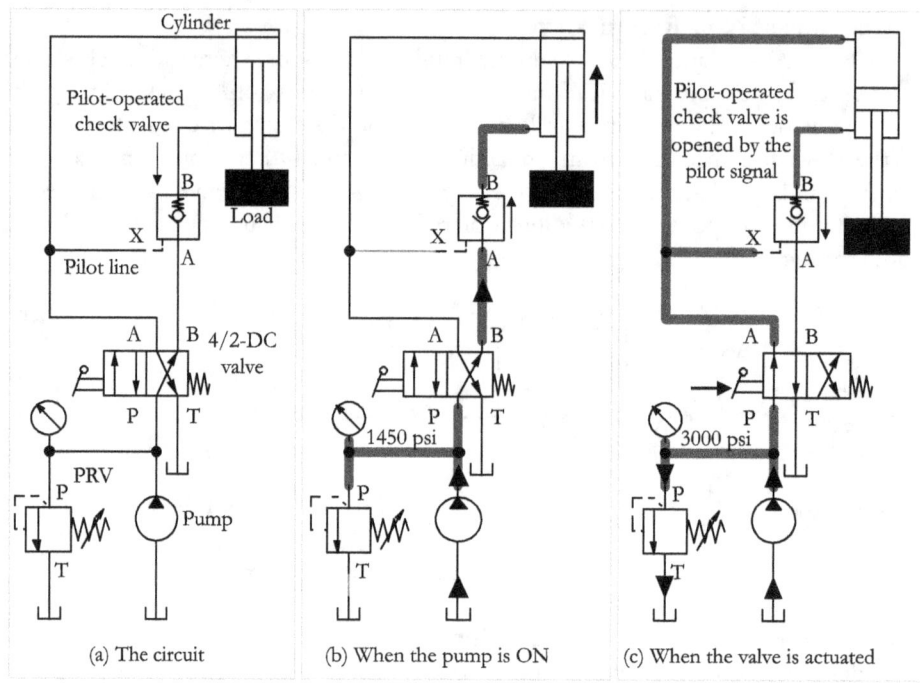

Figure 14.5 | Three positions of a hydraulic circuit for load control (Example 14.1)

When the 4/2-DC valve is shifted to its left envelope, as shown in Figure 14.5(c), the fluid flow is directed toward the piston side of the cylinder, building pressure in this line. The check valve is initially closed until enough pressure is applied in its pilot line. Once sufficient pressure builds, the check valve opens, allowing fluid to escape from the piston-rod side and lowering the load. If the cylinder overruns with the load, the pressure in the pilot line drops. When the pressure falls below a threshold, the check valve closes, blocking the return flow from the cylinder and halting the load's movement. As enough pressure builds up again, the check valve opens and moves the load. This cycle repeats until the cylinder fully descends. This circuit prevents the load from running away. A load-control valve improves operational safety.

Drawback: The pilot-operated check valve is only an economical, on-off, non-modulating device and is not suitable for precise motion control of a free-falling load. The load-control circuit using this valve develops high-pressure spikes as the cylinder extends. As explained, the cylinder tends to run away and then stops cyclically. This runaway-and-stop cycle repeats until the cylinder reaches its final position. Therefore, using this valve to control an overrunning load can cause jerky and severely unstable cylinder motion.

The best way to address runaway load and pressure buildup in the load-holding circuit is to use a counterbalance/brake valve instead of a pilot-operated check valve. A section of Chapter 16 details load-holding and load-locking methods using counterbalance/brake valves.

Example 14.2 | A Hydraulic Circuit for Holding a Heavy Load in a Mid-position

A hydraulic system should be designed with a vertically mounted double-acting hydraulic cylinder for moving heavy loads and, if required, for holding loads in mid positions. Use a 3/2-DC valve, a 4/2-DC valve, and a pilot-operated check valve. Develop a hydraulic circuit to implement the control scheme. The system shall set a max pressure of 3000 psi.

Solution

Figure 14.6(a) shows a hydraulic circuit for controlling a vertically mounted double-acting hydraulic cylinder to move and hold heavy loads in a mid-position against gravity using a pilot-operated check valve, a 4/2-DC valve (V1), and a 3/2-DC valve (V2). The pilot-operated check valve can control flow in both directions to move the cylinder back and forth. The 4/2-DC valve extends the cylinder, and the 3/2-DC valve controls the pilot line of the check valve. A PRV sets the system pressure at 3000 psi.

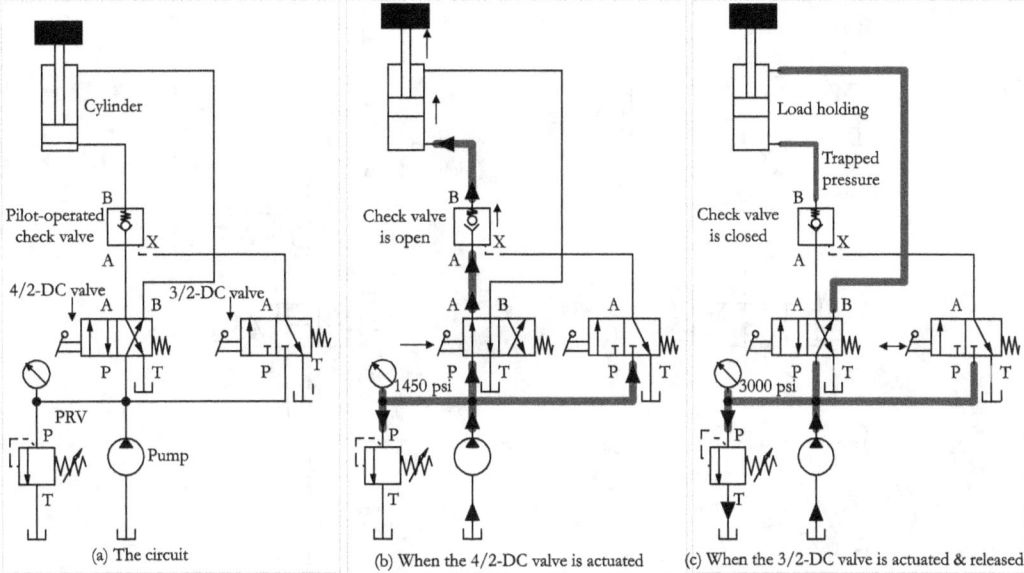

Figure 14.6 | Three positions of a load-holding hydraulic circuit (Example 14.2)

When the 4/2-DC valve is pressed, as shown in Figure 14.6(b), the flow is directed to the piston side of the load-carrying cylinder through the pilot-operated check valve extending the cylinder. As soon as the 4/2-DC valve is released, the flow is directed to the piston-rod side of the cylinder. However, the cylinder's return flow is blocked by the pilot-operated check valve when the pilot pressure is absent. The cylinder then holds the load in that position against gravity.

When the 3/2-DC valve is pressed, as shown in Figure 14.6(c), the flow is directed to the pilot line of the check valve to override the valve's normal blocking state and retract the cylinder. As soon as the 3/2-DC valve is released, the pressure in the pilot line is released. The check valve remains closed, blocking the cylinder return flow. The cylinder then securely holds the load in that position.

Drawback: The cylinder, which holds the load in mid-position during the cylinder's travel, can drift due to leakage through the directional control valve.

Example 14.3 | A Hydraulic Circuit for Holding and Locking a Load in a Vertical Lifting System

A load-carrying double-acting hydraulic cylinder, used in a vertical lifting system, should be able to extend, retract, and stop at any position. The lift must be securely locked in place for an indefinite period without drifting when the pump stops, even if there is a minor leak or a pressure reduction in the main circuit. The system should be capable of reaching a maximum pressure of 3000 psi. Develop a fail-safe control circuit that keeps the lift locked positively in place and prevents it from falling.

Solution

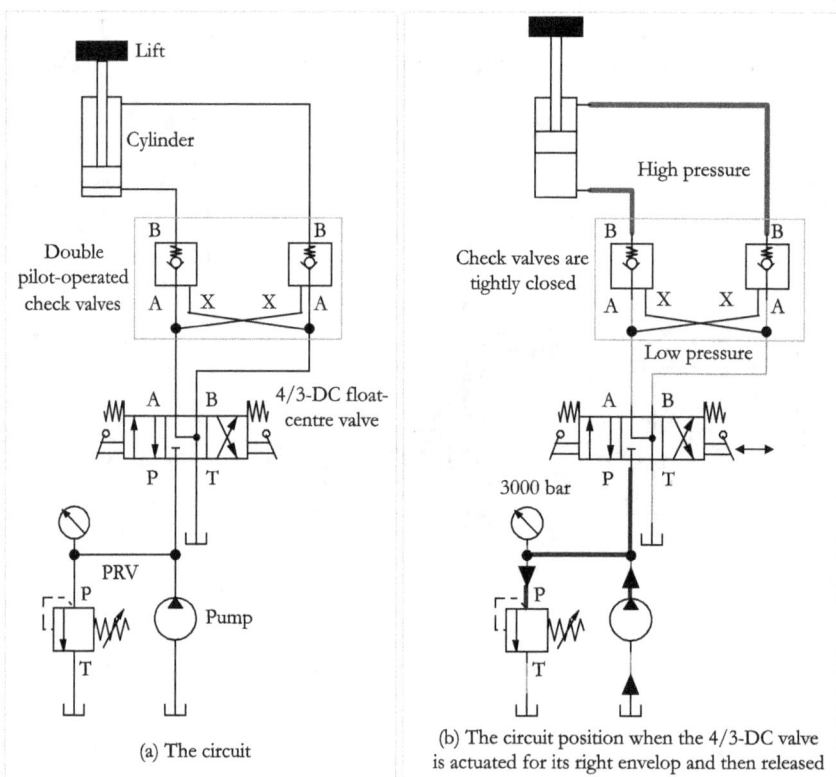

(a) The circuit

(b) The circuit position when the 4/3-DC valve is actuated for its right envelop and then released

Figure 14.7 | Two positions of a load-locking hydraulic circuit (Example 14.3)

Figure 14.7(a) shows the circuit for a vertically mounted hydraulic cylinder with a lifting platform. The circuit is designed with a power pack, double-pilot-operated check valves, and a 4/3-DC float-centre valve to positively lock the load in place. The double pilot-operated check valves are ideal for preventing the associated cylinder from drifting due to gravity or system leaks through the main control valve. The Figure shows that the pilot line of each check valve is cross-connected to the opposite line, enabling automatic pilot operation in either direction of fluid flow. Further, the directional control valve must feature a float-centre spool to vent the pilots, ensuring the check valve poppets can seat tightly.

Figure 14.7(b) shows the circuit position when the 4/3-DC valve is shifted to its right envelope position and then released. When the 4/3-DC valve is actuated, pilot pressure is applied to the normally blocked check valve, enabling it to open. This allows fluid to flow in both directions, retracting the cylinder to lower the elevated load. When the 4/3-DC valve is released and returned to its centre position, the check valves automatically close, trapping the fluid on the actuator side while simultaneously releasing the pressure on the other side through the 4/3-DC valve. This results in the cylinder holding the load in the mid position. The load remains in the mid position and is positively locked by the tightly closed fail-safe check valves. Load-holding and load-locking valves improve the safety and performance of operations.

Objective-type Questions

1. Mark the type of directional control (DC) valve that preferentially permits fluid flow in one direction and blocks the flow in the reverse direction:
   a) 3/2-DC valve
   b) 4/2-DC valve
   c) 4/3-DC valve
   d) Check valve

2. A check valve is essential for:
   a) Load-holding circuit
   b) Energy storage using an accumulator
   c) Pump bypass circuit using an unloading valve
   d) All of the above

3. A critical valve required for a hydraulic load-holding circuit is:
   a) Check valve
   b) Pilot-operated check valve
   c) 2/2-DC valve
   d) 3/2 DC valve

4. Identify the symbol:

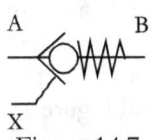

Figure 14.7

   a) Check valve
   b) Orifice valve
   c) Pilot-operated check valve
   d) 2/2-way valve

Review Questions

1) What is an NRV as used in hydraulic circuits? Describe the various types of NRVs.
2) Explain how the primary hydraulic non-return valve works with a neat sketch.
3) Describe the function of a pilot-operated check valve with a neat diagram.
4) Explain the operation of the pilot-operated double check valve.
5) Show the application of a pilot-operated check valve with the help of a simple circuit.
6) Describe the typical applications of check valves, each with the help of a hydraulic circuit.
7) A 4/3-way valve should control a double-acting hydraulic cylinder driving a heavy load. Develop a circuit to hold and lock the load in position for extended periods using pilot-operated check valves.

Objective-type questions - answer key: 1-d, 2-d, 3-b, 4-c

Chapter 15 | Flow Control Valves and Circuits

An important issue in a hydraulic system is how to control the speed of actuators. The flow rate of the system fluid controls the speed of the actuators. The flow rate can be varied using a variable-displacement pump, as in a pump-controlled system, or a fixed-displacement pump in conjunction with flow control valves, as in a valve-controlled system.

Pump-controlled System vs. Valve-controlled System

In the pump-controlled system, the flow rate can be varied by adjusting the pump's swash-plate angle. This system works well for a single-actuator hydraulic circuit. However, for many applications, flow control valves are used to control the flow rate of the system fluid, as they provide a fast response.

Flow Control Valves

An essential function of the flow control valve is to provide hydraulic resistance to fluid flow and, hence, to control the volume of fluid passing through a given point in the system. The flow control function can be realized using a controlled restriction in the fluid flow path. The length of the restriction can be short or long. Accordingly, flow control devices are classified as either orifice or throttle valves. When a flow control valve is to be used in a hydraulic system, the following factors of the valve should be evaluated: (1) the pressure rating, (2) the pressure differential across the valve, (3) the flow rating, and (4) the temperature rating. It is to be noted that the throttling action of a flow control valve is due to the division of the flow between the valve and the system relief valve.

Concept of Flow Division

The circuits of Figures 15.1(a) and (b) explain the concept of the flow division in a hydraulic circuit. Figure 15.1(a) shows a fixed-displacement pump delivering a fixed amount of fluid per unit time (say 13.2 gpm) to the cylinder through a throttle valve. The resulting excess flow cannot go anywhere when the pump delivery is higher than the flow rate through the throttle valve. This excess flow results in a pressure buildup in the line between the pump and the throttle valve until the pump stalls, the seals break, or the line blows. Therefore, it is necessary to bypass the excess flow by installing a pressure relief valve in the flow line, as shown in the modified Figure 15.1(b). That is, a portion of the pump delivery flows through the throttle valve's constant restriction, and any excess delivery, if any, flows through the pressure relief valve. This type of flow division is essential for obtaining flow control in a hydraulic circuit with a fixed-displacement pump.

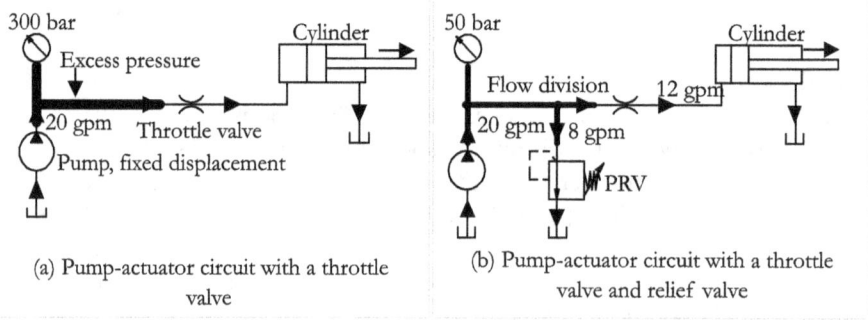

(a) Pump-actuator circuit with a throttle valve

(b) Pump-actuator circuit with a throttle valve and relief valve

Figure 15.1 | Circuit diagrams illustrating the concept of the flow division

Types of Flow Control Valves

Many variants of flow control valves exist to realize additional control functions. An important variant is the one-way flow control valve. Another important consideration in selecting a flow control valve is the effect of variations in the fluid medium's pressure and temperature on the valve. Another class of flow control valves includes flow dividers for dividing a single flow into two or more flows. The flow control valves are also available as line-mounted and cartridge versions. Figure 15.2 shows the symbols for the basic flow control valves and their variants.

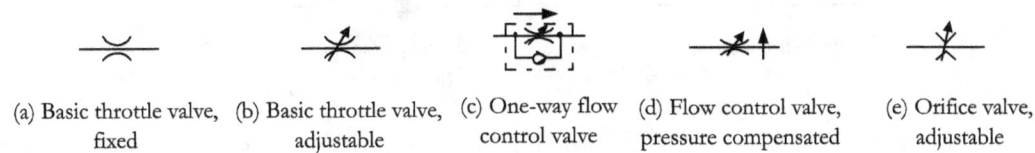

(a) Basic throttle valve, fixed   (b) Basic throttle valve, adjustable   (c) One-way flow control valve   (d) Flow control valve, pressure compensated   (e) Orifice valve, adjustable

Figure 15.2 | Symbolic representations for various types of flow control valves

Typical Applications of Throttle Valves

Throttle valves are used in various applications, including conveyors, food processing machines, and material-handling equipment. Figure 15.3 shows some important applications of the flow control valves. The most important application of a flow control valve is for speed control of a hydraulic actuator. The throttle valve can be installed upstream or downstream of the final control element, as shown in parts (a) and (b) of the figure, respectively. With the integration of a check valve into a flow control valve, the flow in only one direction can be restricted, and the actuator's speed in only one direction can be controlled [See part (c)]. Flow control valves are also frequently used in pressure gauges to dampen pressure surges that may affect them [See part (d)]. The flow control valves can be connected in series with or in parallel to hydraulic motors to control their speeds [See part (e)]. They can also be used as an integral part of other valves, such as directional control valves, counterbalance valves, pressure-reducing valves, and flow dividers.

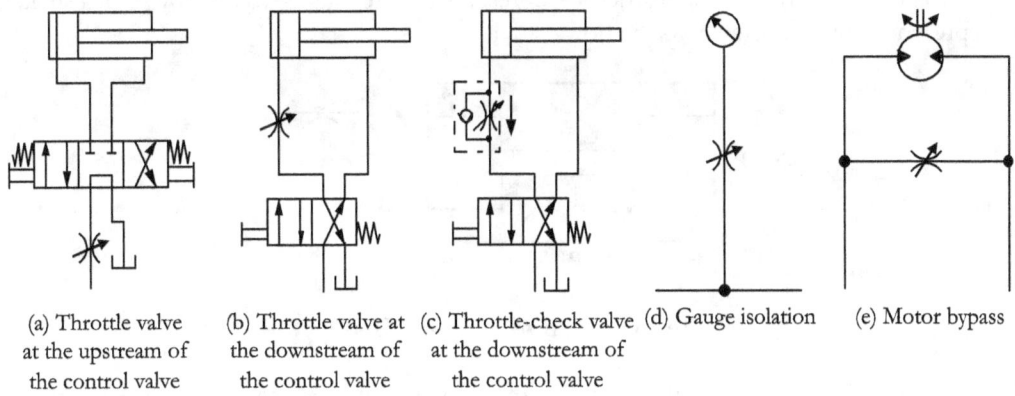

(a) Throttle valve at the upstream of the control valve   (b) Throttle valve at the downstream of the control valve   (c) Throttle-check valve at the downstream of the control valve   (d) Gauge isolation   (e) Motor bypass

Figure 15.3 | Partial hydraulic circuits showing typical applications of throttle valves

Orifice Valve

An orifice valve consists of a housing with a disk, as shown in Figure 15.4. The disk is usually embedded with a sharp-edge or square-edge orifice. The function of the orifice is to control the flow rate through a hydraulic system. The flow rate through the orifice depends on the size of the opening and, hence, the orifice area. As the restriction is of negligible length, the flow rate through the orifice valve is not affected by the fluid's viscosity. Orifice valves can be fixed or variable. The adjustable orifice valve can be used for a simple, infinitely variable-speed control application under constant pressure conditions. An orifice valve permits a finer flow rate adjustment, especially with a triangular inlet opening.

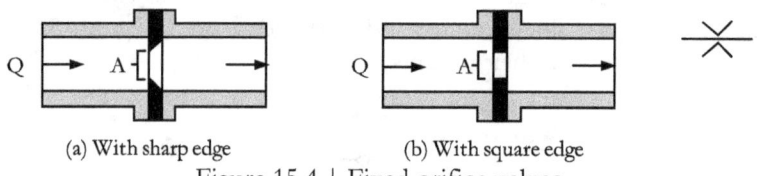

(a) With sharp edge (b) With square edge

Figure 15.4 | Fixed orifice valves

Throttle Valves

A throttle valve is a device with a lengthy restriction that offers resistance to the system fluid flowing through it. Precisely speaking, the throttle valve finely regulates the flow rate of the fluid. According to the type of restriction, there are two types of throttle valves. They are: (1) Fixed type and (2) Adjustable type (Needle valve). In a fixed-type throttle valve, the restriction is fixed, whereas in an adjustable-type throttle valve, the restriction's area can be varied.

There are two basic types of throttle valves based on their dependence on pressure variations. They are: (1) pressure-dependent throttle valves and (2) Pressure-compensated throttle valves. In a pressure-dependent throttle valve, the flow rate varies appreciably with pressure fluctuations across the valve. However, the pressure-compensated throttle valve maintains a constant pressure differential across the valve. The following sections explain these types of throttle valves.

Throttle valve, Fixed-type

A fixed-type throttle valve consists of a housing with a constant restriction of larger length, as shown in Figure 15.5. The valve provides constant hydraulic resistance to fluid flow in either direction. The pressure build-up ahead of the restriction permits a flow division, reducing the flow rate through the system. A broad cross-section of the restriction represents little resistance to the flow, and a small cross-section represents a high resistance to the flow.

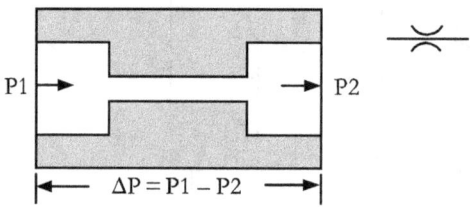

Figure 15.5 | A throttle valve, fixed-type

As the fluid flows through the throttle valve, some of the hydraulic energy is converted to thermal energy due to friction. This energy loss appears as a pressure differential (Δp) across the restriction (or valve). The flow rate through the throttle valve is a function of the following parameters: (1) the cross-section of the restriction, (2) the differential pressure (ΔP) across the valve, and (3) the viscosity of the system fluid. As the pressure drop across the valve fluctuates with variations in the connected load in the system, the flow rate through the fixed-type throttle valve cannot be maintained constant. Therefore, the flow rate through the valve is pressure-dependent.

The fixed-type throttle valve can be used for simple speed-control applications, where the pressure differential across it remains constant and precise speed is not essential. This type of valve controls the speed of a hydraulic cylinder in both directions of its motion.

Adjustable Throttle Valve, Pressure-dependent
A pressure-dependent adjustable throttle valve creates adjustable hydraulic resistance without compensating for pressure variations across the valve. In this type, the flow rate through the valve varies appreciably for the same setting due to the pressure fluctuations across the valve. Figure 15.6 gives the cross-sectional view of the adjustable throttle valve. It consists of an orifice whose cross-section can be controlled by an externally adjustable needle-shaped plunger. The flow through the controlled cross-section can be metered precisely by the pointed needle. However, it can handle only relatively small flows.

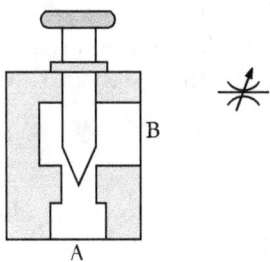

Figure 15.6 | An adjustable throttle valve, pressure-dependent

In a simple throttle valve, the differential pressure across the valve increases as the flow rate increases. Because of this operating characteristic, pressure-dependent flow-control valves cannot be used for precise speed-control applications. This type of throttle valve can be used in applications with minimal pressure fluctuations, such as lifting platforms or clamping fixtures. For precise speed control applications, pressure-compensated throttle valves are most appropriate.

Compensated Flow Controls Valves
Flow control valves can be non-compensated or compensated. In a non-compensated flow control valve, the flow rate across the valve varies with changes in the pressure drop across it. In a compensated type, the valve automatically adjusts to changes in pressure or temperature conditions within the valve to maintain a constant flow rate.

Pressure-compensated Adjustable Throttle Valve
As you know, the flow rate through an elementary throttle valve varies appreciably with the same setting due to the pressure fluctuations across the valve. However, a pressure-compensated throttle valve maintains a constant pressure differential across the valve, irrespective of fluctuations in the connected load or the valve inlet pressure, through a pressure-regulating section.

As shown in Figure 15.7, a pressure-compensated adjustable throttle valve consists of: (1) an adjustable throttling section and (2) a pressure-regulating section. The throttling section controls the flow rate, and the pressure-regulating section regulates the pressure across the valve. The setting of the valve's orifice area determines the flow rate through the valve. The pressure-regulating section consists of the spring-biased compensator spool (piston). Initially, the regulator is fully open. However, the piston adjusts its position as the flow rate varies to maintain a constant flow through the valve. The concept of pressure compensation is proved mathematically in the following paragraphs.

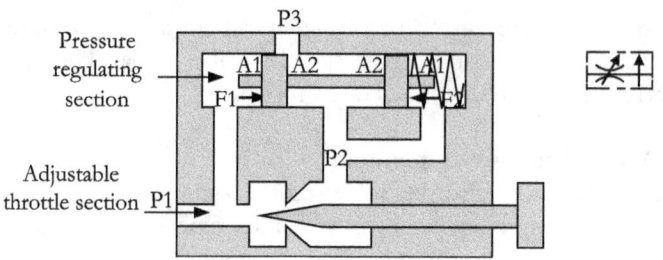

Figure 15.7 | A pressure-compensated flow control valve

Let P1 be the inlet pressure, P2 be the pressure downstream of the throttle section, and P3 be the outlet pressure, as shown in Figure 15.7. Let A1 and A2 be the areas on either side of the pressure-regulating piston, and 'F_S' be the spring force. The regulating piston is loaded from the left by the force F1, which is a function of the pressures P1 and P2, and the piston areas A1 and A2.

That is, F1 = P1 x A1 + P2 x A2

The regulating piston is loaded from the right by a force F2, which is a function of the pressure P2, the piston areas A1 and A2, and the spring force F_S.

That is, F2 = P2 x A2 + P2 x A1 + F_S

Forces F1 and F2 acting on the regulating piston must be in equilibrium. Therefore, F1 = F2.

That is, P1 x A1 + P2 x A2 = P2 x A2 + P2 x A1 + F_S
Or, P1 x A1 = P2 x A1 + F_S
Or, P1 x A1 - P2 x A1 = F_S
Or, (P1 – P2) x A1 = F_S
Or, (P1 – P2) = F_S/ A1
Or, ΔP = F_S/ A1

As F_S and A1 are constant, ΔP must also be constant. When ΔP remains constant irrespective of the pressure fluctuations in the valve, the flow rate through the valve also tends to stay constant.

Temperature Compensation, Throttle Valves
Variations in the fluid temperature in a hydraulic system tend to alter the fluid's viscosity. As the fluid viscosity varies, the flow rate through a throttle valve also changes for a given valve setting. In many hydraulic systems, the fluid flow rate must be maintained constant, regardless of temperature fluctuations. For such applications, temperature-compensated throttle valves can compensate for temperature fluctuations and the consequent changes in flow rates.

In a typical design, the temperature-compensated throttle valve uses a sharp-edged metering orifice, where the flow rate through the orifice is insensitive to variations in fluid viscosity. In some other types of design, the temperature-compensated throttle valve uses an orifice with dissimilar metals to automatically adjust its area in response to variations in system temperature, maintaining a constant flow rate through the valve.

One-way Flow Control Valve

Hydraulic flow control valves, such as orifice valves and elementary throttle valves, as discussed in the previous sections, provide the same flow restriction in either direction of fluid flow. A one-way flow control valve restricts fluid flow in one direction and allows unrestricted flow in the opposite direction.

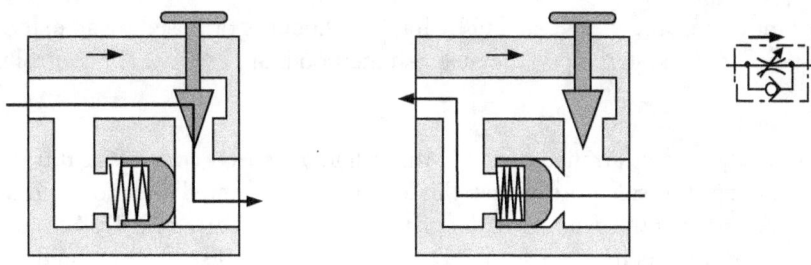

(a) Flow through the throttle section   (b) Flow through the check valve section

Figure 15.8 | Two critical positions of a one-way flow control valve

A one-way flow control valve consists of a throttling section and a check valve section, both integrated into the valve body in parallel. It is also known as a throttle check valve. The annular gap at the throttling point can be controlled by turning the throttling screw. Figure 15.8 gives the schematic diagrams representing the two positions of the one-way flow control valve.

The check valve blocks the system flow in one direction, forcing the fluid to flow through the controlled cross-section, as shown in Figure 15.8(a). Hence, the flow is throttled and its rate is controlled in that direction. In the opposite direction, the fluid flows freely through the opened check valve without any restriction, as shown in Figure 15.8(b). This valve is used when a direction-sensitive speed control of a hydraulic actuator is required, as in clamping applications.

Speed Control of a Cylinder Using a One-way Flow Control Valve

The one-way flow control requirement for an actuator in a hydraulic system can be achieved using a throttle check valve. Usually, this valve is installed in-line between the actuator and its final control element. Depending on how the throttle check valve is placed in the system, there are two methods for direction-sensitive speed control. These are: (1) the meter-in method and (2) the meter-out method.

In addition to these two methods, the bleed-off method can also be used to achieve speed control in hydraulic systems. The following sections further explain these methods.

Meter-in Method

In the meter-in method, the speed of a hydraulic cylinder can be controlled in only one direction by restricting the system flow into the cylinder and allowing unrestricted flow out of it. The meter-in circuit is ideal for applications where an external load opposes the cylinder's motion, providing a definite resistance to flow during the controlled stroke. This method provides finer speed control than other speed control methods. However, this method cannot provide accurate speed control of an actuator under negative load conditions. Some applications that employ the meter-in method of speed control include feeding grinder tables, welding machines, milling machines, and rotary hydraulic motor drives.

Example 15.1 | Speed Control of a Hydraulic Cylinder Using the Meter-in Method
Develop an appropriate circuit to control the speed of a double-acting hydraulic cylinder during its forward stroke under an opposing external load.

Solution
As the forward motion of the double-acting hydraulic cylinder is opposed by an external load offering a positive resistance to the system flow, the meter-in method is appropriate for controlling the cylinder's speed.

Figure 15.9 shows the circuit arrangement for the cylinder's speed control. A throttle check valve is installed between the piston-side port of the cylinder and the associated directional control valve in such a way that the fluid entering the cylinder at the piston-side port is throttled. This control action reduces the cylinder's speed during its forward stroke. During the retraction stroke, the fluid in the cylinder can freely pass into the system reservoir through the open check valve, allowing the cylinder to retract at its standard speed.

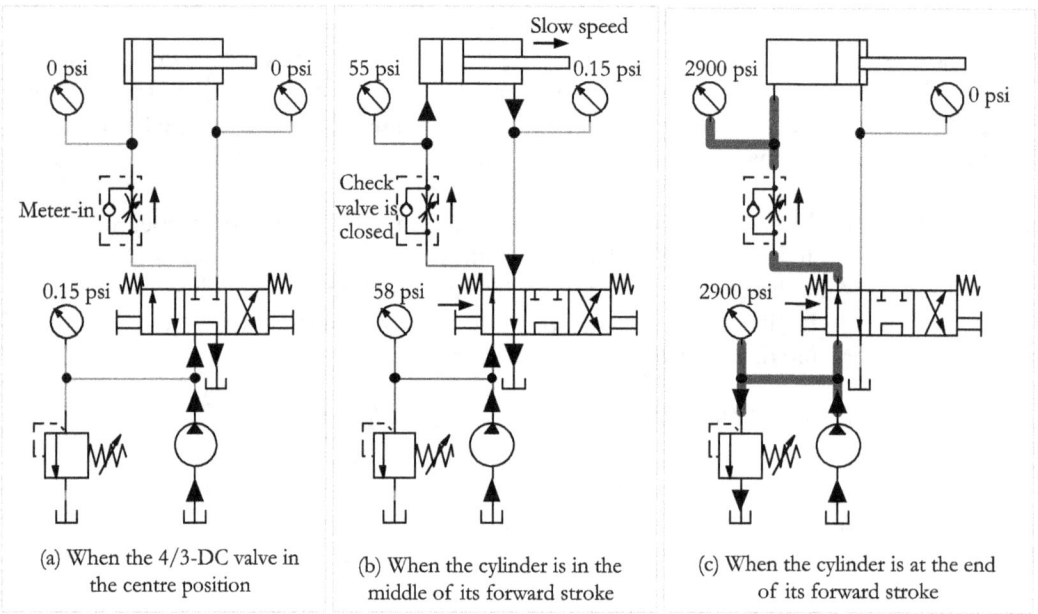

Figure 15.9 | A circuit for controlling the speed of a hydraulic cylinder (Example 15.1)

Meter-out Method
In the meter-out method, the speed of the cylinder in a given direction can be controlled by restricting flow away from the cylinder and allowing unrestricted flow into it. This method maintains constant backpressure while the cylinder moves and prevents its lunging if the load connected to the cylinder drops suddenly. Therefore, this method can be used in actuator systems with overrunning (negative) loads (e.g., crane booms and winches).

However, using the meter-out control for its forward stroke may intensify pressure at the piston rod side of the cylinder. For example, if the pressure on the piston side of a cylinder with an area ratio of 2:1 is 1450 psi, the pressure on the piston rod side can reach 2900 psi.

Example 15.2 | Speed Control of a Hydraulic Cylinder Using the Meter-out Method

Develop an appropriate circuit for speed control during the forward stroke of a vertically mounted double-acting hydraulic cylinder, where a heavy load pulls the piston down. How can speed control be achieved during the return stroke?

Solution

As the weight pulls the piston of the double-acting hydraulic cylinder down, the meter-out method is appropriate for controlling the cylinder's speed to prevent its lunging.

Figure 15.10(a) gives the circuit arrangement for the cylinder's meter-out speed control method. This method installs a throttle-check valve between the cylinder's piston rod-side port and its main directional control valve.

Figure 15.10(b) shows the circuit's position when the 4/3-DC valve is actuated. The fluid entering the cylinder flows freely, and the fluid leaving the cylinder is restricted through the throttle valve to control its speed during its forward stroke.

Figure 15.10(c) shows the circuit's position when the cylinder reached the end of the stroke.

During the return stroke, the fluid from the pump can pass freely through the open check valve to the cylinder, allowing the cylinder to retract at its normal speed.

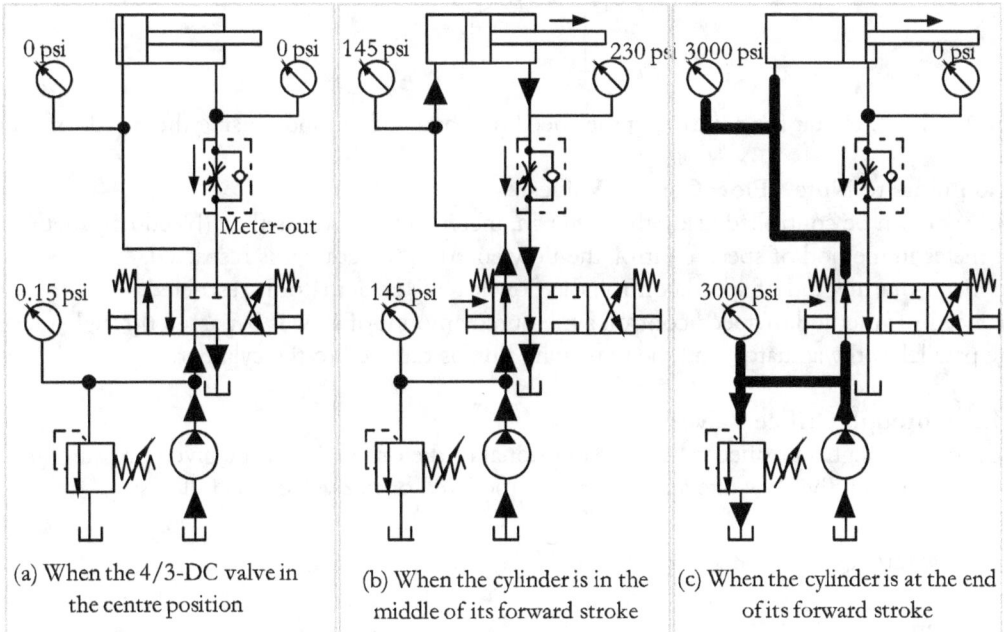

(a) When the 4/3-DC valve in the centre position

(b) When the cylinder is in the middle of its forward stroke

(c) When the cylinder is at the end of its forward stroke

Figure 15.10 | A circuit for controlling the speed of a hydraulic cylinder (Example 15.2)

To achieve speed control during the return stroke of the cylinder, another throttle check valve can be installed in the line between the piston-side port and the main control valve, so that the valve throttles the fluid leaving the cylinder through the piston-side port. (Not shown)

Bleed-Off Method

Another speed-control method for a hydraulic cylinder is bleed-off control (or bypass flow control). In this method, one end of the flow control valve is connected to the cylinder, as shown in Figure 15.11. The other end of the bleed-off valve is directly connected to the system reservoir. A measured part of the fluid intended for the cylinder can be bled off to the reservoir through the throttle valve.

Because the pump pressure is only high enough to move the load, this type of circuit usually generates less heat and is more energy-efficient than the meter-in/meter-out methods. However, the bleed-off control cannot maintain precise speed regulation when the flow in the inlet line changes. Moreover, this speed control method can only control the speed of one cylinder at a time and cannot be used with actuators that have overrunning loads.

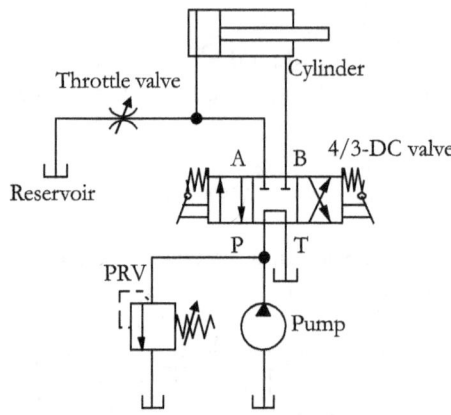

Figure 15.11 | A circuit for controlling the speed of a hydraulic cylinder using the bleed-off method

General Points to Note - Flow Control Valves
The flow rate can be controlled using the meter-in, meter-out, or bypass flow (bleed-off) methods.
-In the meter-in method of speed control, the flow entering an actuator is restricted.
-In the meter-out method of speed control, the flow leaving an actuator is restricted.
-In the bleed-off method of speed control, a measured amount of flow is bypassed through a restricted passage parallel to the actuator, and the remaining fluid is directed to the cylinder.

Flow Rate through Orifice Valves
The flow rate through an orifice valve is proportional to the orifice area for a given pressure drop. The relationship between the flow rate and the area of the fixed orifice valve is as follows:

$$\text{Flow rate, } Q = 38 \times C \times A \times \sqrt{(\Delta P / SG)}$$

Where,
- Q = Flow rate (gpm)
- C = Flow coefficient (= 0.8 for sharp-edged orifice, = 0.6 for square-edged orifice)
- A = Orifice area (in^2)
- ΔP = Pressure drop across the orifice (psi)
- SG = Specific gravity of the fluid (for mineral oil = 0.85 – 0.9)

Example 15.3 | Flow Rate Through an Orifice

A sharp-edged orifice valve having a ¾-inch diameter produces a pressure drop of 80 psi across the valve when fluid with a specific gravity of 0.9 flows through it. Find the flow rate through the valve.

Solution

- Pressure-drop (ΔP) = 80 psi
- Orifice diameter, D = ¾ in
- Orifice area, A = 0.44 in²
- Specific gravity, SG = 0.9
- Capacity coefficient, C = 0.8 [For sharp-edged orifice]

Flow rate, Q = 38 x C x A x $\sqrt{(\Delta P/SG)}$
 = 38 x 0.8 x 0.44 x $\sqrt{(80/0.9)}$
 = 126 gpm

Flow Rate through Throttle Valves

The flow rate through a needle (throttle) valve is substantially influenced by the pressure drop across the valve. As given for the directional control valves, the general equations for the flow rate can be expressed as follows:

Flow rate, Q = Cv x $\sqrt{(\Delta P/SG)}$

Where,
- Q is the flow rate in gpm
- ΔP is the pressure drop developed across the valve in psi
- Kv [Cv] is the capacity coefficient in gpm/$\sqrt{psi}$
- SG is the fluid's specific gravity

The flow coefficient Cv of a flow control valve in its fully open position is determined experimentally. It is listed as the rated Cv in its manufacturer's catalogs.

Example: 15.4

Calculate the capacity coefficient of a flow control valve that experiences a pressure drop of 58 psi for the flow rate of 30 gpm. The hydraulic fluid passing through the valve has a specific gravity of 0.9.

Solution

- Pressure-drop, ΔP = 58 psi
- Flow rate, Q = 25 gpm
- Specific gravity, SG = 0.9

Capacity coefficient, Cv = Q/$\sqrt{(\Delta P/SG)}$
 = 30 /$\sqrt{(58/0.9)}$
 = 3.74 gpm/$\sqrt{psi}$

Regenerative Circuit

In an earlier chapter, we learned that a typical double-acting hydraulic cylinder with piston area 'Ap' and piston rod area 'Ar' produces a higher speed when retracting than when extending, provided the pump flow remains constant. However, many hydraulic circuits require faster extension strokes to reduce cycle times. A regenerative circuit can be employed to increase the speed of a cylinder during its extension stroke. The regenerative circuit is explained using Figure 15.12. The circuit consists of the cylinder powered by a fixed-displacement pump with a flow rate Q_P and controlled by a 3/2-DC valve. The lines to the cylinder's ports are connected in parallel.

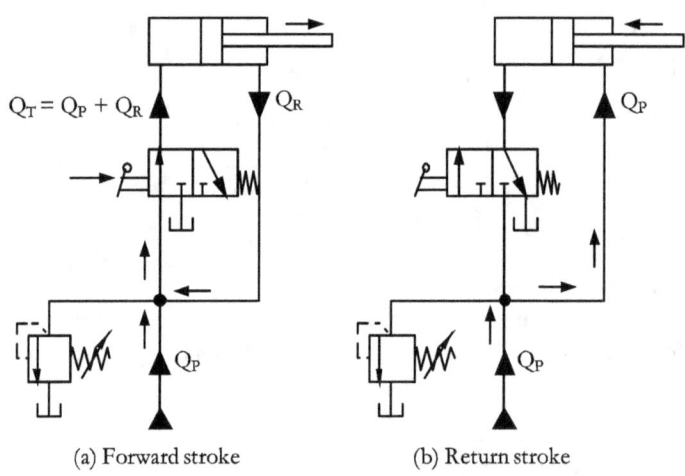

Figure 15.12 | Two critical positions of a regenerative hydraulic circuit

During the cylinder's forward stroke, the fluid from the pump (Q_P) flows to the cap end through the DC valve, and the flow from the piston rod end (Q_R) regenerates with the pump flow. During this stroke, the total flow (Q_T) to the cylinder's cap end is the sum of Q_P and Q_R.

That is, $$Q_T = Q_P + Q_R$$

Therefore, the pump flow rate Q_P is given by

$$Q_P = Q_T - Q_R$$

As we are aware, the total flow rate (Q_T) is the product of the piston area (A_P) and the extension speed (v_{ext}), and the regenerative flow rate (Q_R) is the product of the active area on the piston rod side (that is, $A_P - A_r$) and the extension speed. We have,

$$\begin{aligned} Q_P &= Q_T - Q_R \\ &= A_P \times v_{ext} - (A_P - A_r) \times v_{ext} \\ &= A_r \times v_{ext} \end{aligned}$$

Therefore, $$v_{ext} = Q_P / A_r$$

The equation above suggests that the speed of the extension stroke equals the pump flow rate divided by the piston rod area. Thus, a small rod area provides a significant extension speed.

Ratio of Extension and Retraction Speeds: The ratio of the extension speed (v_{ext}) to the retraction speed (vret) of a hydraulic cylinder is related to the area ratio (A_p/A_r) of the cylinder. The area ratio can be calculated in the following manner:

Retraction speed, $\quad v_{ret} = Q_P / (A_p - A_r)$

Therefore, $\quad v_{ext}/v_{ret} = (A_p - A_r)/A_r$
$\quad v_{ext}/v_{ret} = (A_p/A_r) - 1$

The equation above suggests that the higher the ratio of the piston area to the piston rod area of the cylinder, the higher the ratio of its extending speed to its retraction speed. For a cylinder with a piston area equal to twice its piston rod area, the extension and retraction speeds would remain the same.

Load-carrying Capacity: The load-carrying capacity of the regenerative hydraulic cylinder during its extension stroke is less than that which would have been obtained from a regular double-acting hydraulic cylinder. The load-carrying capacity of the cylinder during its extension stroke equals the system pressure multiplied by the piston rod area, rather than the piston area. This is because the pressure acts on both sides of the piston during the cylinder's extension stroke.

Example 15.5 | Speed and Load-carrying Capacity of a D/A Cylinder in a Regenerative Circuit
A double acting hydraulic cylinder is used in the standard regenerative circuit with a fixed-displacement pump, a pressure relief valve, and a 3/2-DC valve. The system pressure is set at 2900 psi. The piston area is 3.14 in², and the piston-rod area is 1.57 in². The pump discharge is 15.4 in³/s. Find the speed and the load-carrying capacity for: (1) the extension stroke and (2) the retraction stroke of the cylinder.

Solution

Pressure, P $\quad = 2900$ psi
Piston area, A_{ext} $\quad = 3.14$ in²
Rod area, A_{rod} $\quad = 1.57$ in²
Flow rate, Q $\quad = 15.4$ in³/s

Rod_side area, A_{ret} $\quad = A_{ext} - A_{rod}$
$\quad = (3.14 - 1.57) = 1.57$ in²

Extension speed, v_{ext} $\quad = Q/A_{rod}$
$\quad = 15.4/1.57 = 9.8$ in/s

Retraction speed, v_{ret} $\quad = Q/A_{ret}$
$\quad = 15.4/1.57 = 9.8$ in/s

Load carrying capacity, forward stroke $= P \times A_{rod}$
$\quad = 2900 \times 1.57 = 4553$ lb

Load carrying capacity, return stroke $= P \times A_{ret}$
$\quad = 2900 \times 1.57 = 4553$ lb

Flow Divider/Combiner

A flow divider is a hydraulic device that divides a single inlet flow into two or more prescribed outlet flows regardless of the load pressures at the outlet ports. A flow combiner is a hydraulic device that combines two or more flows into a single flow. There are two common types of flow divider/combiner devices. They are: (1) rotary type and (2) sliding-spool type.

Rotary Flow Divider (or Combiner)

A hydraulic rotary flow divider (or combiner) uses two coupled identical hydraulic gear motors/pumps, as shown in Figure 15.13(a). These machines are coupled on the same driveshaft and run at the same speed. These motors control the flow rates through their outlets. The main advantage of the rotary flow dividers is that contaminants or dirt in the fluid have little effect on their performance, thanks to the large clearance between the gears and the housing of each machine. However, their disadvantages are that they are noisy and less accurate due to internal leakages.

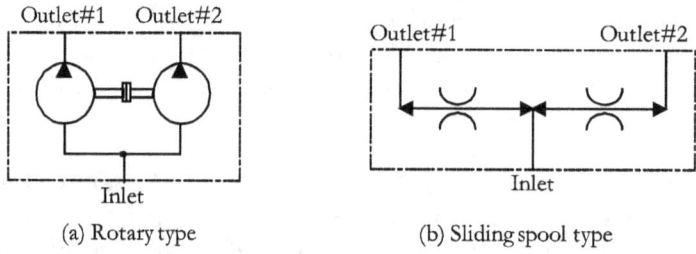

Figure 15.13 | Circuit diagrams showing the basic representations of hydraulic flow dividers

Sliding-Spool Flow Divider

A sliding-spool type flow divider is a pressure-compensated flow control valve that draws fluid from the pump and divides the flow between two functions. Moreover, in the reverse direction, it recombines the flows. It can be designed to split the flow equally between its outlets or to a specified percentage, prioritizing some key components over others. Figure 15.13(b) shows the basic representation of the flow divider with one inlet and two outlets. The inlet flow passes through the orifices, dividing into two streams. The sliding spool dividers are more accurate than the rotary flow dividers. However, they are adversely affected by the contaminants in the system fluid.

The sliding-spool-type hydraulic flow dividers are designed according to priority, proportional priority, and load sensing versions. This type of flow divider is available in a broad range of flow ratings and relief settings. Many sliding-spool flow dividers are also available in adjustable and non-adjustable versions. These extensive options allow a designer to create multiple configurations to meet the requirements of the applications. The load sensing priority valves can provide a dependable flow on demand for getting the load sensing steering, braking, or other priority services while allowing the excess flow to be used for auxiliary functions. Applications include tractors, motor graders, lift trucks, and backhoe/loaders.

Example 15.6 | Circuit for the Synchronized Movement of Two Cylinders Using a Flow Divider
Develop a control circuit to synchronize the movement of two identical cylinders under different load conditions by dividing a fixed pump flow during the forward motion of the cylinders.

Solution

Figure 5.4(a) shows a basic circuit for synchronizing the forward stroke of two identical, unequally loaded double-acting hydraulic cylinders. A constant displacement pump supplies fluid. A 4/3-DC valve is used for the directional control of the cylinders. System pressure can be regulated with the pressure relief valve. Equal fluid streams are directed to the piston side of the cylinders through a flow divider. The flow control valves in the flow divider maintain a constant flow at a given setting, regardless of variations in cylinder load. The cylinders then move synchronously.

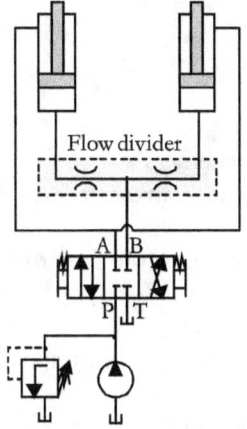

Figure 15.14 | A hydraulic circuit for the synchronized motion of two cylinders (Example 15.6)

Example 15.7 | Control Circuit for Controlling Two Hydraulic Motors Using a Flow Divider

Develop a control circuit for two bi-directional hydraulic motors connected in parallel via a pressure-compensated flow divider. Assume that the load on each motor is the same as that of the other.

Solution

Figure 15.15 gives the circuit for controlling two bidirectional hydraulic motors connected in parallel. A 4/3-DC open-center valve controls each motor. A flow divider divides the pump flow into two and delivers the fluid to the motors.

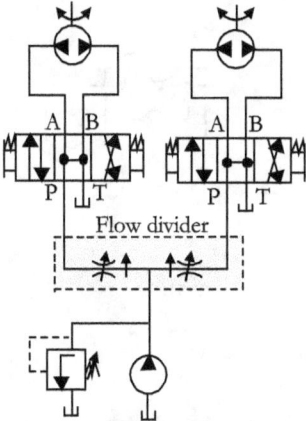

Figure 15.15 | A hydraulic circuit controlling two hydraulic motors (Example 15.7)

Objective-type Questions

1. The flow control in a hydraulic circuit is associated with:
   a) Direction control of actuators
   b) Speed control of actuators
   c) Load sensing
   d) Pressure setting

2. The volume of fluid passing through a hydraulic circuit can be controlled using a:
   a) Poppet
   b) Pilot valve
   c) Metering orifice
   d) Spool

3. Which of the following is the correct statement?
   a) A non-compensated type flow control valve automatically adjusts to the changes in pressure to produce a constant flow rate.
   b) In the meter-out method of speed control, the flow to a cylinder is restricted.
   c) A meter-in speed control circuit is ideal for applications where an external load opposes the motion of the associated hydraulic cylinder.
   d) Bleed-off control is a method of releasing the air trapped in hydraulic systems.

4. What factors are the flow rate through a throttle valve primarily dependent on?
   a) Cross-section of the valve orifice, differential pressure across the valve, and fluid density
   b) Cross-section of the valve orifice, differential pressure across the valve, and fluid viscosity
   c) The shape of the valve orifice, system pressure, and viscosity
   d) Cross-section of the orifice, system pressure, and viscosity

5. In which speed control method is the fluid leaving a hydraulic actuator only throttled?
   a) Meter-in
   b) Meter-out
   c) Bleed-off
   d) Bypass throttling

6. Identify the symbol given below

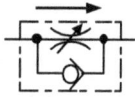

Figure 15.16

   a) Throttle valve
   b) One-way flow control valve
   c) Flow control valve with pressure compensation
   d) Throttle valve with bypass

7. In a conventional double-acting hydraulic cylinder, the speed of the forward stroke is:
   a) less than the speed of the return stroke
   b) equal to the speed of the return stroke
   c) higher than the speed of the return stroke
   d) more uniform than the speed of the return stroke.

8. Identify the symbol given below:

Figure 15.17

   a) Throttle valve
   b) One-way flow control valve
   c) Pressure-compensated flow control valve
   d) Orifice valve

9. For a hydraulic cylinder used in the regenerative circuit, the extension speed would be equal to the retraction speed if the ratio of piston area to piston rod area is
   a) 1:1
   b) 1.5:1
   c) 2:1
   d) 2.5:1

10. A regenerative circuit in a system can
    a) restrict the system flow
    b) divide the system flow into two equal streams
    c) control the speed of the associated cylinder in both directions
    d) increase the speed of the associated cylinder during its extension stroke

Review Questions
1) Explain how the flow rate of fluids in hydraulic systems can be varied.
2) Mention three ways to control the speed of hydraulic actuators. What are their relative advantages and disadvantages?
3) Mention the primary function of flow control valves used in hydraulic systems.
4) Draw the symbols for: (1) Adjustable needle valve, (2) Pressure-compensated throttle valve, (3) One-way flow control valve, and (4) Orifice valve.
5) Explain how flow control is achieved with hydraulic systems.
6) List different types of flow control valves available for hydraulic applications.
7) Which flow control valve is least affected by the fluid viscosity changes? Explain.
8) What does a one-way flow control valve mean in hydraulic systems?
9) Differentiate throttle and orifice valves used in hydraulic systems.
10) Differentiate pressure-dependent and pressure-compensated types of hydraulic flow control valves.
11) List two applications of the flow control valves.
12) What factors affect the flow rate through a throttle valve?
13) Draw the symbol for a pressure-compensated flow control valve.
14) Explain the constructional features of a pressure-compensated flow control valve.
15) How can flow control valves achieve pressure compensation? Explain.

16) Describe the operation of the one-way flow control valves used in hydraulic systems.
17) Differentiate the meter-in and meter-out methods of speed controls used in hydraulic systems.
18) Explain the meter-in method of speed control of a double-acting cylinder with a circuit diagram.
19) Explain a double-acting cylinder's meter-out speed control method with a circuit diagram.
20) Give three methods for controlling the speed of hydraulic double-acting cylinders.
21) What is the meter-in circuit as used in hydraulic systems? What are its limitations?
22) What is the meter-out circuit as used in hydraulic systems? What are its limitations?
23) Explain the function and the use of the bleed-off circuit in hydraulic systems.
24) Contrast the operating characteristics of the meter-in, meter-out, and bleed-off flow control circuits.
25) What is the purpose of the regenerative circuit in hydraulic systems?
26) What kind of circuit do you construct to achieve the same speed for a hydraulic cylinder's forward and return motions? Explain the operation of the circuit.
27) Develop the regenerative control circuit for the rapid forward stroke of a double-acting hydraulic cylinder. A tandem-center 4/3-DC valve can be used as a final control element.
28) What is a flow divider? How can it be used in hydraulic circuits?
29) Develop a hydraulic circuit with a double-acting cylinder for the following requirements: (1) the speeds of the cylinder piston in both directions should be the same, (2) the piston should stop at any desired position, (3) systems pressure should be set to 2000 psi and (4) The system should have facilities for cleaning the hydraulic fluid in at least three locations in the circuit.

Numerical Problems

1) A hydraulic valve has an experimentally determined Cv rating of 8 gpm/$\sqrt{psi}$. The valve is designed to permit 20 gpm of fluid flow. The fluid's specific gravity (SG) can be taken as 8.5. Calculate the anticipated pressure drop across the valve. [Ans: 53.125 psi]

2) A cylinder with an area ratio of 2:1 generates an extension speed of 20 in/s when connected to a hydraulic system. What is the return speed of the cylinder? [Ans: 20 in/s]

3) A double acting hydraulic actuator is used in a regenerative circuit with a pump, pressure relief valve, and a 3/2-DC valve. The system pressure is set at 3046 psi. The piston area is 1256 in^2, and the piston-rod area is 628 in^2. The pump flow is 20 gpm. Find the actuator speed and load-carrying capacity for (1) extending speed and (2) retracting speed. [Ans: Extension stroke: 0.123 in/s, 1912888 lb, Retraction stroke: 0.123 in/s, 1912888 lb]

4) A cylinder with a piston area of 12 in^2 and a piston-rod area of 6 in^2 is connected to a hydraulic system with a pump delivering 75 in^3/s. What speed does the cylinder develop during the extension stroke when connected to a hydraulic system under (1) standard mode and (2) regenerative mode? [Ans: 6.25 in/s, 12.3 in/s]

5) A cylinder with a piston area of 12 in^2 and a piston-rod area of 6 in^2 is connected to a hydraulic system with a pump delivering flow at a maximum pressure of 1523 psi. What is the cylinder's load-carrying capacity during the extension stroke when connected to a hydraulic system under (1) normal mode and (2) regenerative mode? [Ans: 8276 lb, 9138 lb]

6) A d/a hydraulic cylinder is used in a regenerative circuit with a pump, pressure relief valve, and a 3/2-DC valve. The system pressure is set at 1523 psi. The piston area is 12 in^2, and the piston-rod area is 4.8 in^2. The pump flow is 19.5 gpm. Find the cylinder's speed and load-carrying capacity for: (1) extending speed and (2) retracting speed. [Ans: Extension stroke: 15.63 in/s, 7310 lb, Retraction stroke: 10.42 in/s, 10966 lb]

Objective-type questions - answer key: 1-b, 2-c, 3-c, 4-b, 5-b, 6-b, 7-a, 8-c, 9-c, 10-d

Chapter 16 | Pressure Control Valves and Control Circuits

Hydraulic systems use pressure-control valves to perform pressure-related regulation and control tasks. These tasks include limiting the maximum operating pressure in a hydraulic system, reducing pressure in some parts of a hydraulic circuit, unloading pumps, establishing the sequence of movements for actuators, counterbalancing overrunning loads, and braking hydraulic motors while running on inertia. Accordingly, PRVs manifest in hydraulic systems in various forms.

A pressure control valve consists of a spring-biased element, such as a poppet or spool, moving within the valve body. Poppet-type valves provide a fast response, whereas spool-type valves provide consistent pressure control regardless of flow variations in the associated circuit. Pressure control valves are available in cartridge and bodied versions. They find applications in food processing, material handling, and process controls.

Symbolic Representations of Basic Pressure Control Valves

Figure 16.1 shows the basic symbols of pressure control valves. Each symbol mainly consists of a square with a movable line and an arrow. The square merely represents the valve body. The line represents the flow passage within the valve body, and the arrow represents the preferred flow direction through the valve.

A moving part of the valve body controls the flow passage. The moving part is held in its normal position by a spring whose tension can often be adjusted externally. A pressure signal from the system acts on the spring-biased spool, actuating the valve when the system pressure exceeds the valve's set pressure. The pressure control valve is usually pilot-operated to obtain precise control.

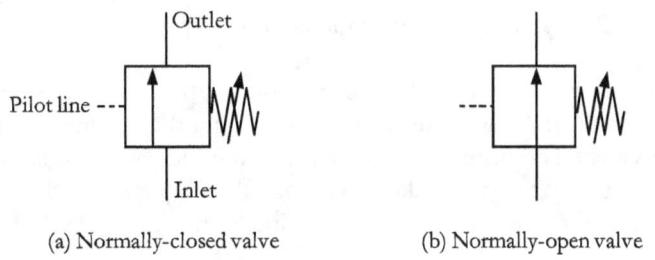

(a) Normally-closed valve   (b) Normally-open valve

Figure 16.1 | Symbolic representations of pressure control valves

The function of an elementary pressure control valve can be interpreted from its symbol. A valve either opens or blocks the internal flow passage in its normal position. Accordingly, pressure control valves can be classified as either normally open or normally closed. The normally closed type pressure control valve, as shown in Figure 16.1(a), prevents flow through the valve when the pilot pressure is below the set pressure and allows flow when the pilot pressure exceeds the set pressure. In contrast, the normally open type pressure control valve, as shown in Figure 16.1(b), allows flow when the pilot pressure is below the valve's set pressure and blocks flow when the pilot pressure exceeds the set value.

Pressure control valves are depicted using variations of basic symbols, providing a standardized representation for easy identification and understanding.

Classification of Pressure Regulating/Control Valves

Based on their function in hydraulic circuits, pressure control valves can be categorized into several basic types. They are: (1) pressure relief valves, (2) pressure-reducing valves, (3) unloading valves, (4) sequence valves, (5) counterbalance valves, and (6) brake valves. Remembering that all these valves work on the same fundamental principle is worthwhile. That is, the spring force in every pressure control valve balances the hydraulic force developed by the static or dynamic pressure prevailing on the inlet or outlet (load) side of the valve. The pressure on the spring-biased moving part produces the hydraulic force, such as a spool. When the hydraulic force exceeds the spring force, the moving part shifts its position to achieve the desired control function. A pressure control valve is provided with a check valve if a return flow passage is required for proper operation. Figure 16.2 gives the symbols for various types of pressure control valves.

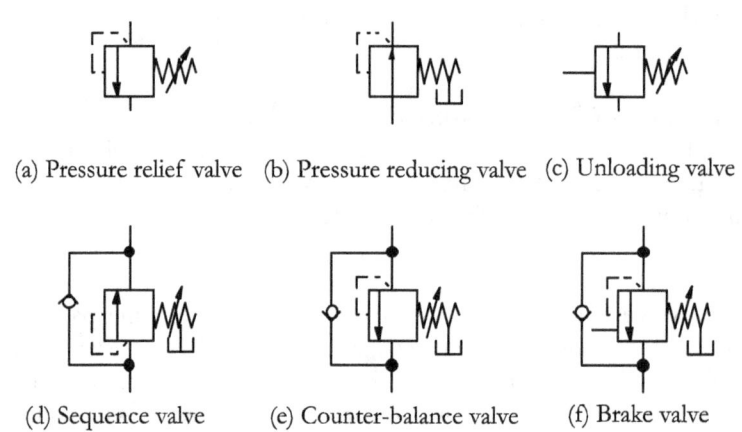

Figure 16.2 | Symbolic representations of some pressure-control valves

The functions of the pressure relief/control valves are also briefly explained below:

Pressure Relief Valve (PRV): A PRV's key function is to limit the maximum system pressure.

- Pressure-reducing valve: The primary role of a pressure-reducing valve is to limit the pressure in some parts of the associated system to a value lower than that required for the rest of the system. It is a normally open valve that throttles or closes the fluid line to maintain the set pressure in the regulated line.

- Unloading Valve: The primary function of the unloading valve is to regulate the pump pressure level by bypassing the fluid to the system reservoir at a low energy level in response to an external pressure signal received from the load section. The valve allows the pump to operate at a minimal load.

- Sequence Valve: The primary function of the sequence valve is to direct the flow from the valve's primary circuit to the secondary circuit in response to a pilot signal to obtain a predetermined sequence of operations.

- Counterbalance Valve: The main function of the counterbalance valve in a hydraulic circuit with a load-carrying actuator is to maintain the preset backpressure in the return line, sufficient to balance the load held by the actuator. An internal pilot passage senses the load pressure.

- Brake Valve (Over-center Valve): The brake valve is an extension of the counterbalance valve with the addition of an external pilot. While the counterbalance valve maintains sufficient backpressure in the circuit's return line, it forces the pump to develop the full set pressure on the counterbalance spring before it allows the associated actuator to move.

Pressure-reducing valves

An important control task in many hydraulic applications is to reduce the pressure in some parts of the circuit. That is, a part of the circuit must work at a pressure lower than that of the primary circuit. For example, excessive force (pressure) applied during clamping in a hydraulic system may damage the workpiece. In such a situation, a pressure-reducing valve can limit the clamping pressure.

A pressure-reducing valve is a normally open type that evaluates and limits the pressure at its outlet. According to how the control pressure acts on the moving element of each valve, there are two types of pressure-reducing valves. They are: (1) Direct-acting type and (2) Pilot-operated type. A pressure-reducing valve can bleed off any leakage fluid through the drain port provided on the valve. A variant of the pressure-reducing valve is a 3-way valve with a tank port. This valve can drain the fluid to the tank when the pressure at the output exceeds the set pressure. A pressure-reducing valve may also be integrated with a check valve to permit unimpeded return flow.

Direct-acting Type Pressure-reducing valve

Figure 16.3 shows a simple direct-acting pressure-reducing valve. It consists of an inlet port A, an outlet port B, a spring-loaded spool, and a pressure-adjusting screw. The valve is provided with an internal control passage to carry the fluid from the outlet (downstream) to act on one end of the spool. This connection can be used to evaluate the outlet pressure. When the outlet pressure remains below the valve setting, the fluid flows from the inlet to the outlet. When the pressure at the valve outlet exceeds its set pressure, the spool shifts to partially or fully block the outlet port. The preset pressure is maintained by the amount of fluid passing through the restricted passage.

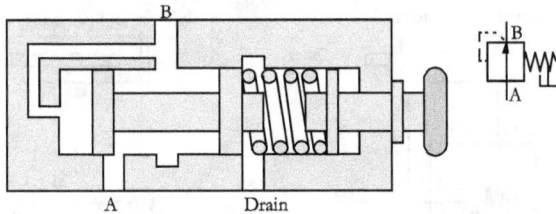

Figure 16.3 | A cross-sectional view of a simple pressure-reducing valve

Pilot-Operated Pressure-reducing Valve

Figure 16.4 shows the cross-sectional view of a pilot-operated pressure-reducing valve. It consists of the main valve section, a pilot section, an inlet port A, an outlet port B, and an internal pilot line for directing fluid from the outlet port to the outlet pressure sensor. The main valve section consists of a spring-biased spool with an axial orifice. Next, the pilot relief valve section consists of a pilot and a spring. The spring biases the spool so that the main valve remains open in its normal position.

Consider that the pressure-reducing valve is connected to a hydraulic system with its output port connected to the low pressure side of the system. When the outlet pressure is below the pilot valve's pressure setting, the main valve spool is hydraulically balanced by the outlet pressure acting on both ends of the main spool. However, when the outlet pressure exceeds the pilot valve setting, the pilot valve opens and relieves the fluid back into the system reservoir. This fluid flow through the pilot valve causes a pressure differential across the main spool, which then shifts the spool toward its closing position against the spring force. The pilot valve relieves just enough fluid to position the main spool so that the flow through the valve equals the flow requirements of the low-pressure side.

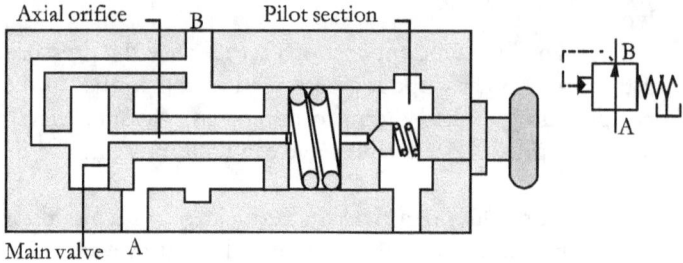

Figure 16.4 | A cross-sectional view of a pilot-operated pressure-reducing valve

Pilot-operated pressure-reducing valves are used for higher flow rates and generally better regulate pressure under those conditions. They also provide good repetitive accuracy. However, fluid contamination can block the flow to the pilot valve section, preventing the main valve from closing correctly. Furthermore, the operation of a pressure-reducing valve in a hydraulic system always generates excess heat due to its throttling effect.

Example 16.1 | Development of a Hydraulic Circuit with a Pressure-reducing Valve
Develop a single-pump hydraulic circuit that employs high pressure in one section of the circuit, say, for a high bending force, and reduced pressure in another part of the circuit, say, for a low clamping force.

Solution

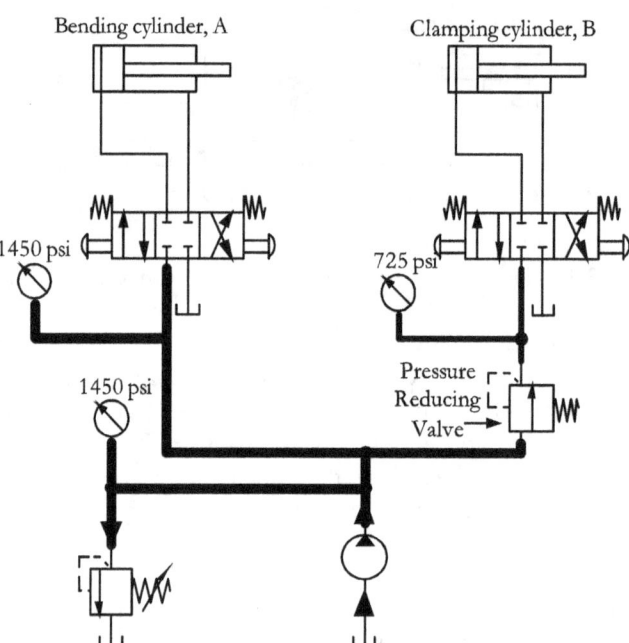

Figure 16.5 | A hydraulic circuit showing the application of a pressure-reducing valve (Example 16.1)

Figure 16.5 shows the hydraulic circuit for the control task of Example 16.1. It consists of a pump, pressure relief valve, pressure-reducing valve, directional control valves, and two cylinders, 'A' and 'B.' The pressure relief valve sets the high pressure (Say 1450 psi) in one circuit section. The pressure-reducing valve sets the reduced pressure (say 725 psi). With this arrangement, the sub-circuit for cylinder A operates at 1450 psi, and the sub-circuit for cylinder B operates at 725 psi.

3-way Pressure-reducing valve

A 3-way pressure-reducing valve maintains constant output pressure regardless of fluctuations in input pressure. When the pressure at the output exceeds the set pressure, the hydraulic fluid is drained off at the tank port. The valve with a relief feature prevents the pressure on the output side from increasing above the set value, e.g., due to an excessive load on the output element.

Unloading Valves

An unloading valve can be used to unload fluid delivery to the associated tank when the critical static pressure at a point in the circuit is reached. Consider a hydraulic system with a pump, reservoir, pressure relief valve, and cylinder. When the cylinder reaches its end-of-stroke position, the pump output can be unloaded to the reservoir at low pressure via the unloading valve rather than dumped at high pressure through the PRV. A hydraulic circuit with an accumulator can also be unloaded when the accumulator is charged fully.

In contrast to the maximum-pressure-limiting control of the pressure relief valve, the unloading valve returns the pump delivery to the reservoir at a low energy level. This valve allows the system to operate cost-effectively by minimizing heat generation.

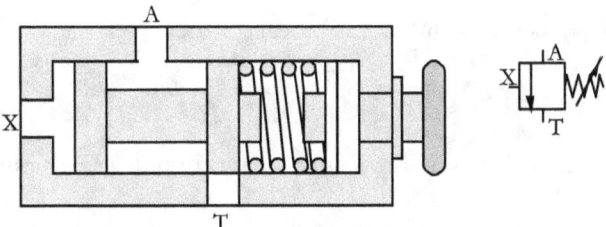

Figure 16.6 | A cross-sectional view of the unloading valve

Figure 16.6 shows the cross-section of an unloading valve. It comprises an inlet port 'A,' tank port 'T,' pilot port 'X,' and a spring-biased spool. In the normal position, the valve remains closed by the spring force. The unloading pressure for the pump can be set by adjusting the spring tension. The pilot port is provided to accept a static pressure signal, which acts on one end of the valve spool.

When the pilot spool receives a sufficient signal, it shifts, diverting the pump delivery to the reservoir at low pressure. The spool remains open until the force from the pilot pressure falls below the spring bias. The spring cavity in the unloading valve can be drained internally or externally. It may be noted that the unloading valve's pressure setting is lower than that of the PRV.

Applications of Unloading Valves: Unloading valves are designed for accumulator circuits, hydraulic motor circuits, and two-pump 'hi-lo' circuits.

In the hi-lo pump circuit, two pumps move a large cylinder at high speed and low pressure. The circuit then shifts to a single pump, which provides high pressure to perform work. This circuit is explained in Example 16.3.

Example 16.2 | Pump Unloading at the End of Cylinder Strokes

A double-acting cylinder is connected to a hydraulic system with a constant-displacement pump and controlled by a 4/3-way closed-center valve. The pump must be unloaded at the end of the forward and return strokes of the cylinder using an unloading valve. Develop a suitable hydraulic circuit.

Solution

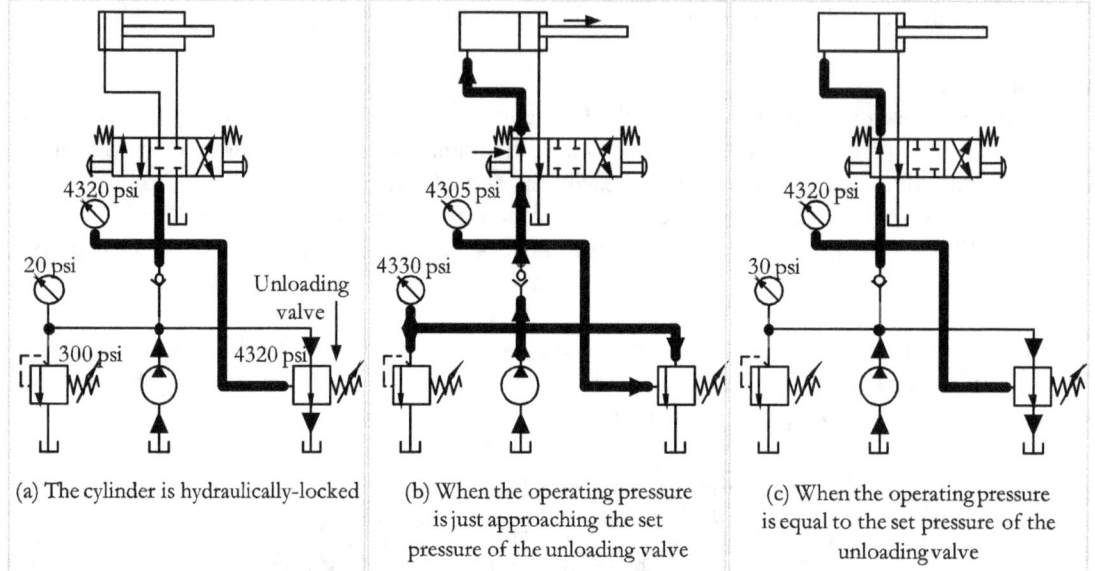

Figure 16.7 | Different positions of a hydraulic circuit for unloading a pump (Example 16.2)

The double-acting hydraulic cylinder can be controlled using the 4/3 DC valve with the closed-center configuration. Figure 16.7 shows three critical positions of the hydraulic circuit for the control task given in Example 16.2.

The cylinder remains hydraulically locked when the 4/3-DC valve is in its center position, as shown in Figure 16.7(a). At the same time, the static pressure acts on the pilot port of the unloading valve, actuating it when the set pressure is reached. The valve then unloads the pump flow to the system reservoir at atmospheric pressure. When the 4/3-DC valve is actuated to the right, as shown in Figure 16.7(b), the flow is directed to the piston-side port of the cylinder, and the cylinder extends. When the cylinder is fully extended, as shown in Figure 16.7(c), the unloading valve unloads the pump flow to the system reservoir at atmospheric pressure.

Two-pump Power Supply

Many hydraulic systems demand high flows at low pressure for rapid tool traverse, and then high pressure at low flow for clamping or feeding operations. This requirement type can be realized using the well-known 'hi-lo' circuit. This circuit uses two hydraulic pumps. One pump, known as a volume pump, is designed to produce a high flow at low pressure. The second pump, known as a pressure pump, is designed to deliver high pressure at low flow. Employing a hi-lo circuit with two pumps of different capacities improves the system's overall efficiency. Moreover, the hi-lo system is less expensive than a single high-pressure, high-flow pump system.

Example 16.3 (Hi-lo Circuit) | A hydraulic system with a hi-lo circuit is to be designed with two pumps (namely, a volume pump and a pressure pump) to provide high flow at low pressure initially and high pressure at low flow subsequently, to control a cylinder. The cylinder must extend rapidly with the high-flow, low-pressure fluid produced by the volume pump until it reaches the work point. At this point, the cylinder must operate with the high-pressure, low-flow fluid delivered by the pressure pump to improve the system's efficiency. Develop a circuit to implement the scheme.

Solution

Figure 16.8 gives two critical positions of the 'hi-lo' circuit for Example 16.3. For easy understanding, assume that the volume pump P_1, rated at 363 psi, delivers 25 gpm, and the pressure pump P_2, rated at 1450 psi, delivers 0.5 gpm. The double-acting cylinder can be controlled using a closed-center 4/3-way valve, a check valve, an unloading valve, and a relief valve.

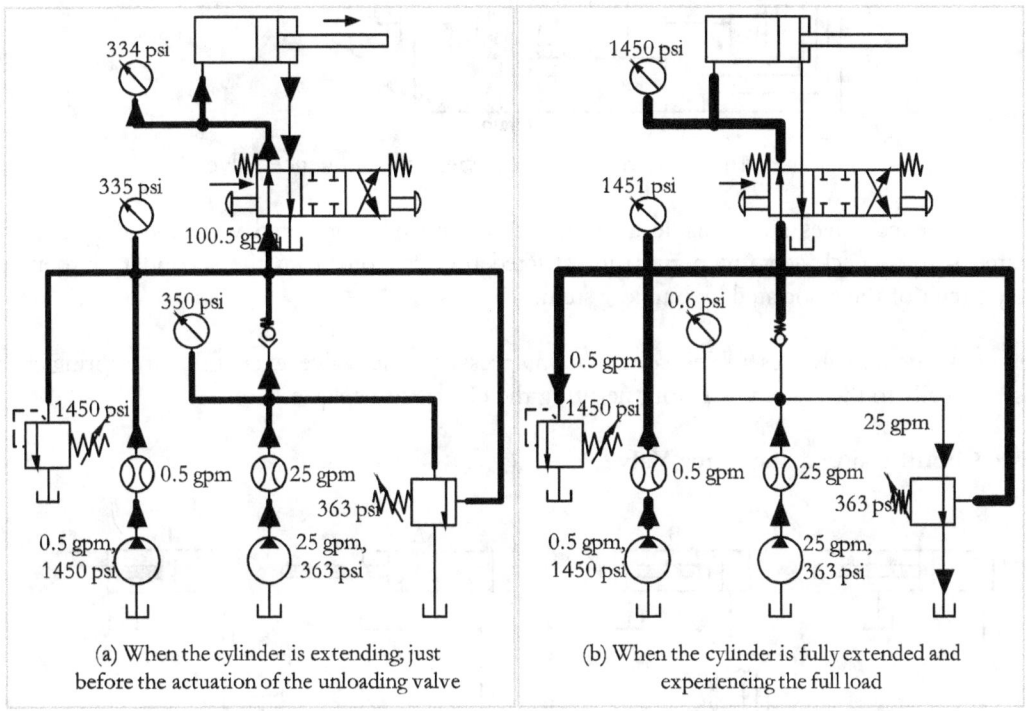

Figure 16.8 | Two positions of a double-pump hydraulic circuit (Example 16.3)

Initially, the combined outputs from pumps P_1 and P_2 give the maximum flow at low pressure to extend the cylinder with a fast approach against a null force, as shown in Figure 16.8(a). When the load pressure increases to the setting of the unloading valve (i.e., 363 psi), the high-flow, low-pressure volume pump (P_1) is bypassed through the unloading valve to the reservoir, allowing the low-flow, high-pressure pump (P_2) to supply all the system requirements, as shown in Figure 16.9(b).

The check valve can isolate the low-pressure side from the high-pressure side, preventing the high pressure developed by pump P_2 from damaging pump P_1. The pressure relief valve limits the maximum pressure (i.e., 1450 psi) that the pressure pump could generate.

Sequence Valves

A pressure sequence valve is a normally closed two-way valve that responds to pressure signals from the associated hydraulic system to permit flow from its input port to its output port when the input pressure reaches a preset value. Figure 16.9 shows the cross-section of a sequence valve. It consists of an inlet port A, an outlet port B, and a spring-biased spool. The externally adjustable spring is provided to set the pressure to the required value. A pilot passage X is provided in the valve. It accepts signals from the inlet and acts on one side of the spool. When sufficient pilot pressure is applied, the spool moves against the spring force and opens the valve, allowing flow through it. The valve is kept open until the pilot sensing pressure exceeds the spring bias.

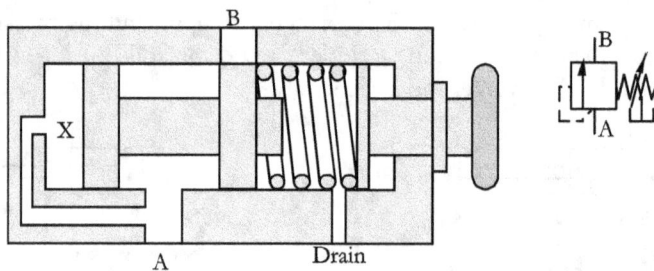

Figure 16.9 | A cross-sectional view of a sequence valve

Pressure sequence valves are available in direct-acting and pilot-operated versions. They are usually integrated with a check valve to permit unrestricted reverse flow from the secondary circuit to the primary circuit of the associated hydraulic system.

A pressure sequence valve resembles a direct-acting pressure relief valve, except that the spring chamber drains externally to the system reservoir, negating the effects of backpressure.

A Basic Circuit Using a Sequence Valve

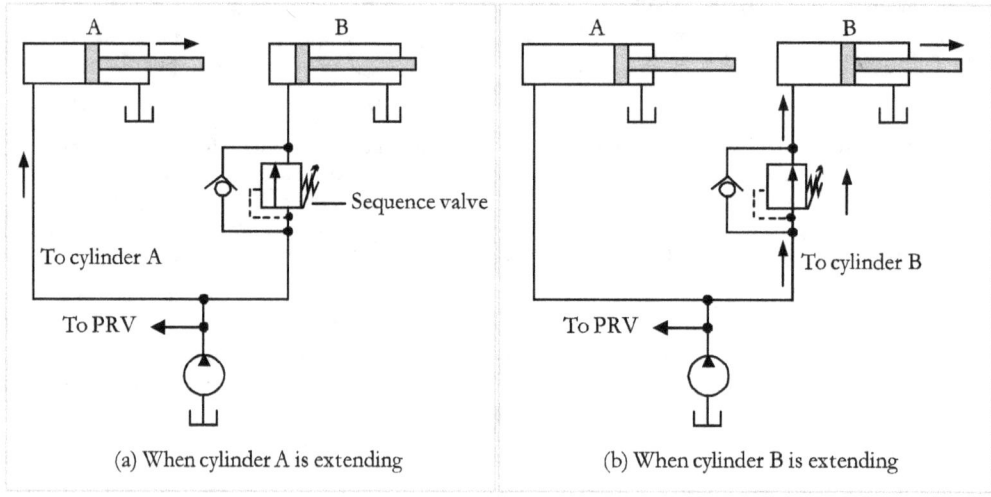

Figure 16.10 | Multiple positions of a circuit for illustrating the function of a sequence valve

Pressure sequence valves can be used to obtain sequential operations of work processes. For example, a pressure sequence valve can be used in a hydraulic system with a clamping cylinder (A) and a drilling cylinder (B) to provide sequence control for these two cylinders. Two circuit positions in Figure 16.10(a) and (b) point to how the sequence valve can be used to control the forward motions of cylinders A and B sequentially for the clamp-and-drill operation. The system pressure increases when cylinder A reaches the end of its forward stroke to clamp a workpiece. When the pressure exceeds the set value of the sequence valve, it opens, allowing the fluid to flow to the drilling cylinder B. Therefore, cylinder B extends and drives the drill unit.

Example: 16.4 | In a clamp-and-cut hydraulic application involving two cylinders A and B, the following sequence of operations is required: First, cylinder A is to extend fully, and then cylinder B is to extend. In the reverse operation, cylinder B must fully retract before cylinder A can retract. Develop a hydraulic circuit using pressure-sequence valves to execute the given sequence of operations.

Solution

Figure 16.11 shows four circuit configurations for achieving the required sequential operation of the two hydraulic cylinders, A and B, using pressure sequence valves SV1 and SV2. A 4/3 DC valve is the main valve for controlling the cylinders.

When the 4/3 DC valve is actuated to the left envelope position, the flow is directed to the piston-side port of cylinder A, and the cylinder extends. When cylinder A reaches the end of its forward stroke, the pressure in the piston-side line of cylinder A increases, and hence, the sequence valve SV2 opens, allowing cylinder B to move forward.

When the 4/3 DC valve is actuated to the right envelope position, the flow is directed to the piston-rod side port of cylinder B, and cylinder B retracts. When cylinder B reaches the end of its return stroke, the pressure in the piston rod-side line of cylinder B increases, and hence, the sequence valve SV1 opens, allowing cylinder A to retract.

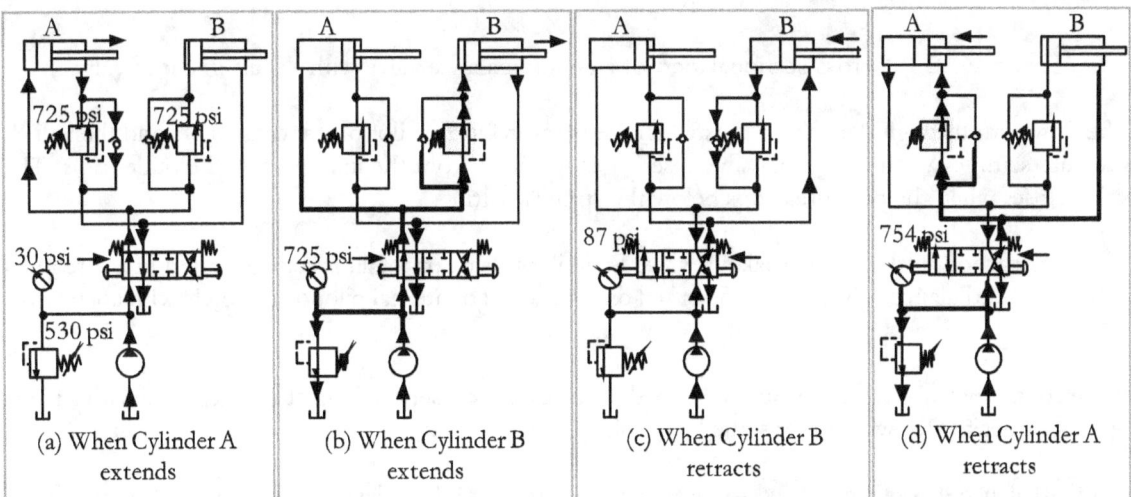

Figure 16.11 | Critical positions of a circuit for the sequence control of two hydraulic cylinders (Example 16.4)

Counterbalancing Overrunning Loads

Vertically mounted cylinders, especially when connected to heavy loads or hydraulic motors on winch drives, are susceptible to the dangers of overrunning loads. Consider a hydraulic cylinder with a heavy load attached. The load forces the cylinder to move faster than its average speed when the associated directional control valve shifts to lower the load. The overrunning load also causes one end of the actuator to starve for fluid, potentially leading to cavitation. Therefore, the actuator with the overrunning load needs some method to retard its downward movement.

Employing a pilot-operated check valve is one way to counterbalance an overrunning load with a manual setting, as explained in Chapter 14. However, the precise setting of the flow rate of this type of control is crucial for its successful operation. Therefore, a counterbalance valve is a better way to control the overrunning load, as the valve dynamically counterbalances the potential free-fall of the connected load. It prevents the actuator from running ahead of the pump due to the load-induced energy. This controlled actuator movement tends to reduce cavitation. However, the dynamic behavior of the counterbalance valve during operation is complex and can often cause oscillations in load movement.

Counterbalance Valve

Figure 16.12 shows the cross-sectional view of the counterbalance valve with a built-in check valve. It is a normally closed valve with an inlet port A and an outlet port B. It also consists of a spring-loaded spool. The externally adjustable spring is provided for setting the pressure. The counterbalance valve must be set at a sufficient pressure to counterbalance the load. A pilot passage is provided to accept a signal from the inlet side of the valve, and the signal acts on one side of the spool. The valve is designed with an internal drain facility.

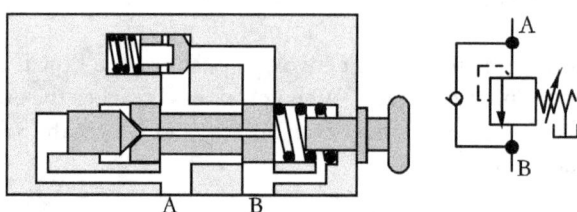

Figure 16.12 | A cross-sectional view of a counterbalance valve with a built-in check valve

The flow path through the valve opens when the pressure at the pilot port increases beyond the valve's pressure setting. When open, the valve discharges the fluid from the inlet port to the outlet port. The valve closes when the pressure drops below the spring's setting.

The construction of the counterbalance valve is similar to that of the indirectly operated pressure relief valve. Counterbalance valves can be operated directly with an internal pilot or remotely with an external pilot.

If the counterbalance valve must hold the load for an extended period, it must be integrated with a pilot-operated check valve with better static load-holding capabilities.

Counterbalance valves are utilized to raise loads and regulate their descent, as seen in tractors, cranes, boom trucks, and elevator systems. Example 16.5 illustrates the use of the counterbalance valve.

Example 16.5 | A Circuit for Controlling a Descending Load Using a Counterbalance Valve

A vertically mounted hydraulic cylinder is coupled to a load that is liable to overrun. The cylinder is controlled by using a 4/3-way open-center DC valve. When the 4/3-way open-center DC valve is shifted to downward load movement, it must be prevented from descending uncontrollably due to the weight of the attached load. Develop a hydraulic circuit to realize the control task using a counterbalance valve.

Solution

Figure 16.13 shows the circuit arrangement for the actuator system with an overrunning load. The primary port of the counterbalance valve is connected to the piston rod end port, and the secondary port is connected to the directional control valve. Assume that the PRV is set at 1450 psi. The counterbalance valve is typically set at a pressure about 5 to 10 percent higher than the load-induced pressure.

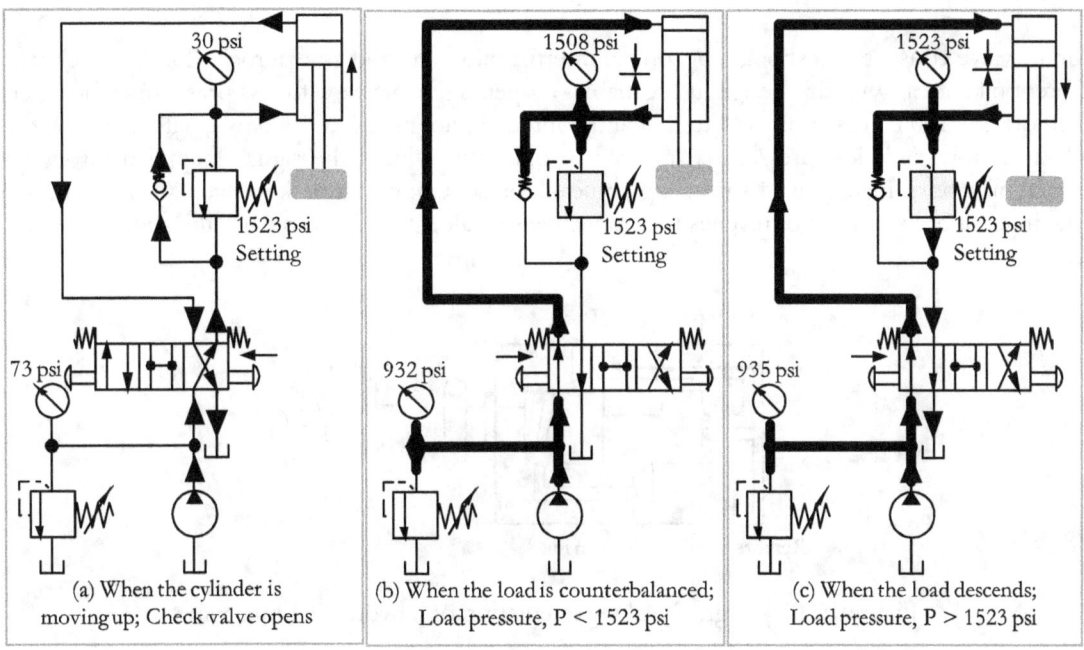

Figure 16.13 | Three critical positions of a hydraulic circuit for illustrating the control action of a counterbalance valve (Example 16.5)

Figure 16.13(a) shows the position of the hydraulic circuit when the 4/3-way valve is brought to its right-hand side envelope to raise the load. As the load is raised, the system fluid is directed to the cylinder through the opened check valve. When the 4/3-way valve is brought to its neutral position, the correctly set counterbalance valve remains closed, preventing the load from free-falling.

Figure 16.13(b) shows the circuit's position when the 4/3-way valve is brought to its left-hand side envelope. The fluid entering the piston-side port of the cylinder tends to move the load with just enough velocity to increase the pressure on the piston rod side. The counterbalance valve senses and evaluates the load pressure.

Figure 16.13(c) shows the circuit's position when the load pressure equals the set pressure. When the load pressure equals or exceeds the set pressure, the counterbalance valve opens, allowing the fluid to flow back to the system reservoir. When the pressure on the piston rod side is relieved, the counterbalance valve closes again.

This dynamic counterbalancing and pressure relief process is repeated until the load reaches its end-of-travel position.

Brake Valves (Over-center Valves)

When abruptly stopped, an overrunning hydraulic motor can develop high pressure, damaging critical system components. The counterbalance valve can reduce backpressure, but the motor continues to run until it coasts to a stop. The pressure drop developed across the counterbalance valve is converted into heat. A specially designed counterbalance valve, called a brake valve (over-center valve), can be used to overcome this disadvantage.

A brake valve consists of a spool, a piston, an internal pilot line, and an external pilot line, each with a different pilot area. With this design, it is capable of opening the valve with less pressure at the external pilot port (say 100 psi) as compared to that at the internal pilot port (say 1000 psi), with an area ratio of 10:1. Therefore, only less pressure is required to open the valve if the signal is applied through the external pilot port. Figure 16.14 shows a graphical representation of a brake valve. When the pressure at the internal or external pilot reaches the respective set value, the piston pushes the spool to open the valve.

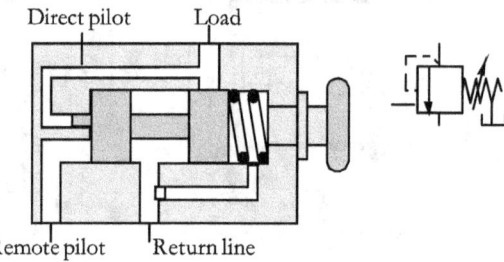

Figure 16.14 | A graphical representation of a hydraulic brake valve

The valve can be used in the circuit of a hydraulic motor, running under high inertia, to stop it quickly, safely, and smoothly. In a typical application for controlling a hydraulic motor with a brake valve, the valve provides static and dynamic load control by evaluating signals from the upstream and downstream sides of the motor. In this way, the valve provides a counterbalancing pressure to the motor.

The brake valves are suitable for load holding, control, and safety. They also provide safety against hose failure and can be used for fast, shock-free deceleration and stoppage of hydraulic motors. Using the brake valve in a hydraulic motor circuit makes the system more efficient, reduces horsepower requirements, and eliminates cavitation and heat generation.

In the event of a hose failure, the brake valve connected to the associated actuator prevents the uncontrolled movement of the load.

Example 16.6 | A Circuit for Controlling a Hydraulic Motor Driving a Loaded Winch

A braking arrangement using a hydraulic valve with an internal and an external pilot is to be incorporated into a hydraulic motor driving a loaded winch. The valve requires 100 psi at the motor inlet to keep it open. The valve closes if the load overruns and the pressure drops below 100 psi. It requires 1000 psi at the internal pilot to open the valve. Develop a hydraulic circuit.

Solution

The load-holding function and the prevention of load overrunning for the hydraulic motor can be realized using a brake valve. Figure 16.15 gives the hydraulic circuit with the hydraulic motor controlled by a 4/3-way valve and the brake valve. When the hydraulic motor runs under normal operating conditions, the external pilot supply opens the brake valve fully at low pressure (e.g.,>100 psi), allowing uninterrupted flow through the valve. In the load control function, the brake valve prevents the motor from running ahead of the pump due to the load-induced energy. If the hydraulic valve tries to run away, the brake valve holds the load until the motor's secondary pressure exceeds the set value of the internal pilot valve (say, 1000 psi).

When the DC valve is moved to its center position to stop the motor, the fluid in the motor lines is recycled because the motor acts as a pump. The pressure at the external pilot line drops below the set value, which is insufficient to keep the brake valve open. The internal pilot now provides a force opposing the spring, which obtains sufficient pressure from the motor (acting as a pump) as it rotates due to load inertia. The force is only sufficient to partially open the valve, reducing flow and creating a braking action.

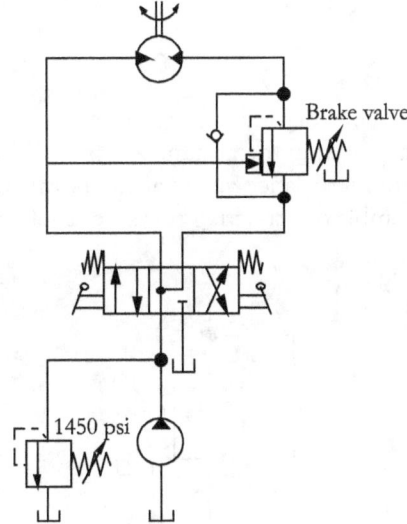

Figure 16.15 | A circuit for braking a hydraulic motor. (Example 16.6)

When stopped, the brake valve (or counterbalance valve) will not entirely keep the motor stationary. No matter how leak-free the valve is, the motor's internal bypass makes it turn slowly. An external braking system is necessary to hold the load against its tendency to creep.

Selection of Pressure Control Valves

When selecting a pressure control valve, the operating pressure range, flow rate capacity, and material compatibility must be considered.

Objective-type Questions

1. Mark the <u>correct</u> statement:
   a) Unloading valves are intended to replace the pressure relief valves in hydraulic systems.
   b) The hi-lo circuit, with one pump delivering the high-flow fluid at low pressure and the other pump delivering the high-pressure fluid at low flow, is more efficient than that using a single high-pressure, high-flow pump.
   c) Meter-out flow control is a better way to control the overrunning load than a counterbalance valve.
   d) A pressure-reducing valve is a normally closed type valve.

2. Identify the valve:

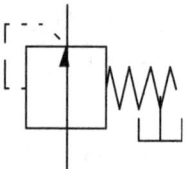

Figure 16.16

   a) Pressure relief valve
   b) Pressure-reducing valve
   c) Unloading valve
   d) Sequence valve

3. Mark the <u>incorrect</u> statement.
   a) Unloading valves are used to control overrunning loads.
   b) Unloading valves are typically used in the accumulator and hi-lo circuits.
   c) Unloading, sequence, and counterbalance valves are all normally closed-type valves.
   d) Sequence valves almost resemble direct-acting pressure relief valves.

4. Identify the valve symbol given below:

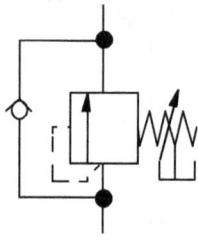

Figure 16.17

   a) Pressure relief valve
   b) Pressure-reducing valve
   c) Unloading valve
   d) Sequence valve

Review Questions

1) Give three functions of the hydraulic pressure control valves with examples.
2) What do the terms 'normally closed' and 'normally open' mean in pressure control valve terminology?
3) Classify the pressure control valves used in hydraulic systems.
4) Give the symbolic representations of a pressure-reducing valve, an unloading valve, a sequence valve, a counterbalance valve, and a brake valve used in hydraulic systems.
5) List any four types of hydraulic pressure control valves.
6) Describe the operation of a pressure-reducing valve meant for hydraulic systems.
7) Draw the graphical symbol of the pressure-reducing valve.
8) Explain the difference between hydraulic pressure relief and pressure-reducing valves used in hydraulic systems, giving their graphical symbols.
9) Which type of hydraulic pressure control valve is normally open?
10) Briefly explain the function of pressure-reducing valves used in hydraulic systems.
11) Mention the function of an unloading valve used in hydraulic systems.
12) What are the different ways to unload a hydraulic pump economically?
13) Explain the operation of an unloading valve with a neat sketch.
14) What is the main difference between the unloading and pressure relief valves?
15) Explain two typical hydraulic applications of the unloading valves.
16) Explain the 'hi-lo circuit' for the control of a hydraulic cylinder with a two-pump power supply.
17) Mention the function of a sequence valve in hydraulic systems.
18) Give the ISO symbol for the unloading valve used in hydraulic systems.
19) Explain the operation of a sequence valve in a hydraulic system with a neat sketch.
20) Explain a typical hydraulic application of the sequence valves.
21) Explain how sequential forward strokes of two hydraulic cylinders can be achieved using the sequence valves with the help of a circuit diagram.
22) Mention the purpose of the counterbalance valve in a hydraulic system.
23) Explain the operation of the counterbalance valve used in a hydraulic system with a neat sketch.
24) Explain typical hydraulic applications of the counterbalance valves.
25) Design a hydraulic circuit using a counterbalance valve to prevent a vertically mounted cylinder from descending due to the weight of its load.
26) Give the symbol of the following hydraulic pressure control valves: (1) Pressure-reducing valve, (2) Unloading valve, (3) Sequence valve, and (4) Counterbalance valve
27) List any four types of hydraulic pressure control valves, mentioning their functions.
28) Compare the pressure-reducing, unloading, sequence, and counterbalance valves.
29) What factors should be considered while selecting a pressure control valve?

Numerical Problems

1. A hydraulic system uses a fixed-displacement piston pump, which delivers the system fluid at 10.5 gpm. The operating pressure is 3046 psi. (1) What is the power loss in the system if the system employs only a pressure relief valve (PRV)? (2) What is the power loss if the system adds an unloading valve to discharge the flow at a pressure of 50 psi?

Objective-type questions - Answer key: 1-b, 2-b, 3-a, 4-d

Chapter 17 | Hydraulic Accumulators and Circuits

An accumulator is a device used to absorb shock pressures and store energy in a hydraulic system. It mainly consists of a vessel containing a hydraulic fluid, which is held under pressure by a raised weight, a spring, or a volume of compressed gas. It is, thus, possible to store potential energy in the accumulator when the associated system pressure remains higher than that of the accumulator. The accumulator can release the stored energy back into the system to perform useful hydraulic tasks when the system pressure falls below the accumulator's pressure. Figure 17.1 shows the realistic view of a gas-charged accumulator.

Figure 17.1 | Graphic view of a hydraulic accumulator with a safety shut-off valve [Tobul Make]
Courtesy: **Tobul Accumulator, USA**

Accumulators are employed in hydraulic systems to realize many functions. The essential functions are presented in the following sections.

Generation of Shocks in Hydraulic Systems
Shock pressures are generated in a hydraulic system when flow is abruptly blocked or when the flow direction is suddenly reversed as a directional valve is shifted. Shocks may also be caused by the jerkiness of a load attached to a cylinder or motor in the system.

Shock Absorbing Function of Hydraulic Accumulators
Two circuits in Figure 17.2 demonstrate the effect of shock pressures on a hydraulic system with a pump, valve, actuator, and interconnecting lines, and the elimination of shock pressures by adding an accumulator.

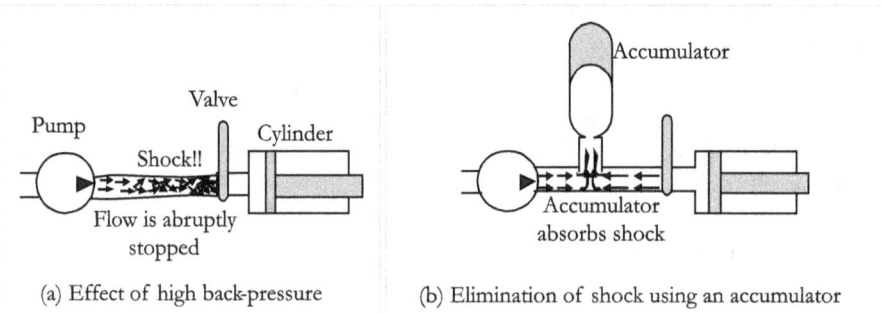

Figure 17.2 | Illustrating the development of shock pressure in a hydraulic system and its elimination

Figure 17.2(a) represents the movement of a column of high-energy fluid through the hydraulic circuit without an accumulator. When the flow is abruptly blocked, for example, by the rapid closure of a valve, pressure waves are generated that travel back and forth through the system. The resulting hydraulic shock and vibration can develop peak pressures several times greater than normal. They may rip tubes, blow seals, produce fatigue failures, jar parts, and split the pump housing.

Installing an appropriately sized accumulator in the system, as shown in Figure 17.2(b), can prevent the damaging effects of hydraulic shock. The accumulator can absorb the kinetic energy of the fluid's moving column and suppress the resulting hydraulic shock. This action prevents system parts such as pumps, valves, hoses, and fittings from being damaged by pressure spikes.

Other Functions of Hydraulic Accumulators
Accumulators can perform many other functions besides storing energy and absorbing shocks, such as dampening pressure pulsations, storing energy during periods of low demand and releasing it during periods of high demand, and compensating for leakages.

Pulsation Dampening
Delivery fluctuations from the system pump, irregularities in fluid flow, thermal variations, or excessive loads are usually the cause of pressure pulsations in a hydraulic system. These pressure fluctuations can cause variations in the actuator's speed and inferior system performance. Adding a properly located and correctly sized accumulator cushions the pressure pulsations and keeps the system pressure relatively constant.

Energy Storage and Release
A typical machine cycle during the operation of a hydraulic system consists of extended periods (say, 80% of cycle time) of little flow and short periods (say, 20% of cycle time) of high-volume flow, as shown graphically in Figure 17.3.

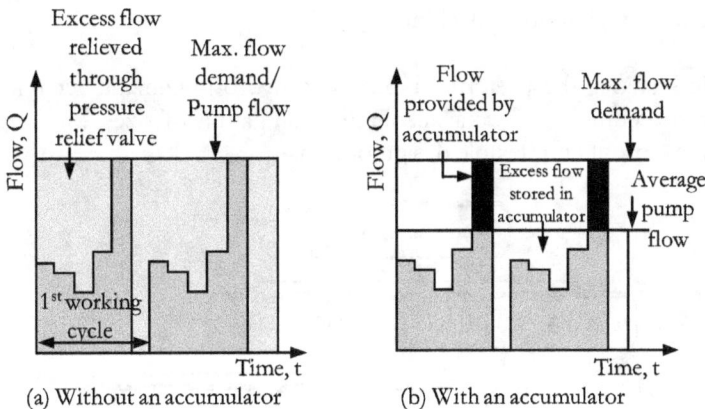

Figure 17.3 | Schematic diagrams showing typical machine cycles in hydraulic systems

If a fixed-displacement pump with an intermittent duty cycle supplies the fluid, the system pressure relief valve bypasses the fluid for most of the time (say 80%). If the pump operates continuously, the pressure relief valve's discharge of pressurized fluid represents a significant power loss, as shown in Figure 17.3(a).

With the addition of a correctly sized accumulator, energy can be stored during periods of low demand and then added to the pump flow during short periods of high demand. Therefore, the pump unit can be resized to match the average power requirement of the machine's operating cycle for high-pressure fluid discharge, as shown in Figure 17.3(b).

Cost Reduction
Employing an accumulator in a high-performance hydraulic system with an intermittent duty cycle allows the use of a smaller-capacity pump.

Leak Compensation
On hydraulic presses used for molding, bonding, etc., constant pressure on the work is necessary during long curing periods. However, pressure changes can occur in such systems due to fluid leakage. A charged hydraulic accumulator can compensate for the pressure changes by supplying additional fluid. At the same time, the pump remains unloaded during an extended system idle period to compensate for fluid loss.

Auxiliary Power Source
Energy stored in a fully charged accumulator can be used in a hydraulic system to meet a sudden, high-power demand for a comparatively short time to complete the cycle, or as a standby power source during power failures. It can also be used for the independent operation of pilot circuits in a hydraulic system when the pump flow is required to perform the main operating movements.

Classification of Accumulators
According to the construction method, accumulators are classified into three basic types. They are: (1) weight-loaded accumulators, (2) Spring-loaded accumulators, and (3) hydro-pneumatic (gas-charged) accumulators. The classification chart of Figure 17.4 shows all these types and their sub-types.

Weight-loaded types are no longer used in modern hydraulic applications, and spring-loaded types are now virtually confined to small mobile machinery.

In modern hydraulic systems, the preferred type is the hydro-pneumatic accumulator. The bladder, diaphragm, and piston accumulators are available in a full gamut of sizes, pressure ratings, materials, and port configurations, providing flexible design options.

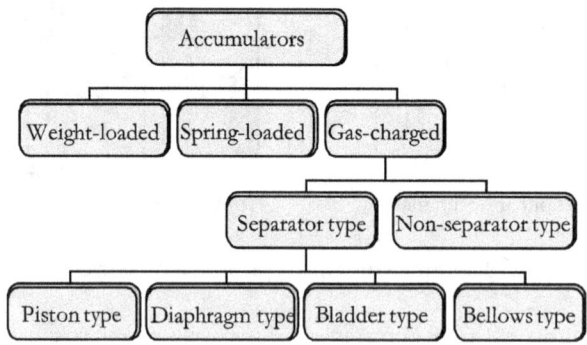

Figure 17.4 | Classification of hydraulic accumulators

Accumulator Symbols

The standard symbols, shown in Figure 17.5, represent the different types of accumulators according to ISO 1219.

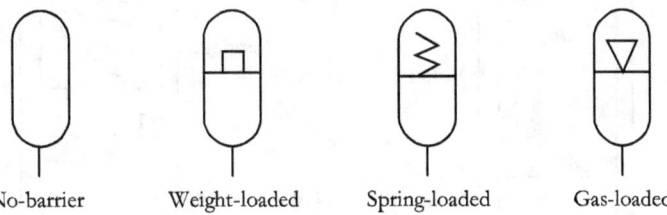

Figure 17.5 | Symbolic representations of hydraulic accumulators

General Constructional Features of Accumulators

Figure 17.6 shows simplified cutaway views of the different types of hydraulic accumulators. Although these are not complete representations of accumulators, they illustrate their general constructional features.

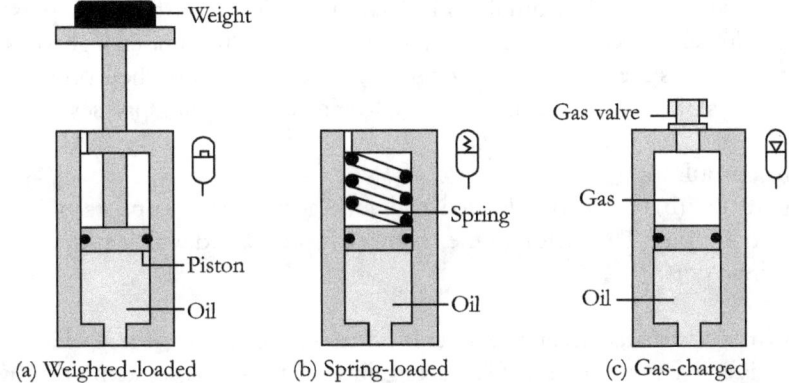

Figure 17.6 | Cross-sectional views of different types of hydraulic accumulators

A hydraulic accumulator usually consists of two chambers separated by a piston, diaphragm, or bladder (bag). One chamber admits fluid from the associated system, and the second chamber maintains a weight, spring, or volume of pressurized gas. Hydraulic energy is stored when the system fluid under pressure acts against a weight-loaded or spring-charged piston, a gas-charged diaphragm, or a bladder.

Weight-loaded Accumulator

The weight-loaded hydraulic accumulator consists of a vertically mounted, thick-walled steel cylinder with a piston. A deadweight or a series of dead weights is placed on the top of the piston, as shown schematically in Figure 17.7(a). For the weight-loaded, the non-pressure side usually has a drain port that connects to the system reservoir to relieve any leakage fluid and avoid any potential backpressure on the piston.

When the fluid under pressure works against them, the piston and dead weight are raised. At the same time, gravity exerts a downward force on the piston, thereby energizing the fluid in the cylinder.

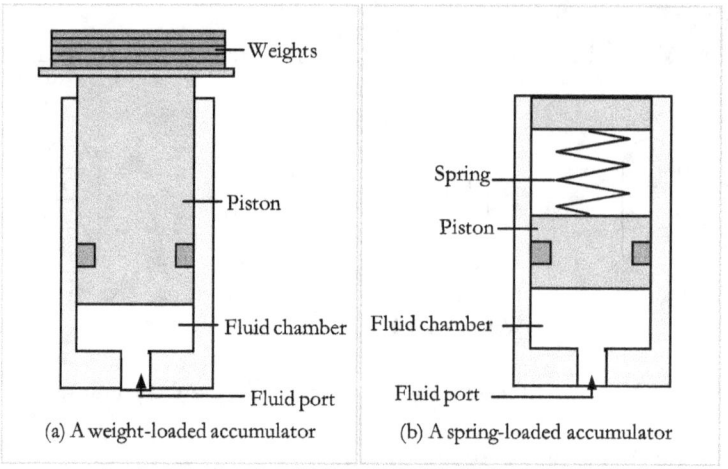

Figure 17.7 | Types of accumulators

An advantage of the weight-loaded accumulator is that it maintains a constant fluid pressure throughout the unit's output volume. It can also supply a large volume of fluid under high pressure. However, weight-loaded accumulators are bulky, cumbersome, and expensive, and their portability is out of the question. Accumulators of this type are used in large forging and molding presses.

Spring-loaded Accumulator
As shown in Figure 17.7(b), the spring-loaded hydraulic accumulator consists of a cylinder body, a movable piston, and a spring. The piston in the chamber is pre-loaded with a spring. The non-pressure side usually has a drain port.

As pressure from the associated hydraulic circuit enters the chamber, the spring compresses. The compressed spring is an energy source that can act against the piston. As system pressure decreases, the fluid from the accumulator is forced into the circuit by the charged spring.

Spring-loaded accumulators are usually smaller and less expensive than weight-loaded accumulators. Their mounting is straightforward and can be fitted directly to the power unit. The pressure generated by this accumulator type depends on the spring's size and pre-loading. The accumulator pressure reaches its peak as the spring is compressed to its maximum, and it drops to its lowest level as it approaches its free length. Therefore, the pressure does not remain constant throughout the spring-loaded accumulator's volume output.

Spring-loaded accumulators are typically suitable for low-pressure, low-volume hydraulic applications and cannot be used for high-cycle-rate applications.

Gas-charged Accumulators
The low compressibility of the hydraulic oil (fluid) makes it difficult to store hydraulic energy in small volumes. Still, it can transfer considerable force to a load surface. On the other hand, gas is a highly compressible medium that can store large amounts of energy in a small volume. Gas-charged accumulators use these two properties to perform better than other types of accumulators. Hence, they are, by far, the most commonly used type of accumulator.

Moreover, they offer a good dynamic response during their operation. Remember that neither air nor oxygen should be used to charge the accumulator to avoid the danger of an explosion. The gas-charged accumulators fall into two basic categories. They are: (1) Non-separator type and (2) Separator type.

Non-separator type Accumulator
Some of the earlier accumulators used in hydraulic systems were non-separator (or direct-contact) fluid-gas containers. Approximately half of this type of accumulator was filled with the fluid, and the other half was filled with nitrogen gas, with no physical separation between them. The shut-off valve at the fluid port must be closed before stopping the system to prevent fluid and gas from the accumulator from escaping. In this type of accumulator, foaming and gas absorption into the fluid medium were observed. Therefore, non-separator-type accumulators are not used in modern hydraulic systems, but many older units may still be in service.

Separator type Accumulators
A commonly used type of gas-charged accumulator is the separator type. This accumulator type has a physical barrier between the gas and fluid media, such as a diaphragm, bag, or floating piston. This barrier efficiently exploits the gas's compressibility. A port is provided for the hydraulic connection. A fill port is provided to supply nitrogen gas. The separator type of accumulators can be classified into the following three types:

- Piston Type
- Diaphragm type
- Bladder type

Piston Accumulator
As shown in Figure 17.8, a piston accumulator consists of a cylinder (or shell) with a finely finished internal surface and a freely floating, lightweight metal piston (typically aluminum). The cylinder is manufactured from a seamless, homogeneous tube. The movable piston with O-ring seals effectively separates the cylinder into fluid and gas sections. The gas section is pre-charged with dry nitrogen gas. The fluid part is connected to the hydraulic system so that the piston accumulator draws fluid from the system when the system pressure rises, thereby compressing the gas. When the system pressure drops, the gas expands, forcing the fluid back into the system.

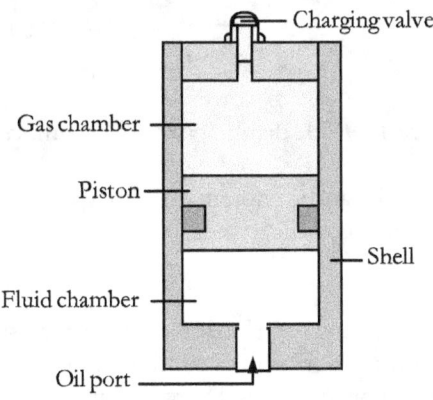

Figure 17.8 | A piston accumulator

Piston accumulators are compact devices. They can provide higher flow rates than other gas accumulators for the same accumulator volume. Several accumulators can be connected in parallel to achieve this higher flow capacity. Other advantages of piston accumulators are their portability and ability to handle a broad range of temperatures.

The main disadvantages of this type of accumulator are its susceptibility to fluid contamination and its hysteresis caused by seal friction.

Because of the piston's inertia and the friction of the piston seals, it cannot be used as a pressure pulsation dampener or a shock absorber.

Diaphragm Accumulator

A diaphragm (membrane) accumulator consists of two steel hemispheres (shells) and a flexible synthetic diaphragm secured to the shells, as shown schematically in Figure 17.9. It also consists of a gas chamber and a fluid chamber. The diaphragm separates the gas and fluid sections. The gas chamber is pre-charged to a definite pressure, and the fluid chamber is connected to the hydraulic system. A valve plate is set into the base of the diaphragm. This shuts off the hydraulic outlet when the accumulator is empty, preventing diaphragm damage.

The diaphragm accumulator draws fluid from the system and compresses the gas when the pressure increases. When the system pressure decreases, the stored energy in the compressed gas is released into the system.

The diaphragm is typically made of Buna-N, Viton, Butyl, PTFE, or Hydrin. Remember that the material used for an accumulator must be compatible with the system fluid. If an accumulator discharges rapidly, the nitrogen gas may cool below the permitted temperature, leading to diaphragm cracking.

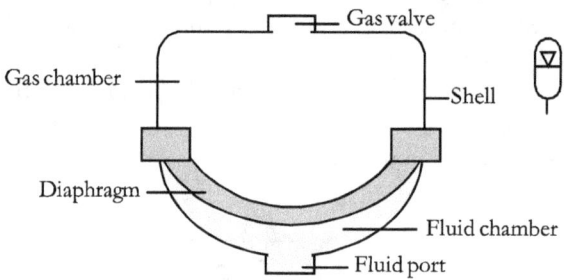

Figure 17.9 | A diaphragm-type accumulator

By design, a diaphragm accumulator can be mounted in any position. However, in a system where contamination is a severe problem, a vertical mount with the accumulator's fluid port oriented downward is usually preferred.

Diaphragm accumulators are fast-acting and exhibit no hysteresis. They are not affected by contamination and behave consistently under similar conditions. The speed of a diaphragm accumulator is governed by the gas, as there is no piston mass. Therefore, it reacts quickly to changes in the system pressure. Hence, it is an excellent choice for pressure-pulsation damping.

Bladder Accumulator

Figure 17.10 shows a schematic diagram of a bladder (or bag) accumulator. It comprises a seamless cylindrical pressure vessel (shell) and an internal elastomeric bladder (bag). Next, the bladder divides the shell into two chambers: the fluid chamber on the system side and the gas chamber inside the bladder. The fluid chamber is connected to a hydraulic system, and the gas chamber is pre-charged to a specified pressure.

When the system pressure increases, fluid from the system is drawn into the fluid chamber, and the gas is compressed. Any pressure drop in the system causes the bag to expand, forcing the stored fluid from the accumulator back into the system.

A poppet valve prevents the bag from being pulled into the downstream tube if it overexpands.

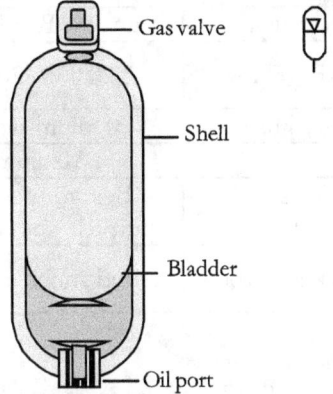

Figure 17.10 | A bladder accumulator

The bladder accumulator shells are made from a seamless carbon steel tube, which is the standard material. The tube is typically heat-treated to ensure excellent mechanical properties. For operation with chemically aggressive media, corrosion protection can be provided by nickel plating the accumulator shell. If this is not enough, stainless steel hydraulic accumulators must be used.

The bladder is typically composed of Acrylonitrile butadiene rubber (NBR), Ethylene oxide epichlorohydrin rubber (ECO), Butyl rubber, or Fluorine rubber (FKM). The bladder material should be chosen based on the operating medium or temperature.

Depending on the application requirements, bladder accumulators can be installed vertically, horizontally, or at any angle.

Like diaphragm accumulators, bladder accumulators are fast-acting and exhibit no hysteresis. They are an excellent option for pressure pulsation damping and shock suppression applications. A bladder accumulator is not susceptible to contamination and provides consistent behavior under similar conditions.

The main limitation of bladder accumulators is their larger size compared to other comparable accumulators.

Comparison of Gas-charged Accumulators

Hydro-pneumatic accumulators come in various designs, constructions, operations, applications, and pressure ratings. The choice of an accumulator depends on specific system requirements, including pressure levels, space constraints, and performance expectations. Table 17.1 compares diaphragm, bladder, and piston hydraulic accumulators across various parameters, including size, pressure rating, flow rate, seal materials, temperature limits, weight, cost, applications, and shock-absorption capability.

Table 17.1 | Comparison of accumulators

Parameter / property	Diaphragm	Bladder	Piston
Size	1 gallon	15 gallons	350 gallons
Working pressure	3600 psi	10000 psi	36000 psi
Flow rate	40 gpm	240 gpm	240 gpm
Seal materials	Buna-N, Butyl	Buna-N, Butyl, Viton	Buna-N, Viton
Temperature limits (Typical)	-40°F to 176°F	-20°F to 200°F	-20°F to 165°F
Heaviness	Light-weight	Medium-weight	Heavy
Cost	Low	Medium	High
Application	Suitable for small volumes and flow rates	Best for general-purpose applications	Best for large volumes or high flow rates
Shock suppression ability	Good shock absorber	Good shock absorber	Not suitable for suppressing shocks

Accumulator Pre-charging

A hydro-pneumatic accumulator must be filled with dry inert gas, such as nitrogen gas, while no fluid is in the fluid chamber. The pre-charge level of the gas medium is an essential parameter for the gas accumulator, as the pre-charge pressure, along with the accumulator volume, determines the maximum amount of hydraulic energy that can be stored.

Generally, a gas accumulator is pre-charged to a certain percentage of the minimum system pressure (P_1), depending upon the accumulator type and application, and according to the manufacturer's recommendation.

For energy storage applications, the pre-charge pressure P_0 typically is 80 to 90 percent of the system's minimum working (operating) pressure P_1.

The pre-charge pressure P_0 for a pulsation compensator or a shock absorber can be 65 to 80 percent of the minimum operating pressure P_1.

The gas pre-charge pressure is further limited by the maximum pressure ratio $P_2:P_0$, which should not exceed 4:1.

That is, $P_2:P_0 \leq 4:1$

Where P_2 is the maximum operating pressure.

Safety Requirements for Hydraulic Accumulators

A hydraulic accumulator is a pressure vessel that stores a large amount of potential energy for subsequent release to perform useful hydraulic functions. Accumulators can be dangerous to personnel and property if they discharge the stored pressure inadvertently. Moreover, they are subject to regulations applicable to the place of their installation. Therefore, it is necessary to isolate the accumulators from the associated systems and discharge their pressure during maintenance or emergency periods. Safety devices must be incorporated into an accumulator to provide a shut-off facility and pressure-limiting and relief features. So, a safety-and-shut-off block, in conjunction with an accumulator, is recommended to protect the system and personnel from the hazardous stored energy.

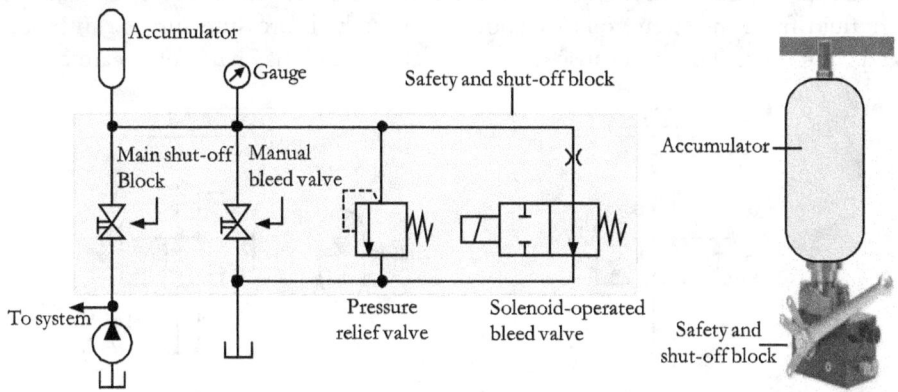

Figure 17.11 | A circuit layout of a safety-and-shut-off block for a hydraulic accumulator

Safety-and-Shut-off Block

Figure 17.11 shows the circuit diagram of a safety-and-shut-off block. It is a multifunctional valve system connected to a hydraulic accumulator. Further, it consists of a shut-off valve, a manual bleed valve, a pressure relief valve, an optional 2-way solenoid-operated bleed valve, and a pressure gauge. The block can be configured modularly with a host of connection options, enabling versatile mounting.

The shut-off valve instantly isolates the accumulator from the hydraulic system for maintenance or emergency use. Once isolated, the accumulator can be safely discharged to the reservoir through the manual bleed valve.

If used, the optional solenoid-operated bleed valve automatically releases the stored energy in the accumulator during an emergency shutdown or loss of electrical power.

A pressure relief valve protects the accumulator from over-pressurization. Qualified personnel must commission and maintain the safety block and associated equipment.

Application Areas of Accumulators

Accumulators are widely used in industrial and mobile applications to absorb shocks, store energy, and compensate for leakage. They are used in large hydraulic presses, vehicles, construction equipment, offshore equipment, and mining equipment. Accumulators can supplement the pump flow in hydraulic applications, such as aircraft landing gear, which require a considerable volume of fluid.

Basic Hydraulic Circuits with Accumulators

Hydraulic accumulators perform many functions in hydraulic systems. They are mostly employed for shock absorption and energy storage.

An accumulator as a Hydraulic Shock Absorber

Figure 17.12(a) shows a simple hydraulic circuit for controlling the direction of a double-acting cylinder using a 4/3–way, closed-center directional control valve.

As we know, high-pressure surges or shock waves are generated in the system when the valve is rapidly shifted to its neutral position. These shock waves are caused by the associated PRV failing to drain the high-pressure fluid from the circuit quickly enough. These high-pressure surges can be dangerous to personnel and equipment. They can cause noise and damage and may harm upstream components.

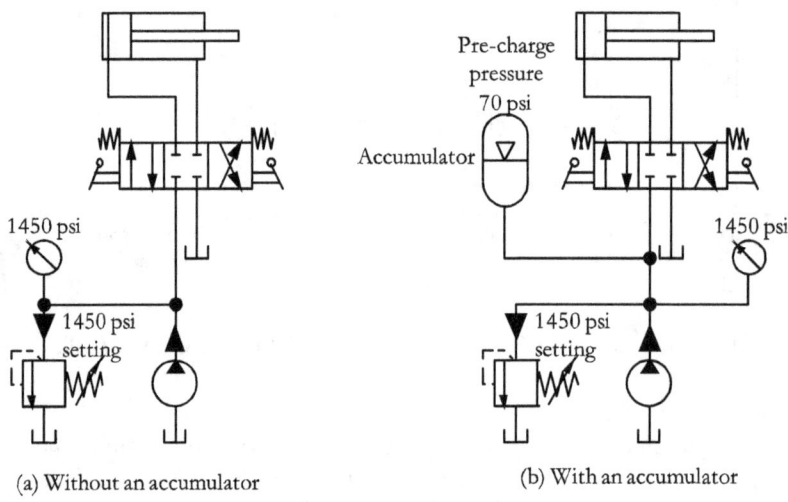

Figure 17.12 | Basic hydraulic circuits

As shown in Figure 17.12 (b), a properly sized accumulator must be connected to the circuit to suppress shock waves. The accumulator compresses the gas in the chamber to absorb the pressure surge and eliminate the shock. It is important to install the accumulator as close as possible to the potential source of shock.

The accumulator absorbs shock waves whenever they appear in the circuit. While piston accumulators can be used, bladder accumulators act more quickly and are often the better choice. To best absorb shock, the accumulator should have a large fluid port.

However, this circuit cannot store the excess energy in the accumulator, as the pressure relief valve slowly drains the excess fluid back to the system reservoir.

Additional components, such as a check valve, an unloading valve, and a throttle valve, can serve additional functions, including storing energy, unloading the pump when the system pressure is reached, and adjusting the discharge flow rate from the accumulator to meet system requirements. The following sections explain the circuits for realizing these accumulator functions.

An Accumulator as an Auxiliary Power Source

Figure 17.13 shows different positions of a hydraulic circuit, necessarily with a fixed-displacement pump, reservoir, unloading valve, and accumulator. The accumulator should be sized appropriately to store energy during periods of low demand and to provide auxiliary power during periods of high demand.

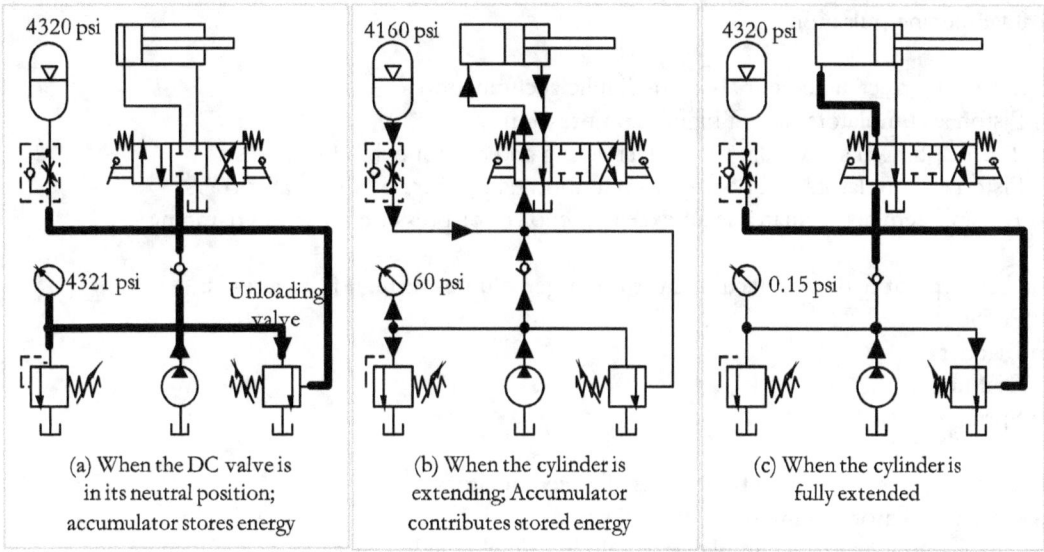

Figure 17.13 | Three positions of a circuit with an accumulator acting as an auxiliary power source

Figure 17.13(a) shows the position of the circuit when the 4/3-way, closed-center DC valve is pulled to its neutral position. In this position, the pump charges the accumulator and builds pressure through the check valve. The unloading valve remains closed until the set pressure is reached. The unloading valve opens and unloads the pump, allowing the flow to return to the reservoir at low pressure when the valve's pressure setting is reached. The unloading of the pump flow protects the system against over-pressurization. The isolation check valve prevents the accumulator fluid from flowing back to the pump and traps the pressure in the accumulator.

Figure 17.13(b) shows the position when the 4/3-way DC valve is shifted to its left envelope for the forward stroke of the cylinder. During this position, the pump flow is supplemented by the accumulator, which releases the stored energy. Remember that there is a volume of fluid under elevated pressure in the accumulator that can discharge almost instantaneously into the system; therefore, when the accumulator releases stored energy, a throttle valve is required to set the flow rate within the system's requirements.

Figure 17.13(c) shows the circuit's position when the cylinder is fully extended and the accumulator is fully charged. When the set pressure is reached, the unloading valve opens, unloading the pump to the reservoir at low pressure.

When the 4/3-way DC valve is shifted to its right envelope for the cylinder's return stroke, the accumulator and the unloading valve can expect similar actions, as explained in the previous paragraphs.

Objective-type Questions

1. A function of hydraulic accumulators is:
   a) Limiting pressure
   b) Controlling speed
   c) Fluid conditioning
   d) Dampening pulsations

2. Mark the <u>incorrect</u> statement about hydraulic accumulators.
   a) Piston accumulators have a higher compression ratio
   b) The diaphragm accumulators are fast-acting and do not exhibit hysteresis
   c) Piston accumulators can be used as pulsation dampeners or shock absorbers
   d) A diaphragm accumulator is an excellent choice for pressure pulsation damping

3. The best type of hydraulic accumulator for large volumes or high flow rates is:
   a) Piston
   b) Bladder
   c) Diaphragm
   d) Spring-loaded

4. Mark the <u>correct</u> statement about hydraulic accumulators.
   a) An accumulator can intensify the pressure.
   b) A hydro-pneumatic accumulator should be pre-charged before use.
   c) A piston accumulator is used for low pressure, low-flow applications.
   d) An accumulator's safety and control block is an optional component.

Review Questions

1) Define a hydraulic accumulator.
2) Explain the essential operation of a hydraulic accumulator using a simple circuit.
3) List four essential functions of hydraulic accumulators.
4) Explain how the pressure pulsations in a hydraulic rock drilling operation can be suppressed.
5) Describe the function of a hydraulic accumulator used as a shock absorber.
6) What adverse effect can happen in a primary hydraulic loader system with a 2-ton front-loading bucket suddenly stopped? How can this effect be overcome?
7) Explain the function of a hydraulic accumulator used as a pump supplement.
8) Explain how the capacity requirements of the pump, utilized in a hydraulic system delivering only a short period of high-volume flow during a given cycle, can be reduced.
9) Develop a hydraulic circuit to carry out a work process in which significant fluid flows for a short period in each cycle.
10) Write briefly about the three most common applications of hydraulic accumulators.
11) Classify hydraulic accumulators.
12) Briefly explain three basic types of accumulators used in hydraulic systems.
13) Draw the symbols for the weight-loaded, spring-loaded, and gas-charged hydraulic accumulators.
14) Explain the principle of operation of hydraulic accumulators.
15) Briefly explain the operation of a weight-loaded hydraulic accumulator with a simple sketch.
16) Give some advantages and disadvantages of the weight-loaded hydraulic accumulators.
17) Briefly explain the operation of a spring-loaded hydraulic accumulator with a simple sketch.

18) Give some benefits and disadvantages of the spring-loaded hydraulic accumulators.
19) Name three major classifications of gas-charged hydraulic accumulators. Write one advantage of each accumulator.
20) Explain the fundamental operating principle of hydro-pneumatic accumulators.
21) What are the advantages of gas-charged accumulators compared to other hydraulic accumulators?
22) What gas is used in hydro-pneumatic accumulators? Why can oxygen and air not be used?
23) Why is the non-separator-type gas-charged accumulator not used in modern hydraulic systems?
24) Briefly explain the operation of a piston-type hydraulic accumulator with a simple sketch.
25) Briefly explain the constructional features of piston-type hydraulic accumulators
26) Give one advantage and one disadvantage of the piston-type hydraulic accumulators.
27) Briefly explain the principle of operation of a hydraulic diaphragm accumulator with a simple sketch.
28) Briefly describe the constructional features of the hydraulic diaphragm-type accumulators
29) Give one advantage and one disadvantage of the hydraulic diaphragm accumulators.
30) Briefly explain the operation of a hydraulic bladder accumulator with a simple sketch.
31) Explain the function and purpose of a gas-filled accumulator.
32) Briefly explain the constructional features of the hydraulic bladder-type accumulators
33) Give one advantage and one disadvantage of the hydraulic bladder accumulator.
34) Compare the diaphragm, bladder, and piston accumulators for typical sizes, pressures, flow rates, and suitability for shock suppression.
35) Give Short notes on (a) pre-charging the gas-loaded hydraulic accumulators and (b) dump valves in hydraulic accumulator circuits.
36) Explain how a load-induced shock can be reduced in a hydraulic system with a circuit diagram.
37) Explain the safety requirements of accumulators in hydraulic circuits.
38) Describe the functions of the safety-and-shut-off block associated with hydraulic accumulators.
39) What are the three important accumulator pressure conditions?
40) Draw a hydraulic circuit for extending and retracting a double-acting hydraulic cylinder with provision for holding pressure, leakage compensation, and power savings. The circuit may include the following essential components: a pump, a check valve, an accumulator, a 4/3-DC valve, and a double-acting cylinder.
41) A large hydraulic cylinder should clamp a workpiece. The cylinder requires a high volume of fluid while it extends rapidly. However, the circuit needs no additional fluid while the cylinder is clamping. Assume that the circuit has an extended dwell time. Develop a circuit capable of storing energy at times of low demand and using the stored energy when a large volume of fluid is required for a short period.
42) A diesel engine is started using a hydraulic motor. The maximum engine power requirement is only for a short time when the start signal is given with a 2/2-DC valve and the interval between two starting operations is long. An accumulator is used to supplement the pump's power during start-up. The pump should unload during idle time. Develop a hydraulic circuit to implement the scheme.

Objective-type questions - Answer key: *1-d, 2-c, 3-a, 4-b*

Chapter 18 | Hydraulic Seals

Seals (or packings) are deformable materials sandwiched between the mating surfaces to close small gaps. They are critical parts, as they prevent system fluid from leaking past the clearances, retain fluid under pressure within the system, and keep foreign matter out. They also prevent the metal-to-metal contact of sliding surfaces. They are made from polymer materials.

Polymerization
A single molecule consisting of a group of similar or dissimilar atoms is known as a mer or monomer. When conditions are right, many monomers can combine chemically to form long, chain-like structures. The resulting macromolecules, incorporating thousands of monomers, are known as polymers. Polyethylene, rubber, and plastics are examples of polymer-based materials. Depending on how the molecules are arranged in polymers, they can be amorphous or crystalline. Amorphous polymers are composed of long and twisted molecular chains that are non-symmetrical. The intermolecular forces are weak in amorphous polymers. It may be noted that all rubbers are amorphous at room temperature. When a compressive load or stretching force is applied to the elastomeric material, the entangled molecular chains uncoil and straighten. When the stress is removed, the chains tend to coil up again, reverting to their normal state of entanglement.

Crystalline Polymers
Crystalline polymers are composed of orderly arranged molecules. The intermolecular forces are strong in crystalline polymers. The orderly arrangement of polymer molecules is rigid. Most plastics are either crystalline or semi-crystalline.

Polymer Additives
Seals can encounter potentially harmful service and environmental conditions, such as extreme pressure, heat, or cold. Other ingredients must be added to enhance their physical and chemical properties. These ingredients may include fillers (to reinforce the material), cure activators and accelerators (to increase cure speed), plasticizers (to aid flexibility), and pigments (for colorization)

Properties of Polymers
The chain length, and thus the molecular weight, significantly impacts the polymer's physical properties. Polymers with high molecular weight are essential for formulating tough materials for applications under severe operating conditions. When a polymer is heated, its intermolecular forces decrease. The seal material must be chemically compatible with the system fluid.

Measurement of Hardness
Hardness is measured with a portable instrument called the 'Shore durometer.' It uses a cone indenter, loaded by a calibrated spring, to measure the test specimen's resistance to indentation. The penetration depth of the indenter under the load determines the hardness of the specimen. The 'Shore A' scale is used for testing soft elastomers, whereas the 'Shore D' scale is used for testing hard elastomers. The Shore A loading force is 822 grams, using a lighter spring and a 35-degree angle indenter point. The Shore D loading force is 4536 grams, using a stiffer spring and a sharp 30-degree indenter.

When the durometer is pressed against a flat elastomer, the indenter point is forced back against the spring. The force is reflected on the gauge, which has an arbitrary scale of 0 to 100. Harder substances generate large durometer numbers. The most common hardness range for seal materials for the hydraulic system is from 50 to 80 Shore A. The Shore D durometer accurately measures the hardness of harder materials than 90 Shore A.

Terms and Definitions
Resilience: It is the seal material's ability to return to its original size and shape after deformation.

Compression set: It is the amount by which the seal material remains short of its original shape after being released from a compressive load. This factor is usually expressed as a percentage of its original dimension. Due to compression set, a seal often hardens and assumes the shape of the gland.

Extrusion: It refers to the flow of a part of an O-ring into the clearance between two mating metal parts when subjected to high pressure. The extruded portion of the seal is liable to be nibbled away from the low-pressure side. Continuous biting away of the extruded part can lead to complete seal failure. Anti-extrusion devices can be used to avoid extrusion.

Factors Affecting Seal Performance
The performance of hydraulic seals is influenced by factors such as system pressure, fluid temperature, speed of moving parts, component surface finish, and exposure to oxygen, ozone, and sunlight. These are briefly explained below.

Pressure
Seals are subjected to operating and shock pressures. Excessive pressure causes seal extrusion. Next, pressure spikes can deform the seals. Repeated deformations result in premature wear of the seal materials.

Temperature
Lower temperatures may harden the seals and make them brittle. At higher temperatures, the seal materials may become too soft to withstand the applied pressure and are susceptible to extrusion. Table A10.1 in Appendix 10 provides details on seal materials and their temperature ranges.

Speed
As the speed of a moving part at the seal contact surface increases, the fluid film between the sealing surfaces breaks down, increasing friction.

Surface Finish
The performance of a seal is significantly affected by the finish of the surface it moves across. The surface finish has a major impact on friction, wear, and sealing performance. A smoother finish reduces wear on the seal and improves its ability to create a better seal with the mating surface. Rough or poorly finished metal surfaces increase friction. Extremely smooth surfaces lack the necessary indentations to retain lubricating fluids. All surface finishes aim to minimize wear on seals. The ideal average roughness, Ra, for working surfaces is 0.5 to 0.6 microns (1.9685×10^{-5} to 2.3622×10^{-5} inches).

Oxygen, Ozone & Sunlight
Oxygen, especially along with heat, causes the seal to harden. Ozone and sunlight can also break the polymer chains.

Requirements for Seals

- Seals must have good mechanical strength, high resistance to abrasion, stability of form, low compression set, high tear strength, low water absorption property, and good anti-extrusion behavior.
- Seals must be capable of withstanding high pressures and dynamic forces within the system. Softer materials may be perfectly fine for low-pressure seals, but seal materials must be harder, stronger, and extrusion-resistant at higher operating pressures.
- The seal contact surfaces should be appropriately finished to meet the conflicting requirements of reducing friction and improving their lubrication-holding ability.
- Seals must retain most of their original properties when subjected to elevated temperatures.
- Hydraulic seals must have good chemical stability as they may be exposed to petroleum or synthetic fluids and harmful agents, such as acids, heat, sunlight, oxygen, ozone, and weather.
- Seals must have enough tolerance to side loads and vibration.

Classification of Hydraulic Seals

A static seal seals high-pressure fluid between stationary parts in hydraulic devices. A dynamic seal is used for reciprocating or rotary motion. In reciprocating seal applications, a seal slides back and forth within its gland. In rotary seal applications, a seal moves radially within its groove.

Static Seals

The O-ring is probably the most commonly used static seal. In high-pressure systems, static seals may be configured with backup rings to prevent excessive compression and seal extrusion.

Dynamic Seals

Hydraulic seals for oscillatory or slow-rotation applications with surface speeds of less than 50 feet per minute are usually classified as dynamic seals. Seals for applications involving high-speed rotation, with surface speeds exceeding 50 feet per minute, are classified as high-speed rotary seals.

Seal Materials

Elastomer Group
Acrylo Nitrile or Butadiene (NBR)
- Based on butadiene and acrylonitrile copolymer
- Excellent abrasion resistance, high tensile strength, and high resilience
- Limited resistance to heat
- Excellent compatibility with petroleum-based fluids

Viton (Fluorocarbon Rubber or FKM)
- Carbon backboned polymers, highly fluorinated
- Excellent heat resistance, with thermal stability up to 262°C
- Low compression set and excellent aging characteristics
- Compatible with a broad range of fluids

Silicon Rubber
- Made from silicon, oxygen, and carbon
- Offers excellent resistance to compression set at high temperatures
- Highly resistant to sunlight, ozone, oxygen, and moisture
- Mainly used as static seals

Ethylene Propylene Rubber (EPR)
- Offers excellent resistance to heat, ozone, and UV light
- Used for ester-based fluids, such as Sky-drol
- Not suitable for petroleum-based fluids

Plastic Group
The plastic group includes polyurethane and nylon.

Polyurethane
- Formulated from copolymers of ether or ester-based urethanes
- Good mechanical properties, such as high resilience and high tensile strength
- Good resistance to extrusion, abrasion, tear, oxidation, and oil swell
- An excellent choice for petroleum-based fluids

Nylon
- Formulated from synthetic rubber and fluorine
- Very high heat resistance and excellent mechanical properties

P T F E (Teflon) Group
PTFE (poly tetra fluoro ethylene) or Teflon is a fluoro-plastic distinguished by excellent resistance to chemicals

Virgin P T F E
- Offers excellent resistance to most chemicals
- Very low coefficient of friction
- Operates over a broad range of temperatures

15% Glass-filled and 60% Bronze-filled PTFE
A virgin PTFE is fortified with glass fiber and bronze to retain its toughness and flexibility, reduce thermal expansion, and improve wear strength.
- Excellent chemical inertness and high heat resistance
- Higher resistance to extrusion as compared to virgin PTFE
- Preferred for high-power hydraulic applications
- Used for making piston seals, rod seals, and wipers

Objective-type Questions
1. Which of the following is not a polymer additive?
    a) Fillers
    b) Demulsifiers
    c) Plasticizers
    d) Pigments

2. Which instrument is used to measure the hardness of hydraulic seals?
    a) Durometer
    b) Viscometer
    c) Spectrometer
    d) None

3. Mark the term for measuring the indentation resistance of the seal material.
   a) Elasticity
   b) Hardness
   c) Modulus
   d) Flex resistance

4. Which seal materials offer high heat resistance and chemical inertness?
   a) Silicon rubber
   b) Polyurethane
   c) Nylon
   d) PTFE

5. Mark the incorrect statement.
   a) The molecules in amorphous materials are branched and non-symmetrical
   b) All rubbers and elastomers are crystalline at room temperature
   c) Crystalline seal materials are linear and symmetrical materials
   d) In plastic materials, intermolecular van der Waals forces are high

Review Questions
1) What are the functions of hydraulic seals?
2) What are polymers?
3) Briefly explain the classification of polymers.
4) What are the basic types of hydraulic seal materials?
5) Describe the process of polymerization.
6) Distinguish between the amorphous and the crystalline polymers.
7) What is the difference between rubber and plastic in their molecular arrangement?
8) What is the effect of increased chain length for a polymer material used in a hydraulic seal?
9) What is the reason for the resilient nature of hydraulic seals?
10) Why are additives added to the base elastomeric materials used in hydraulic systems?
11) Enumerate five essential properties of hydraulic seal materials.
12) How is the hardness of a seal material measured?
13) Differentiate the Shore A and the Shore D scales of the hardness measurement of seals.
14) Define the following: (1) Compression set, (2) Seal extrusion, and (3) Tear resistance.
15) What are the critical factors affecting the hydraulic seal performance? Explain briefly.
16) How do extreme temperatures and friction affect hydraulic seals?
17) Explain the impact of the rough finish of the contact surfaces of hydraulic components on seals.
18) State four essential requirements of seals used in hydraulic systems
19) Describe the functions of the static and the dynamic seals used in hydraulic systems.
20) Differentiate between the static and the dynamic seals used in hydraulic devices.
21) List and describe two major dynamic seals used in hydraulic systems.
22) List two hydraulic seal materials for use in extreme temperature conditions.
23) List two types of materials, each for the piston seals and wipers used in hydraulic cylinders.
24) List some commonly used hydraulic seal materials.

Objective-type questions - Answer key: 1-b, 2-a, 3-b, 4-d, 5-b

Chapter 19 | Fluid Conductors

Various components are assembled in a conventional hydraulic system through a conductor system. The conductor system is a network of conductors that connect to the components through fittings to deliver fluid effectively. Pipes, tubes, and hoses are the three basic fluid conductors used in hydraulic systems to transfer energy.

A conductor is a pressure-tight vessel conveying enough pressurized fluid through a leak-free system. It must have smooth interiors to reduce friction in the sliding parts of system components and turbulence in fluid flow. Moreover, it must have sufficient wall thickness to withstand high operating and shock pressures. Further, it must withstand high system and ambient temperatures. It must also be compatible with the type of fluid used.

Critical considerations for selecting fluid conductors include their construction, sizing, installation, routing, and applicable standards. The following sections present a concise and structured account of various aspects of fluid conductors.

Terms and Definitions, Fluid Conductor
The fluid power industry uses many conductor-related terms to specify the performance levels of fluid power systems. The following sections briefly cover some commonly used terms and definitions for fluid conductors.

Diametrical Size
The diametrical size of a conductor is specified by its inside diameter, outside diameter, or nominal size.

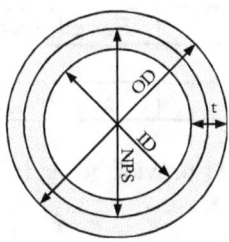

Figure 19.1 | Size specifications of a fluid conductor

Inside Diameter, D_i
It is the smallest cross-sectional diameter of a conductor, as shown in Figure 19.1.

Outside Diameter, D_o
It is the largest cross-sectional diameter of a conductor, as shown in Figure 19.1.

Nominal Size
The nominal size of a pipe (NPS), as shown in Figure 19.1, may correspond to the inside diameter (ID), outside diameter (OD), or a value in between its OD and ID, depending on the size, type, and applicable standard (ISO, ANSI, or SAE) of the pipe. In ANSI and SAE standards, for example, the size of a pipe is specified in terms of nominal pipe size (NPS), and in the SI system, it is specified in terms of Nominal Diameter (DN).

Wall Thickness, t

The wall thickness of a pipe or tube determines the maximum pressure it can withstand. It is expressed as a schedule number in ANSI and SAE standards or metric units. It is given by:

Wall thickness, $t = (D_o - D_i) / 2$

Example 19.1 | Determine the inside diameter of a pipe with an OD of 1.9 inches and a wall thickness of 0.145 inches.

Solution

Pipe OD	= 1.9 inch
Wall thickness	= 0.145 inch
Inside diameter of the pipe	= 1.9 – 2 x 0.145
	= 1.61 inch

Schedule Number

In ANSI/SAE systems, the wall thickness of a pipe is usually described in terms of a 'schedule number.' The schedule numbers vary from 5 to 160 in a graded manner. Altogether, there are eleven different schedule numbers. They are: 5, 10, 20, 30, 40, 60, 80, 100, 120, 140, and 160. A larger schedule number for a given pipe size indicates heavier wall thickness. The outer diameter remains the same for a given pipe size, but the inside diameter decreases as the schedule number increases. For example, Table 19.1 lists the wall thicknesses corresponding to schedules 40, 80, and 160 for a pipe of nominal size ½.

Table 19.1 | Wall thickness of pipes in terms of schedule numbers corresponding to ½" NPS

Nominal Pipe Size	Outside Diameter	Wall thickness		
		Schedule 40	Schedule 80	Schedule 160
--	inch	inch	inch	inch
½	0.500	0.840	0.109	0.147	0.188

Tables A11.1 and A11.2 of Appendix 11 provide general specifications for steel pipes and tubes.

Hoop Stress

Hoop stress, as shown in Figure 19.2, of a pipe or tube is the circumferential stress acting on the wall of the conductor trying to split it. It is the maximum pressure that the material of the conductor is capable of withstanding before pulling apart. Consider a thin-walled conductor of the inside diameter (D_i), wall thickness (t, where $t < 0.1 \times D_i$), and length (L). The conductor is subjected to operating pressure, P.

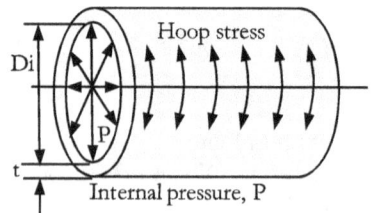

Figure 19.2 | Hoop stress

The operating pressure, acting normal to the inside surface of the pipe, induces a circumferential force in the conductor that tends to split the conductor into two halves. The normal surface area can be the projected area ($D_i \times L$) of one-half of the pipe. The circumferential force (or burst force) is given by:

$$\text{Circumferential force} = P \times D_i \times L$$

The tensile force, which is trying to resist the splitting of the conductor, acts on the cross-sectional area (tL) of each wall. Therefore, the resistive force is given by:

$$\text{Resistive force} = 2tL \times \text{Hoop stress}.$$

Equating the circumferential force to the resistive force, we get,

$$P \times D_i \times L = 2tL \times \text{Hoop stress}$$

Therefore,

$$\text{Hoop stress} = P \times D_i / 2t$$

The conductor must have sufficient tensile strength to prevent it from bursting under excessive hoop stress.

Example 19.2 | Determine the hoop stress developed in a pipe with an outside diameter of 2.375 inches and a wall thickness of 0.218 inches when the pipe is subjected to a pressure of 1000 psi.

Solution

Outside diameter of the pipe, D_o = 2.375 inches
Wall thickness of the pipe, t = 0.218 inch
Pressure, P = 1000 psi

Outside diameter of the pipe, D_i = 2.375 − (2 × 0.218) inch = 1.939 inch

Hoop stress developed = $P \times D_i / 2t$
= 1000 × 1.939 / (2 × 0.218) psi
= 4447 psi

Burst Pressure

It is the internal pressure inside a fluid conductor that causes it to burst or rupture. A ruptured tube is shown in Figure 19.3. The fluid conductor bursts when the hoop stress exerted on the conductor exceeds the tensile strength (S) of the conductor material. Barlow's formula is commonly used to predict the burst pressures in ductile thin-wall tubes ($t < 0.1 \times D_i$)

$$\text{Burst pressure (BP)} = 2tS / D_i$$

Tensile stress across the wall thickness is not uniform in thick-walled pipes. Therefore, the following formula must account for this non-uniform tensile stress.

$$\text{Burst pressure (BP)} = 2tS / (D_i + 1.2t)$$

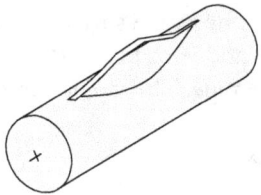

Figure 19.3 | A ruptured tube

Working Pressure

The working pressure of a conductor is the maximum sustained pressure it can withstand. It is calculated by dividing the conductor's burst pressure by a safety factor.

$$\text{Working pressure (WP)} = \frac{\text{Burst pressure (BP)}}{\text{Safety factor (SF)}}$$

- A safety factor of 4:1 is used for hydraulic applications where shock and mechanical strain are not considerable.
- A safety factor of 6:1 should be used where considerable shock and mechanical strain are expected.
- A safety factor of 8:1 should be used where severe hydraulic shock and mechanical strain are expected.

Example 19.3 | Determine the maximum pressure rating of a thick-wall pipe of OD of 1.315 inches with a wall thickness of 0.25 inches. The pipe material is carbon steel with a tensile strength of 50000 psi. Assume a safety factor of 4.

Solution

Pipe OD, D_o	= 1.315 inch
Wall thickness, t	= 0.25 inch
Tensile strength, S	= 50000 psi
Safety factor, SF	= 4
Pipe ID, D_i	= $D_o - 2 \times t$
	= 1.315 − (2 × 0.25) inch
	= 0.815 inch
Burst factor	= $2ts / (D_i + 1.2t)$
	= 2 × 0.25 × 50000 / (0.815 + 1.2 × 0.25)
	= 22421 psi
Pressure rating of the pipe	= Burst pressure / Safety factor
	= 22421 / 4 = 5605 psi

Minimum Bend Radius

It is the smallest radius of the curved section of a tube beyond which it should not be bent without flattening, kinking, or wrinkling, as shown in Figure 19.4. Bending the tube beyond the limit causes severe backpressure and damages the conductor internally, leading to its premature failure.

The bend radius is typically measured along the tube's centerline as the distance from the center of curvature to the centerline. A rule of thumb suggests a minimum bend radius of three times the outside diameter.

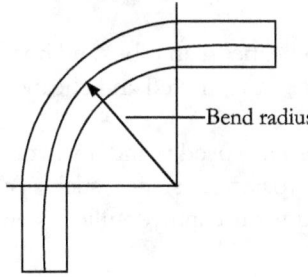

Figure 19.4 | Minimum bend radius of the tube

Pipe and Tube Materials
The pipe and tube materials suitable for high-pressure industrial hydraulic service are: (1) Cold-drawn seamless carbon steel and (2) Cold-drawn seamless stainless steel. Carbon steel is generally used for pipes and tubes employed in indoor hydraulic applications. Stainless steel pipes and tubes are used in applications that require corrosion resistance, such as in chemical equipment or marine vessels. Hot-rolled pipes are not recommended for hydraulic services because they have scales on the inside and outside. The inside scales affect the cleanliness level of the fluid.

Steel is an alloy of iron and carbon. Four types of steel exist based on their physical properties and unique chemical compositions, which depend on the amount of carbon and other alloying elements. They are carbon steel, alloy steel, stainless steel, and tool steel.

Conductor Sizing
Proper conductor sizing for various hydraulic system components results in an optimal balance of efficiency and cost-effectiveness. A conductor that is too small can lead to high fluid velocity, high pressure drops, and increased heat generation. Too-large conductors increase system cost. Thus, optimum conductor sizing is critical.
- Calculate the minimum I D from the flow-rate requirements
- Select the standard-size conductor
- Determine the wall thickness
- Determine the burst pressure
- Determine the working pressure

Selection of Conductor
The selection of conductors for a fluid power system is critical to its efficient, trouble-free operation. For a hydraulic application, the conductor system involves selecting its type, size, and material, and matching the fittings to the system.

Pipes
Pipes are rigid conductors used to contain and convey hydraulic fluids. They are highly resistant to bending. It is difficult to shape rigid pipes into the desired configuration. Remember, configuring a piping system is more labor-intensive. Many fittings, such as elbows and tees, must be used while routing a piping system. They are liable to transmit shock and vibration between components. All piping segments should be secured with clamps, preferably with damped ones, to absorb the shock and vibration and prevent their propagation.

Constructional Details, Pipes
Pipes have a thicker wall but are much cheaper than tubes and hoses. Hence, they are generally employed in applications where conductor size is not restricted and cheaper conductor systems are preferred.

Carbon steel is generally used for pipes employed in indoor hydraulic applications. Stainless steel pipes are used in applications that require corrosion resistance, such as chemical equipment or marine vessels. A galvanized pipe is not recommended for use in hydraulic systems.

Advantages and Disadvantages, Pipes
Pipes have many advantages and disadvantages. Some advantages and disadvantages are listed below:
- A conductor system with steel pipes is the least expensive way to assemble a hydraulic system with low to medium-pressure ratings
- As pipes are made of inflexible material and have a large wall thickness, they are difficult to form into the desired configuration and install
- They cannot withstand high surge pressures

Basic Requirements, Pipes
- Pipes must have sufficient cross-sectional areas to satisfy the flow rate requirements without producing excessive pressure drops
- They must be strong enough to withstand the working pressure, shock pressures, and vibration
- They should have smooth interiors to reduce the friction and flow turbulence
- They must be compatible with the type of fluid used
- They must withstand high operating temperatures
- They must be supported by damped mountings to absorb both shock and vibration

Size Specifications, Pipes
A pipe should have sufficient ID and a smooth inside surface to reduce frictional forces. The wall thickness of the pipe decides its pressure rating. The optimal pipe size should be determined by minimizing the sum of energy and piping costs.

Pipe sizes are standardized to reduce the number of sizes. In current practice, pipe size is defined with two sets of numbers:
1) Nominal pipe size (NPS), Nominal diameter (DN), or Nominal bore
2) Pipe schedule (wall thickness).

Wall Thickness – Schedule Numbers
Manufacturers offer standard and non-standard sizes. Schedule numbers 40, 80, and 160 are most commonly used to specify the wall thickness of pipes for hydraulic systems.
- Schedule number 40 conforms to the 'standard' wall thickness for low pressures.
- Schedule number 80 conforms to the 'extra heavy' wall thickness for high pressures.
- Schedule number 160 conforms to the 'double extra heavy' wall thickness.

Wall Thickness – in Metric Units
Wall thickness is specified in mm.

Nominal Pipe Size (NPS)

It is a size standard established by the American National Standards Institute (ANSI). It is the number that defines the pipe's size. For a 6 NPS pipe, the nominal size is 6". Pipe nominal sizes do not always correspond to their inside diameter or outside diameter. Hydraulic pipes come in nominal sizes from 1/8" to 42". Each size is available in a variety of wall thicknesses.

Table 19.2 | Nominal pipe sizes

Nominal Pipe Size		Outside Diameter	Wall thickness		
			Schedule 40	Schedule 80	Schedule 160
--	inch	inch	inch	inch	inch
1/8	0.125	0.405	0.068	0.095	--
1/4	0.250	0.540	0.088	0.119	--
3/8	0.375	0.675	0.091	0.126	--
1/2	0.500	0.840	0.109	0.147	0.188
3/4	0.750	1.050	0.113	0.154	0.219
1	1.000	1.315	0.133	0.179	0.250
1 1/4	1.250	1.660	0.140	0.191	0.250
1 1/2	1.500	1.900	0.145	0.200	0.281
2	2.000	2.375	0.154	0.218	0.344
2 1/2	2.500	2.875	0.203	0.276	0.375
3	3.000	3.500	0.216	0.300	0.438
3 1/2	3.500	4.000	0.226	0.318	--
4	4.000	4.500	0.237	0.337	0.531
5	5.000	5.563	0.258	0.375	0.625
6	6.000	6.625	0.280	0.432	0.719
8	8.000	8.625	0.322	0.500	0.906
10	10.00	10.75	0.365	0.594	1.125
12	12.00	12.75	0.406	0.688	1.312
14	14.00	14.00	0.438	0.750	1.406
16	16.00	16.00	0.500	0.844	1.594
18	18.00	18.00	0.562	0.938	1.781

Table 19.2 shows that for a pipe with a nominal size between 1/8" and 12", its internal diameter (ID) differs from the nominal size, depending on the schedule number. For a pipe with a nominal size of 14" or above, the pipe OD corresponds to the nominal pipe size. Any increase in wall thickness decreases the pipe's inside diameter.

Example 19.4 | Refer to Table 19.2. Calculate the IDs of schedule 40 pipes corresponding to the following nominal pipe sizes: (1) 2" and (2) 14."

Solution

(1)

NPS	= 2 in
OD	= 2.375 in
Wall thickness (t)	= 0.154

ID	= OD – 2 x t
	= 2.375 – (2 x 0.154) in
	= 2.067 in

(2)

NPS	= 14 in
OD	= 14 in
Wall thickness (t)	= 0.438

ID	= OD – 2 x t
	= 14 – (2 x 0.438) in
	= 13.124 in

Fittings, Pipe

Pipe connections are coupled using welded, flanged, or threaded joints. The type of jointing technology is selected based on working pressure, pipe size, pipe material, fitting standards, and other conditions, such as potential pressure surges in the system and environmental conditions. Figure 19.5 shows an assortment of pipe fittings.

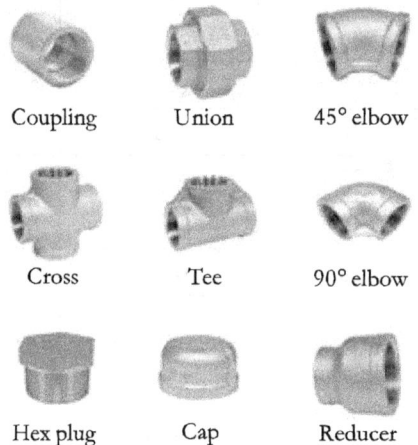

Figure 19.5 | An assortment of pipe fittings

Pipe fittings, such as sleeves, elbows, tees, and bends, can join sections of pipes together. The threaded connections are most common and are used in applications with pressures up to 2500 psi.

Thread Joints for Pipes

Thread joints are used for hydraulic service to produce a leak-proof metal-to-metal seal. They are either tapered or straight, as shown in Figure 19.6. Next, the pipe threads are made pressure-tight by sealing them. Pipe threads used in hydraulic piping can be divided into two types: (1) Standard pipe threads and (2) Dry-seal pipe threads.

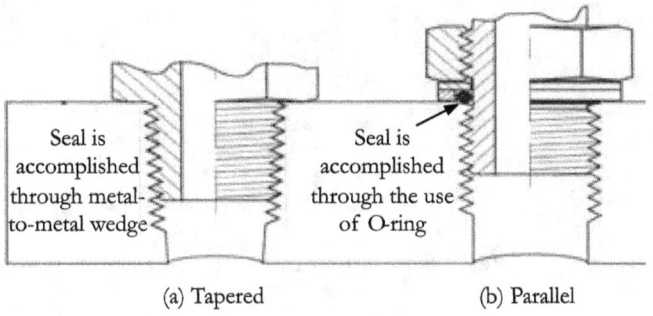

(a) Tapered   (b) Parallel

Figure 19.6 | Types of thread joints

Standard Pipe thread

As shown in Figure 19.7(b), the standard pipe thread has tapered threads that produce a metal-to-metal seal. The taper is 1/16 of an inch. The connection is made pressure-tight by sealing the threads. This type of thread leaves a clearance spirally as the pipe is tightened. Pipe threads require sealants, such as Teflon tape or joint compound, to fill gaps between the threads and make the joint leakproof. However, these threads often develop leaks that are difficult to repair in the field.

Tapered pipe threads are used for standard hydraulic services (except for toxic or corrosive applications) at pressures up to 1500 psi and temperatures up to 900°F. They can be used for hydraulic services without pressure limitations for connection to equipment only, such as pumps, valves, cylinders, accumulators, gauges, and hoses.

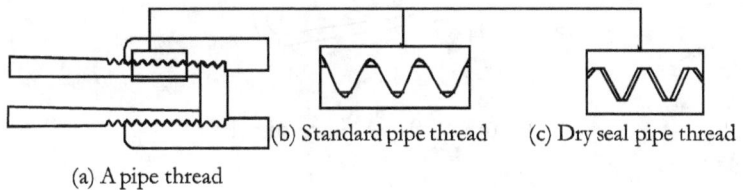

(a) A pipe thread   (b) Standard pipe thread   (c) Dry seal pipe thread

Figure 19.7 | Pipe threads

Dry-seal Pipe Thread

When larger pipes are used, dry-seal pipe threads are most appropriate. In this type of thread, pressure-tight joints are not made on the threads. Both threads are parallel, and sealing is achieved by compressing a soft material onto the external thread, as shown in Figure 19.7(c). When tightened, the dry-seal thread eliminates the radial clearance. This type of thread form tends to minimize thread leaks.

Straight-thread with 'O' Ring

A straight-thread with an O-ring-type fitting, as shown in Figure 19.8, may be used for connections to equipment without pressure or service limitations. However, this type may not be used for joining pipe sections.

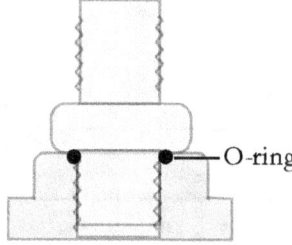

Figure 19.8 | Straight thread with O-ring

Hydraulic Tubes

The tube is the most widely used type of hydraulic conductor. Remember, a tube system meets the same general requirements as the piping system. The tube is generally a small-diameter, thin-walled pipe. It can be bent into almost any shape, thus reducing the number of fittings. It is easier to handle a tube system. However, it is usually more expensive than the piping system due to its tighter manufacturing tolerances.

Tube Construction

Seamless tubes, as shown in Figure 19.9, are formed by cold drawing a billet over a piercing rod. Welded tubes are made by forming a piece of cold-rolled steel into a tube and then joining it along its longitudinal seam through a material fusion process. Tubes must be pre-formed or bent before installation. Bending them requires careful planning and skill. When installed properly, they appear neat.

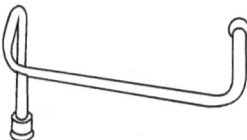

Figure 19.9 | A piece of tube

Tube Material

The tube, constructed of dead soft, cold-drawn carbon steel, has become the accepted standard for hydraulics because it has the mechanical properties required to withstand high pressures (tensile strength of 55000 psi). If greater strength is required, the tube material can be AISI 4130 steel, which has a tensile strength of 72000 psi.

Specifications of Tube

Essential specifications for hydraulic tubes include their size, pressure ratings, and minimum bend radius. Tube size is always specified by its outside diameter (OD). Available tube sizes in metric units range from 4 mm to 42 mm. Available sizes in English units include 1/16-in increments from 1/8-in outside diameter to 3/8-in. From 3/8-in to 1-in, the increments are 1/8 in, and for sizes beyond 1-in, the increments are ¼-in. A dash number represents the outside diameter (OD) of the tube expressed in terms of sixteenths of an inch. Table 19.4 gives a size chart for a seamless cold-drawn steel tube.

Typical Size and Pressure Chart for Carbon Steel hydraulic tubing (Inch Sizes)
Table 19.3 gives the size chart for carbon steel tubing.

Table 19.3 | Typical size chart for carbon steel tubing (inch sizes)

Tube OD (inch)	Wall thickness (inch)	Max. Working Pressure @ 6:1 SF (psi)	Burst Pressure (psi)
3/16	0.035	3422	20533
1/4	0.035	2567	15400
1/4	0.049	3593	21560
5/16	0.049	2875	17248
5/16	0.065	3813	22880
3/8	0.049	2396	14373
3/8	0.065	3178	19067
1/2	0.049	1797	10780
1/2	0.065	2383	14300
1/2	0.083	3043	18260
5/8	0.065	1907	11440
5/8	0.095	2787	16720
3/4	0.049	1198	7187
3/4	0.065	1589	9533
3/4	0.095	2322	13933
3/4	0.109	2664	15987
1	0.065	1192	7150
1	0.095	1742	10450
1	0.120	2200	13200
1-1/4	0.095	1393	8360
1-1/4	0.120	1760	10560

Pressure Rating, Tubing
Burst pressure is the pressure at which a tube ruptures. Working pressure is the value considered safe when operating the system under normal circumstances. The safety factor is the ratio of the working pressure to the burst pressure. Generally, a safety factor of at least 4:1 is essential for hydraulic systems.

Example 19.5 | Determine the inside diameter of the tubing section with an OD of 0.5 inches and a wall thickness of 0.049 inches.

Solution

Pipe OD, D_o = 0.5 inch
Wall thickness, t = 0.049 inch
Inside diameter of the pipe = $D_o - 2t$
= 0.5 − (2 x 0.049)
= 0.402 inch

Example 19.6 | Determine the hoop stress developed in a tube with an outside diameter of ½ inch and a wall thickness of 0.083 inches when the tubing is subjected to a pressure of 3000 psi.

Solution

Outside diameter of the tubing, D_i	= 0.5 inch
Wall thickness of the tubing, t	= 0.083 inch
Pressure, P	= 3000 psi
Internal Diameter	= 0.5 − (2 x 0.083) = 0.334 inch
Hoop stress developed	= P x D_i / 2t
	= 3000 x 0.334 / (2 x 0.083) psi
	= 6036 psi

Example 19.7 | Determine the maximum pressure rating of a tube of OD of ¾" with a wall thickness of 0.049". The tube material is carbon steel with a tensile strength of 50000 psi. Assume a safety factor of 6.

Solution

Tube OD, Do	= 0.75 inch
Wall thickness, t	= 0.049 inch
Tensile strength, S	= 50000 psi
Safety factor, SF	= 6
Tube ID, Di	= Do − 2 x t
	= 0.75 − (2 x 0.049) inch
	= 0.652 inch
Burst pressure	= 2ts / Di
	= 2 x 0.049 x 50000 / 0.652
	=7515 psi
Pressure rating of the tube	= Burst Pressure / Safety factor
	= 1569 / 6 = 1252 psi

Minimum Bend Radius
It is the smallest radius of the curved section of a tube beyond which it cannot be bent without flattening, kinking, or wrinkling.

Tube Bending Process
Hydraulic tubes can be bent by hand or using power-bending equipment. Steel tubes can be bent by unique methods, such as roll forming, press forming, mandrel bending, or table forming. In a tube-bending process, pressure is applied to bend the tube around a correctly sized die to form the required radius. A tube has a standard bend radius to which it must be bent without flattening, kinking, or wrinkling.

Selection of Tube
Proper tube size and material are critical for the efficient and trouble-free operation of the associated fluid power system. The selection of the proper tube involves choosing the right tube material and determining the optimum tube size (O.D. and wall thickness). Proper tube sizing for various hydraulic system components results in an optimal balance of efficiency and cost-effectiveness.

Advantages and Disadvantages of Tubes

The main advantage of a tube is that it can be bent into shape, requiring fewer fittings. Fewer connections generally mean a lower risk of leaks. Tubes are also known for their ability to absorb vibration and for a smooth interior finish, which promotes easy fluid flow.

Fittings, Tube

Since the wall sections of a tube are relatively thin, threading cannot be used to seal the tube connections. Therefore, fittings are to be used to make tube connections. A variety of tube fittings are available for hydraulic applications. Tubes can be joined with flaring, brazing, or couplings. Flared or flareless fittings are used for tube-end connections.

Flared Fitting, Tube

It comprises a nut, a sleeve over the flared tube, and a body, as shown in Figure 19.10. When forming flares, it is necessary to prepare the tube: cut square, file smoothly, and remove burrs. The tube must then be flared by inserting a flaring tool, typically a mandrel or rolling cone. The most critical step in making a flare tube fitting is forming the flare without galling, over-thinning, or splitting the tube end. The sleeve and nut are pushed smoothly over the tube end. The sleeve prevents the nut from twisting when the nut is tightened. When the nut is screwed onto the body, it draws the sleeve and the flare against the body, thus forming a seal.

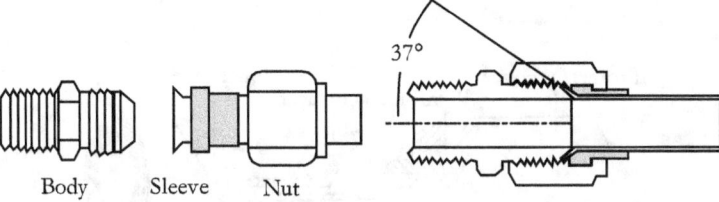

Figure 19.10 | Flare fitting

Flared tube fittings are suitable for connecting thin to medium-diameter wall tubes. They are preferred when a durable, leak-proof connection is needed and the tube can be flared. These fittings offer a high degree of long-term reliability and are ideal for joining hydraulic tube and hose systems. They are particularly effective at preventing leaks and are suitable for critical aerospace and automotive braking system applications. The fittings are manufactured to the SAE J514 standard with a 37° flare, commonly called JIC (Joint Industry Council). Flared fittings offer several benefits, including easy assembly, leak resistance, and compatibility with thin-walled tubes.

Flareless (Compression) Fitting, Tube

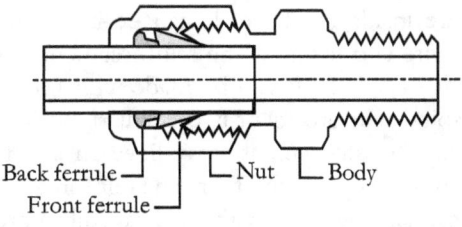

Figure 19.11 | Compression fitting

It consists of a body, ferrule(s), and a nut, as shown in Figure 19.11. First, the ferrules and the nut are slipped over the tube. The tube is inserted into the body, where it butts up against the shoulder. When the nut is screwed onto the body, the ferrule bites into the tube's skin to provide holding strength for the connection. This tight connection provides a positive seal. Compression fittings are used on medium and heavy-wall tubes or when the tube cannot be flared. Flareless fittings are suitable for systems with high vibration and dynamic loading. They offer reliability in vibration-prone environments and a reduced risk of leaks.

Hoses

Hoses are the most flexible and versatile type of conductor. They can bend and flex easily and better accommodate vibration and pulsations than tubes. They are selected when rigid pipes or semi-rigid tubes cannot be used, as in applications with components that move relative to each other or when excessive vibration and recurrent pressure pulsations exist. A hose assembly consists of a hose and end fittings that connect directly to adjoining pipework or fittings. A hose assembly must have correct end-fitting configurations. It is easy to install a hose assembly with a well-thought-out routing layout. Given its superior routing advantage, a hose is generally preferred over a metal tube.

Hose Construction

As shown in Figure 19.12, hydraulic hoses have three parts: an inner tube, a reinforcement layer, and a protective cover.

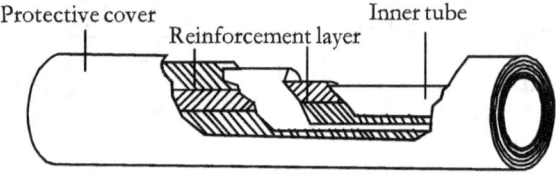

Figure 19.12 | Cut-section of a hose

Inner Tube

The inner tube is the hose's lining that comes into direct contact with the fluid. The choice of inner tube material is determined by factors such as the type of hydraulic fluid used, temperature range, pressure rating, compatibility, and environmental conditions. The inner tube must resist chemical corrosion from the fluid and endure extreme variations in fluid temperature. Inner tubes are typically made from synthetic rubber, thermoplastics, or PTFE (polytetrafluoroethylene).

Reinforcement Layer

The Reinforcement Layer provides the strength to withstand internal and external pressures and determines the working pressure inside the hose. It can be made of steel wires, textiles, synthetic materials, or a combination of wire and textiles. The reinforcement layer is constructed with a single layer or multiple layers of braids or spirals. It can be made with four or six layers to meet the most demanding applications. This spiral reinforcement is particularly well-suited to high-pressure impulse applications. A hose with multiple reinforcements may have an anti-friction layer between them to prevent the steel wires from rubbing against each other. The reinforcement layer of the hose, connected to the suction side of a pump, can also be made with a helical coil to prevent the hose from collapsing under a partial vacuum during fluid suction. This type of construction also prevents the hose from collapsing from tight bends. Braided reinforcement can be made of steel wire, textile, or a combination of wire and textile. Depending on the hose's intended use, the reinforcement layer may be single- or multi-layered.

Protective Layer
The primary purpose of the cover is to protect the tube and reinforcement from abrasion, corrosion, extreme temperatures, UV light, and ozone. The cover can be made from synthetic rubber, fiber braids, or both, depending on the application. Hoses with synthetic rubber covers are generally preferred over those with textile-braid covers because they are more abrasion-resistant.

Types of Hoses by Operating Pressures
Another way to classify hydraulic hoses is by pressure rating. The classification is presented below:

- **Low pressure Hoses**: They are designed for various applications with operating pressures below 300 psi. Their reinforcement is usually textile.
- **Medium-pressure Hoses:** They are used for hydraulic applications requiring 300 to 3,000 psi operating pressures. A medium-pressure hose is constructed with a single-wire braid, multiple-wire braid, and/or textile braid.
- **High-pressure Hoses:** They are frequently found in high-pressure hydraulic applications, such as construction equipment requiring 3,000 to 6,000 psi operating pressures. These hoses are often called 'two-wire' braid hoses because each one has a reinforcement of high-tensile strength two-wire steel braid.

Suction Hose: A hose connected to the suction service is subjected to crushing forces because the atmospheric pressure outside the hose is higher than the internal pressure. They also tend to restrict the flow through it. Therefore, hoses connected to the pump suction line must withstand the pressure differential across them. The best way to prevent the hose from collapsing is to reinforce the hose with a helical wire. The size and spacing of the helical wire reinforcement depend on the hose size and the pressure differential.

Very High-Pressure Hoses: They are used for off-highway equipment and heavy-duty machinery that experience steep pressure surges. The oil-resistant synthetic tubes in these hoses are reinforced with four or six layers of spiraled, high-tensile steel wire over a layer of yarn braid.

Hose Size Specifications
Choose a hose with an adequate inside diameter to minimize pressure loss and avoid hose damage from heat generated by excessive fluid turbulence. An undersized hose increases pressure loss, and an oversized hose adds unnecessary cost, weight, and bulk. The essential specifications of a hose include the inside diameter (ID), wall thickness (t), pressure ratings, and minimum bend radius.

Table A11.3 of Appendix 11 gives Dash numbers and the corresponding hose IDs.

Inside Diameter: The size of a hose is specified by its inside diameter (ID). The ID must provide the proper volume of fluid for the specific application. The ID is specified in dash sizes or metric units. As mentioned in a previous section, a dash number indicates the hose size in 1/16-inch increments. For example, a hose with a 1/4" (=4/16") inside diameter indicates that it has four numbers of 1/16" segments. Therefore, its dash size would be a -4 (dash 4).

Pressure Rating: The pressure rating of a hose is determined by its construction, which is governed by its number of layers, materials, and construction method. The more reinforcement layers a hose has, the more pressure it can withstand. The rated pressure of a hose must be higher than the normal system pressure and any pressure surges it will encounter.

Minimum Bend Radius: This is an essential consideration in hose design and selection. Table 19.5 presents typical hose parameters, and Table A11.4 in Appendix 11 provides hose specifications in metric and inch sizes.

Table 19.4 | Typical parameters of hoses

Dash Number	ID inch	Work pressure psi	Min. Burst pressure psi	Min. bend radius inch
-2	1/8	3000	16000	
-3	3/16	3000	16000	
-4	1/4	3000	16000	1.5
-5	5/16	3000	16000	
-6	3/8	3000	16000	2.5
-8	1/2	3000	16000	2.9
-10	5/8	3000	16000	3.3
-12	3/4	3000	16000	4.0
-14	7/8	3000	16000	
-16	1	3000	16000	5.0
-20	1 1/4	3000	16000	12.0
-24	1 1/2	3000	12000	14.0
-32	2	3000	12000	
-36	2 1/4	3000	12000	
-40	2 1/2	3000	12000	
-48	3	3000	12000	
-56	3 1/2	3000	12000	
-64	4	3000	12000	
-72	4 1/2	3000	12000	

Types of Hose Motions
Various types of hose motions are shown in Figure 19.13.

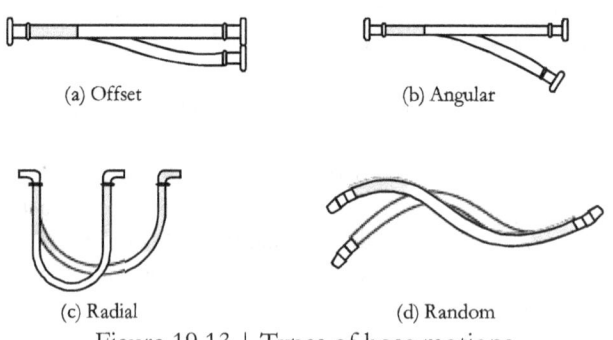

(a) Offset (b) Angular (c) Radial (d) Random

Figure 19.13 | Types of hose motions

Offset motion occurs when one end of the hose is moved in a plane perpendicular to its longitudinal axis while the other ends remain parallel.
- The angular motion of the hose occurs when one end of the hose is moved in a simple bend, with its ends not remaining parallel
- The radial motion of the hose occurs when the hose is bent in a circular arc
- The random motion of the hose occurs in random planes

Standards of Hoses
ISO and SAE specify standards for hoses that define dimensional and performance parameters.
- ISO 1436-1, wire braid reinforced
- ISO 4079-1, textile reinforced
- ISO 3949, thermoplastic textile reinforced
- ISO 3862-1, spiral wire reinforced

The SAE J517 (US) standard includes 100R numbers (SAE 100R1 to SAE 100R18) that specify hose construction, dimensions, pressure, and temperature. Extracts of the specifications of SAE 100 R2 and SAE 100 R18 are presented below:

SAE 100 R2 Specifications:
The hose shall consist of an inner tube of oil-resistant synthetic rubber, steel wire reinforcement, and an oil—and weather-resistant synthetic rubber cover. A braid or ply of suitable material may be used over the inner tube, the wire reinforcement, or both to anchor the synthetic rubber to the wire.

SAE 100 R18 Specifications:
The hose shall consist of a thermoplastic inner tube resistant to hydraulic fluids, suitable synthetic fiber reinforcement, hydraulic fluid, and a weather-resistant thermoplastic cover.

Table A11.5 of Appendix 11 gives various standards relevant to hydraulic fluid conductors.

Selection of Hose
Selecting the right hose for a hydraulic system is the first step toward its safe operation and long service life. It begins by selecting the right components, including hoses, couplings, crimping equipment, and accessories. Hoses must meet the size, pressure, bend radius, and routing requirements. The equipment type, working and impulse pressures, fluid to be used, and bend radius must also be known.

Applications of Hoses
Hoses are used when rigid or semi-rigid pipes or tubes cannot be used, as in applications involving the movement of machine parts.

Advantages and Disadvantages, Hoses
Hoses offer many advantages. They are flexible and portable. Next, they absorb and dampen pressure surges and vibrations. Routing hoses is easier and faster than other types of conductors, even around obstacles. They require no brazing or specialized bending.

However, mixing and matching couplings from one manufacturer with hoses from another can lead to premature or catastrophic assembly failure. Hoses are susceptible to abuse, misapplication, and improper plumbing.

Fittings, Hose

Hose fittings can be either permanent or reusable. Permanent hose fittings are installed on the hose by crimping and cannot be disassembled, and reusable hose fittings are screwed or clamped on the hose end. The big difference between the skive and the no-skive hose is in the thickness of the outer cover. The thicker cover requires a different fitting shell.

Quick Couplings (or Disconnects)

They are used for convenience, as they can be installed and removed by hand when repeated connections and disconnections of the lines are needed. A quick disconnect (QD) coupling has a male side and a female coupler. Quick couplings can be poppet or flat face.

In the poppet type, the male poppet (nipple) is depressed when it engages with the coupler. This action opens the valve to allow hydraulic fluid to flow. Poppet-type quick couplers fall into ISO A or ISO B styles.

In the flat-face couplers, both coupling sides have flat surfaces. A flat-face male nipple will mate with a female flat-face coupler. The back end of flat-face couplers can come with NPT, JIC, ORFS, or straight O-ring threads.

Based on the coupling's valving, hydraulic couplings generally fall into two groups: double-shutoff and straight-through.

Double Shut-Off Couplings

They are extensively used when it is essential to minimize fluid loss upon disconnection. Both halves of the coupler, the body, and the nipple, contain shut-off valves, as shown in Figure 19.14(a). These valves open automatically when the body and nipple are connected and close automatically when the two halves are disconnected—keeping fluid loss to a minimum.

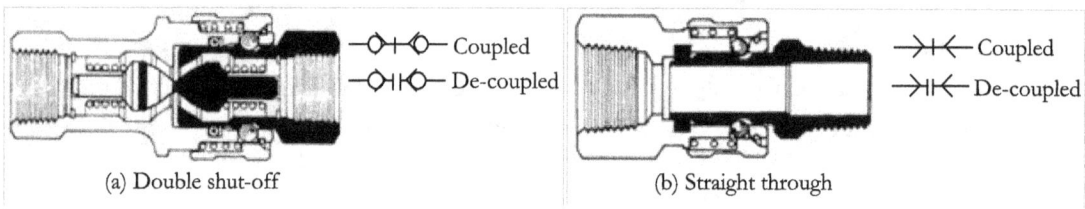

(a) Double shut-off   (b) Straight through

Figure 19.14 | Cross-sectional views of hydraulic couplings

Straight-Thru Couplings

Figure 19.14(b) shows the cross-sectional view of a double straight-through coupling. It has no valve in either half and is ideal for a maximum flow application. Its smooth, open bore offers the lowest pressure drop compared to any other type of quick disconnect coupling and allows it to be thoroughly cleaned. Since no valve is in either half, the fluid flow should be shut off before disconnecting the coupling. A straight-through coupling is used where flow must be unrestricted.

Every manufacturer etches a part number on each coupler for proper identification. It is better to consult the manufacturer's reference guide to ensure that QDs match correctly.

Objective-type Questions
1. The schedule number is associated with:
   a) Viscosity index
   b) The wall thickness of a pipe
   c) The hardness of a seal material
   d) The acidity level of a hydraulic fluid

2. The dash number is used to specify:
   a) Wall thickness of the tube
   b) The elasticity of seal materials
   c) Inside diameter (ID) of a hose
   d) Contamination concentration level

3. Mark the <u>incorrect</u> statement
   a) Hoop stress of a given length of pipe is the circumferential stress acting on the wall of the pipe under the operating pressure
   b) Bending the hose/tube to a smaller radius than its rated minimum bend radius may result in the premature failure of the hose/tube
   c) Flared or flareless-type fittings can be used for tube end-connections
   d) Hose length must be exactly equal to the actual distance between their end connections

Review Questions
1) What is the function of fluid conductors in hydraulic systems?
2) List three types of fluid conductors used in hydraulic systems. Briefly explain their flexibility.
3) What are the requirements for the satisfactory function of fluid conductors in hydraulic systems?
4) State the reason for the energy losses in the fluid conductors used in hydraulic systems.
5) What are the reasons for the fluid leakages in a hydraulic distribution system?
6) What are the ways to minimize fluid leakage in a hydraulic fluid distribution system?
7) What is the definition of 'schedule number' when referring to piping and fittings?
8) What variables determine the wall thickness and the factor of safety of a fluid conductor?
9) Briefly explain the terms about fluid conductors: (1) Bend radius, (2) Tensile stress, (3) Burst pressure, and (4) Working pressure.
10) What factors determine the pressure rating of a fluid conductor?
11) List out the procedure to calculate the size of a fluid conductor for a hydraulic system.
12) Explain the purpose of hydraulic pipes briefly.
13) State two disadvantages of using pipes in hydraulic systems
14) How are the pipes used as fluid conductors in hydraulic systems specified?
15) What is the schedule number of a pipe used in hydraulic systems?
16) Describe the methods of coupling pipes in hydraulic systems.
17) What are the functions of pipe threads as used in hydraulic systems?
18) State some common materials used in the manufacturing of hydraulic pipes.
19) State the significant disadvantages of using steel pipes in hydraulic systems.
20) What are the two types of thread configurations used in the piping systems for hydraulic systems?
21) Name two common types of pipe joints used in hydraulic systems.
22) What is the disadvantage of threaded fittings for hydraulic systems?
23) Briefly explain the use of tubes in hydraulic systems.
24) Why are steel tubes more commonly used than steel pipes in hydraulic systems?

25) State the common materials used for manufacturing hydraulic tubes.
26) How do you specify hydraulic tubes?
27) What factors determine the size of a tube?
28) Mention one advantage and one disadvantage of hydraulic tubes.
29) Mention two advantages of the hydraulic tubes over the pipes
30) Describe any two methods of coupling the tubes in hydraulic systems.
31) Describe different types of tube fittings as used in hydraulic systems.
32) Briefly explain the correct methods for bending hydraulic tubes.
33) Briefly explain the correct methods for flaring hydraulic tubes.
34) What is a flare fitting, as used in hydraulic systems? Why is flaring needed, and how is it done?
35) What is the difference between the flared and compression fittings?
36) List the parts of a flared tube-fitting assembly as used in hydraulic systems.
37) List the parts of a flareless tube-fitting assembly used in a hydraulic system.
38) Briefly explain the use of hoses in hydraulic conductor systems.
39) Mention three essential hose elements as used in hydraulic systems.
40) Describe the basic constructional features of hoses as used in hydraulic systems.
41) What is the purpose of providing a protective outer layer for a hydraulic hose?
42) Under what conditions hoses are used in hydraulic systems?
43) How are the hoses used as fluid conductors in hydraulic systems specified?
44) What does the dash number of a hydraulic hose refer to?
45) What determines the pressure rating of a hydraulic hose?
46) Explain how the pressure rating of hydraulic hoses is increased.
47) What are the advantages of hydraulic hoses?
48) Mention three factors to consider when selecting hydraulic hoses.
49) Briefly explain five essential parameters when selecting a hydraulic hose.
50) Give a brief note: (1) Hose routing in hydraulic systems and (2) Applications of hydraulic hoses.
51) Explain the purpose of the quick disconnect coupling, as used in hydraulic systems.

Numerical Problems

1) Calculate the minimum inside diameter of the suction pipe in a hydraulic system to handle the flow rate of 10.6 gpm with an average fluid velocity not exceeding 1.6 ft/s. [Ans: 1.65 in]
2) Determine the size of the pressure line of a hydraulic system with a 10 gpm positive-displacement pump. The recommended flow velocity through the pressure line is 16.4 ft/s. [Ans:0.5 in]
3) Determine the size of the return line of a hydraulic system with a 16.25 gpm positive-displacement pump. The recommended return flow velocity is 8.2 ft/s. [Ans: 0.9 in]
4) A hydraulic system should permit a flow rate of 10.6 gpm with an average fluid velocity not exceeding 13 ft/s. Calculate the minimum inside diameter of the pressure conductor in the system. [Ans: 0.575 in]
5) A hydraulic system should permit a flow rate of 10.6 gpm with an average fluid velocity not exceeding 6.5 ft/s. Calculate the minimum inside diameter of a return-line conductor. [Ans: 0.814 in]
6) Find the schedule number of a steel pipe for a hydraulic system at the estimated working pressure of 1625 psi. The allowable stress is 58000 psi.[Ans: 30]
7) Calculate the burst pressure of a seamless cold-drawn steel tubing of outside diameter 0.98 in and wall thickness 0.098. The tubing has a tensile strength of 58000 psi. [Ans: 12889 psi]
8) Determine the safe working pressure for a steel tube with a burst pressure of 13000 psi, assuming a safety factor of 8. [Ans: 1625 psi]

Objective-type questions - answer key: 1-b, 2-c, 3-d

Chapter 20 | Hydraulic Applications

Hydraulic power transmission systems generate and transmit large amounts of power to hydraulic machinery employed in various applications.

Why is Hydraulics the Preferred Technology for Industrial, Mobile, and Other Applications?
Conventional hydraulic systems can handle large forces with simple, compact components. They eliminate the need for mechanical linkages. They are capable of withstanding the rigorous duty cycles and severe operating conditions of applications. They offer excellent resistance to shock loads and provide accurate and reliable control of work operations. The electro-hydraulic techniques complement the outstanding advantages of conventional hydraulics.

Moreover, electrohydraulic systems can easily be interfaced with PLCs. Cartridge valves can be used to develop simple, flexible, and versatile hydraulic systems. All these advantages have made hydraulics the preferred technology for stationary and mobile hydraulic applications.

Evolution of Hydraulic Applications
The expansion of hydraulic applications has evolved over time. In the twentieth century, hydraulic systems were developed to drive and control ships and aircraft systems. Military applications, such as radars, missile launchers, and vehicles, preferred hydraulics. Because of their early success, the list grew to include various areas in the last few decades. Hydraulically powered tools, machines, and systems are used in industrial, mobile, aerospace, marine, construction, mining, defense, and entertainment fields. Let us go into some fine details.

Stationary Industrial Applications
Hydraulic technology provides muscle power in many stationary industrial applications. Electrical connectivity to these applications makes it easy to use electric motors as prime movers for hydraulic power packs. This technology can be employed in normal and harsh industrial applications.

Hydraulic technology can be used in typical stationary applications to obtain various work operations such as feed motions, spindle drive operations, clamping, pressing, and drilling. The end applications for these work operations may include CNC lathes, turning centers, machine tools, and material handling. Further, this technology is well-suited for harsh industrial applications involving large forces and extreme heat, such as steel mills, foundry plants, metal-cutting and forming machines, and plastic injection molding machines.

Mobile Applications
'Mobile hydraulics' encompasses a large sector of off-road vehicles, including forklifts, tractors, bulldozers, excavators, dump trucks, highway trucks, breakers, crushers, and crawler drills. Many infrastructure/construction/development projects make use of mobile hydraulic technology.

Electric connectivity is extremely difficult in mobile applications, so internal combustion (IC) engines are invariably used as the prime movers.

Hydrostatic transmission systems drive various work operations, such as excavation, lifting, and material loading. They are also used in mobile equipment for precise, remote control of speed and torque/force.

Aerospace Applications
Hydraulic power is used in commercial and military aircraft, spacecraft, and support equipment. Aerospace hydraulic applications include landing gear, brakes, flight controls, and cargo-handling equipment.

Marine Applications
Marine applications are characterized by severe operating conditions, such as corrosion, pitting, and extreme temperatures. Stainless steel cylinders must be used to handle harsh environments. Hydraulic systems are also used in dockyard cranes and container handlers.

Mining Applications
Hydraulic technology is the preferred choice for highly mechanized mining equipment used to excavate ores. Large earth-moving equipment, rock drill rigs, and blast-hole drills are used to extract precious minerals in mining applications. Hydraulic devices can handle harsh environments and rigorous duty cycles in mining applications.

Defense Applications
The most demanding applications of defense weapon systems use the power of hydraulic components. They are used in radar antenna positioners and missile launchers.

Entertainment Sector
Hydraulic systems provide a natural look and feel to creature motions in amusement or theme parks. They can also accurately control high masses and the inertia developed on rides.

Typical Applications of Hydraulic Technology
Let us now examine typical hydraulic applications, beginning with a hydraulic press.

Hydraulic Press
The primary function of a hydraulic press, as shown in Figure 20.1, is to convey forces and movements to its tool and die section for the forming or blanking operation of workpieces. Modern presses can configure work parameters, such as travel distance and force. The press uses a large-volume cylinder to drive the work operations. It is controlled by fast-acting valves for high performance. Further, it incorporates interlocking and barrier guards to safeguard equipment and personnel. The capacity of hydraulic presses typically ranges from 5 to 5000 Tons.

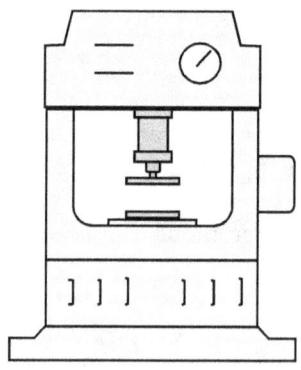

Figure 20.1 | Hydraulic Press

Hydraulic Work-holding Systems

Next, we will examine the essential features of modern work-holding systems that employ high-pressure hydraulics. These systems are used in fixtures for positioning and clamping even complex, irregularly shaped parts. They find applications with complexities ranging from simple machining or welding fixtures to robot-assisted machining centers. They are suitable for systems that require holding critical tolerances, such as when machining a surface with tight tolerances.

A workholding system uses many short-stroke conventional cylinders and swing clamp cylinders, as shown in Figure 20.2. A swing clamp cylinder, attached to a clamping arm, can extend and retract like standard cylinders. Next, the clamping arm can swing by a certain angle. This feature allows easy loading and unloading of parts. Workholding systems are generally preferable in high-volume applications. Hydraulic clamping is less time-consuming but costlier than mechanical clamping.

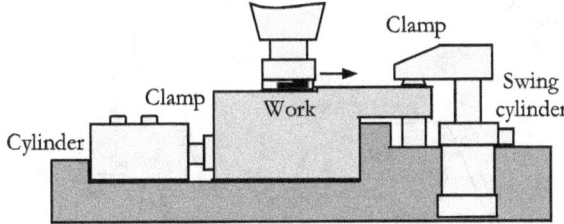

Figure 20.2 | Workholding system

Plastic Injection Molding Machine

A plastic injection molding machine has a mold section and an injection section, as shown in Figure 20.3. The mold section performs the clamping and molding operations. The mold clamping arrangement has moving and fixed platens. These platens, driven by a cylinder, hold the mold together under pressure. The injection section consists of a barrel, a screw, a heater, a nozzle, and hydraulic actuators. A screw, driven by a low-speed, high-torque (LSHT) hydraulic motor and rotating inside the barrel, shears and heats the granular raw plastic, softening it to a plastic or gently flowing state. The plastic is then injected into the mold by a pressurized cylinder. The mold can be opened after the set time.

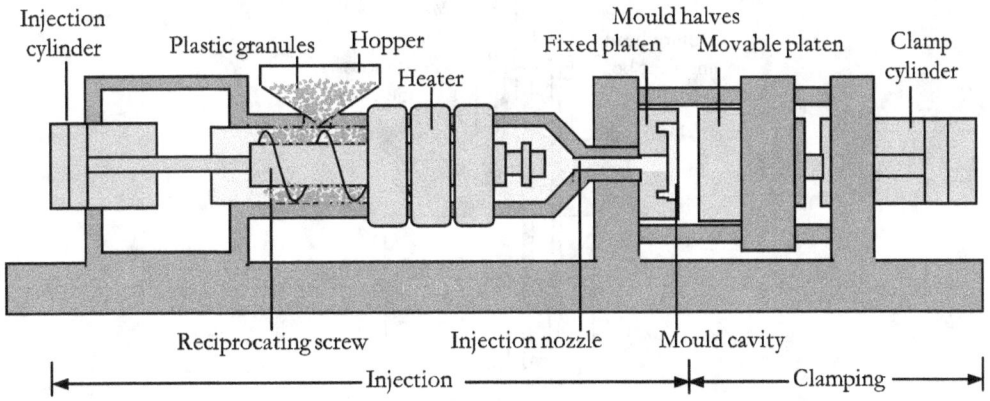

Figure 20.3 | Plastic injection molding machine

Hydraulic Excavator

The hydraulic excavator is a piece of heavy construction equipment operating like a combined human arm and hand. It is used to dig and move large objects. It consists of three parts: a set of working elements, an upper structure, and a lower structure, as shown in Figure 20.4. The working parts consist of a boom, an arm, and a bucket, powered by hydraulic cylinders. The boom provides the arm's up-and-down movements. The powerful arm with the bucket attachment is designed for digging, lifting, and loading soil. The hydraulic system in the excavator typically consists of three engine-driven pumps. Two pumps supply fluid to the cylinders and motors at high pressures, typically up to 5000 psi. The third pump supplies fluid to its pilot control circuits at low pressures, typically up to 725 psi. Joysticks are provided to control the movement of the bucket and boom assembly, as well as the rotation of the upper structure. A boom suspension circuit is incorporated into the wheeled excavator to smooth the bounce of the loaded bucket during transport through rugged terrain. Many of the control solutions for the excavator can be easily achieved through cartridge valves and manifolds. The power rating typically ranges from 12.5 to 4425 hp.

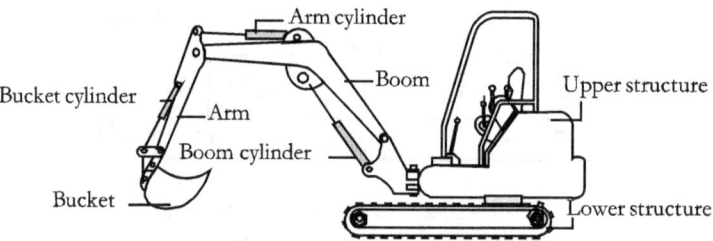

Figure 20.4 | Excavator

Drill Rigs

A rig is a machine used to drill onshore or offshore wells for oil or mineral exploration. It consists of drilling, feeding, and propulsion systems, as shown in Figure 20.5. The rig is mounted on an undercarriage. The drilling system makes holes. Next, the feed unit moves the drill unit up and down. A propulsion system runs the undercarriage. The required movements of the drill unit, feed unit, and propulsion system are hydraulically actuated and controlled.

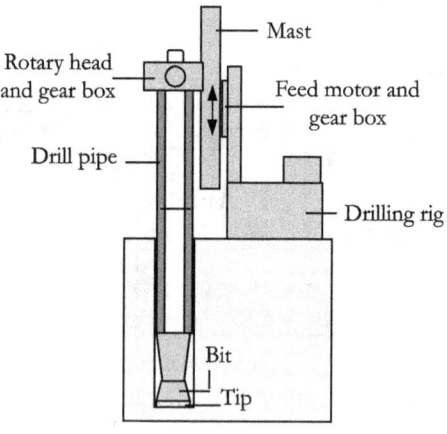

Figure 20.5 | Drill rig

Objective-type Questions

1. A hydraulic component used in a workholding system is:
   a) Swing cylinder
   b) Hydraulic motor
   c) Semi-rotary actuator
   d) Long-stroke cylinder

2. An excavator uses:
   a) Joystick
   b) Cylinders
   c) Cartridge valves
   d) All of the above

Review Questions
List five advantages of hydraulic systems.
List three disadvantages of hydraulic systems.
What are the benefits of a hydraulic power transmission system?
Why is hydraulic technology the preferred choice for industrial and mobile applications?
Explain the development of hydraulic applications.
What are the features of stationary hydraulic applications?
What are the end applications of hydraulic technology in stationary systems?
What are the features of mobile hydraulic applications?
What are the end applications of hydraulic technology in mobile systems?
Give a brief overview of aerospace hydraulic applications.
Give a brief overview of marine hydraulic applications.
Give a brief overview of defense hydraulic applications.
How does the entertainment sector make use of hydraulic power?
Describe the hydraulic applications in presses.
How can safety be incorporated into hydraulic presses?
Describe the hydraulic applications in workholding systems.
What advantages do workholding systems offer compared to mechanical clamping?
Describe the hydraulic applications in plastic injection molding systems.
What are the main blocks of a plastic injection molding machine?
What hydraulic components can drive work operations in plastic injection molding machines?
Describe the hydraulic applications in excavators.
What working parts in an excavator are powered by hydraulic cylinders?
What is the purpose of a bucket cylinder in an excavator?
What is the purpose of a boom cylinder in an excavator?
What is the purpose of an arm cylinder in an excavator?
What is the purpose of a boom suspension circuit in an excavator?
Describe the hydraulic applications in drill rigs.
List five industrial applications where hydraulic components are used.
List five mobile applications where hydraulic components are used.

Objective-type questions - answer key: 1-a, 2-d

Chapter 21 | Maintenance, Troubleshooting, and Safety of Hydraulic Systems

Generally, the term 'maintenance' of a system covers a broad range of routine maintenance and repair activities intended to keep the system in a satisfactory working condition. The same broad definition also applies to the specialised area of hydraulic machines and systems.

Therefore, a hydraulic technician should have the necessary maintenance and troubleshooting skills, as well as knowledge of the physical laws of hydraulics and the functions and symbols of hydraulic elements, to perform successful maintenance and troubleshooting.

Adequate safeguards are also essential to prevent personal injury and damage to a hydraulic machine during maintenance and troubleshooting activities.

Classification of Maintenance
Maintenance can be classified into the following two basic categories:
- Preventive (or proactive) maintenance
- Corrective (or breakdown or reactive) maintenance

Preventive maintenance is undertaken when a machine is operating correctly to prevent potential failure. It is performed regularly, usually according to a schedule or checklist, to ensure the smooth, efficient operation of all machine components at all times. The purpose of performing preventive maintenance on a machine is to prolong its useful service life.

On the other hand, corrective maintenance is performed on a machine after it fails. This activity consists of finding the fault and repairing the machine. Preventive maintenance is the most effective maintenance strategy, and one has to focus on preventing failure rather than troubleshooting the machine.

Definitions of Maintenance Activities
Maintenance generally involves closely related activities, such as inspection, servicing, examination, and overhaul. The following paragraphs explain the meanings of these maintenance-related activities for a machine.

Inspection: This term refers to the maintenance activity that comprises the careful observation/scrutiny of the machine, usually without dismantling it. This activity usually includes visual and operational checks.
Servicing: This term refers to the cleaning, adjustment, lubrication, and other servicing functions of the machine without dismantling it.
Examination: This term refers to the inspection of the machine, including necessary dismantling, measurements, and non-destructive testing to obtain useful information on the condition of its components/subassemblies.
Overhaul: This term refers to the extensive work done to repair or replace worn-out or defective parts of the machine. The parts are dismantled, partly or wholly. The components that are worn beyond the acceptable limit are replaced. The assembly is followed by functional checks and measurements to ensure the machine operates satisfactorily.

Requirements for Preventive Maintenance

The primary objective of any preventive maintenance activity on a machine is to prevent its failure or breakdown. The most general requirements to achieve this goal are as follows:

Know the Machine

The first step in preventative maintenance is to become familiar with the machine. For that, a maintenance technician should know the layout, line routing, the functioning of all its components, and the operation of the whole circuit. Once you know the machine, regular maintenance becomes easy.

Understand and Follow the Best Maintenance Practices

Understanding proper maintenance procedures and the knack for following them are prerequisites for performing effective preventive maintenance.

Compile a Maintenance Checklist

Following the best maintenance practices, developing and implementing a maintenance checklist or schedule for the machine is also necessary. Lay down the intervals (say daily, weekly, and monthly) at which inspection and servicing operations are to be carried out.

Follow the Instruction Manual

Trained maintenance personnel should carefully study the 'Instruction manual of the installation, operation, and maintenance' of the machine supplied by its manufacturer to compile its maintenance checklist. The machine manufacturer provides only general guidelines, which cannot be taken as the exact maintenance program for meeting the local requirements.

Ensure the Safety

It is imperative to ensure the safety of personnel and equipment during machine preventive maintenance. Therefore, the maintenance person must always follow safe practices during maintenance activities.

Stock Spares

Spares are crucial for maintenance duty and should be readily available. Hence, it is good practice to stock the essential spare parts for the machine, with proper inventory control, to facilitate fault servicing with minimal interruption to production.

Consequences of Poor Maintenance in Hydraulic Systems

Lack of regular maintenance in a hydraulic system may lead to undesirable consequences. These consequences may include the following: (1) the accumulation of dust, (2) the presence of air and water in the system, (3) the formation of sludge and rust, (4) the premature wear of moving parts, (5) internal or external leakages, (6) the loss of fluid, (7) decomposed packing, (8) cavitation, and (9) mechanical damage to parts.

These effects eventually result in increased system downtime and production shortfalls. Therefore, preventive maintenance must consider the individual components of the system.

The following sections explain the regular maintenance activities on power packs, fluid media, valves, accumulators, cylinders, seals, and piping in hydraulic systems.

Preventive Maintenance of Hydraulic Systems

A hydraulic system failure can be attributed to the following: (1) contaminated system fluid, (2) insufficient fluid, (3) wrong fluid, (4) dirty filters, (5) loose connections, (6) leakage, (7) excessive heat, (8) excessive pressure, and (9) cavitation.

However, carrying out maintenance in a planned manner can eliminate most of the hydraulic system failures. In general, the list of preventive maintenance activities for a hydraulic system could be:

- Checking the fluid level in the system reservoir and, if necessary, topping it up
- Cleaning of the reservoir, lines, fittings, and other parts of the system
- Maintaining the system fluid in a clean state
- Cleaning and repairing/replacing the hydraulic filters in the system
- Plugging the internal and external leaks in the system
- Setting the pressure and the flow rate as per the requirements of the system
- Restricting the system heat, turbulence, vibration, and noise within limits
- Checking and maintaining the system components, such as pumps, valves, actuators, and lines
- Making sure that the new parts of the system are clean
- Keeping the exposed parts protected using covers before their assembly in the system
- Filtering the new hydraulic fluid before filling the system
- Monitoring and reducing cavitation in the system
- Obtaining the fluid sample from the system and carrying out the fluid analysis
- Periodic checking and recording of the values of pressure, flow, and temperature in the system

Maintenance of Power Packs

The power pack must be mounted in a clean and dry location. Periodic maintenance is essential for keeping the entire unit in good working condition at all times. Proper maintenance of the power pack is as good as the maintenance of its constituent parts. The following sections elaborate on the maintenance activities specific to each power-pack component.

Maintenance of Reservoirs

A hydraulic reservoir should be serviced regularly. Monitoring fluid temperature and controlling fluid turbulence in the reservoir are essential. The fluid turbulence can be controlled by adjusting the velocity of the return flow to the reservoir. The reservoir cap should not be removed while the system is in operation, as the hot fluid inside the reservoir may cause injury to personnel. The following maintenance activities on the reservoir should be carried out regularly:

- Clean the exterior surfaces of the reservoir to remove dust, dirt, and leakage,
- Examine the reservoir externally for any signs of rust or corrosion,
- Paint the corroded areas of the reservoir,
- Check the reservoir's fluid level regularly,
- Clean the interior surfaces of the reservoir through the access provided,
- Protect the reservoir from the ingress of water, air, or other contaminants,
- Clean the magnets inside the reservoir frequently,
- Check for the loose inlet pipe on the suction line and the drop tube on the return line of the reservoir, and
- Inspect the screwed connections for leaks and tighten these connections if necessary.

Maintenance of Hydraulic Pumps

Under normal operating conditions, hydraulic pumps require little attention. However, if due care is not taken, they may fail prematurely. A hydraulic pump's failure may be attributable to dirty system fluid, corrosion, insufficient fluid in the reservoir, cavitation, improper assembly and alignment, loose bolting and connections, and overload.

The primary maintenance consideration for the pump is keeping the system fluid clean. The gears/vanes/pistons in the pump can be abraded by the contaminants, resulting in inefficient pumping. A leak check around the pump seals is required. It is also necessary to periodically check the pump's drive shaft for misalignment or damage. Check for the wear and improper tensioning of V-belts of belt-driven pumps. Verify the pump for unusual noise or vibration, or both. It is essential to reduce cavitation in the pump. The pump may require proper priming to avoid its possible failures.

Maintenance of Hydraulic Fluids

An essential concern in a well-designed hydraulic system is maintaining its fluid medium in a clean state. Remember, the fluids used in modern hydraulic systems are liable to be stressed and degraded by contaminants and the severe working conditions imposed on them.

A proactive maintenance programme should be followed to properly control hydraulic fluid contamination. Two important proactive maintenance activities for hydraulic fluids are filtration and fluid analysis. The proactive maintenance activities in a hydraulic system ensure (1) maximum operating performance and long-term reliability, (2) reduced downtime, (3) reduced overall costs, and (4) enhanced service life.

General Maintenance Activities for Hydraulic Fluids

The following activities are to be carried out on hydraulic fluids at regular intervals to keep contamination out of the fluids:

- Clean areas around reservoirs, coolers, filters, cylinder piston-rods, fill-plugs, and dipsticks
- Clean hoses, tubes, and piping during their installation
- Check pumps, reservoirs, cylinders, and lines for leaks
- Check the fluid level in the reservoirs and keep them filled
- Filter fluids when filling
- Keep hoses capped and plugged when they are removed or opened
- Keep the new components covered until they are ready for installation
- Change the filters regularly
- Check the appearance and smell of hydraulic fluids
- Check for excess heat development in fluids
- Take the fluid samples regularly
- Send the fluid samples for fluid analysis
- Monitor the particulate level in fluids
- Check for excess water content and air in fluids
- Check for the symptoms of oxidation in fluids
- Measure the viscosities of fluids
- Replace the fluids at the appropriate time

Monitoring Hydraulic Fluids in Service
The following paragraphs describe some of the fluid monitoring activities.

Check Appearance and Smell
Very often, the appearance or smell of hydraulic fluid alone indicates its suitability for a given application. The colour of the clean hydraulic fluid is amber. A frothy or milky fluid indicates excess air. A hazy or dark-coloured hydraulic fluid indicates a high rate of oxidation.

Some fluids have a bland, oily smell, while others have no smell. A marked change in the smell of a given volume of hydraulic fluid shows a chemical breakdown of the fluid. If a distinct change in the look or smell of the fluid is detected, a chemical analysis of the fluid is appropriate.

Monitor Particulate Level
After a fluid is selected and added to a piece of hydraulic machinery, the next essential step is to install correctly sized filters at appropriate locations in the associated hydraulic circuit. This provision removes particulates and tends to achieve the specified target cleanliness level in the circuit.

Maintain Filters
The effectiveness of filters decreases when contaminants clog them. Therefore, they must be replaced regularly, as per the manufacturer's recommendations or as needed.

Check Excess Water Content
The water content of a fluid used in a hydraulic system should not exceed 100 ppm [0.01%]. Excessive water indicates an ineffective breather filter or heat exchanger in the system. Water can be eliminated from the system fluid by adsorption, absorption, centrifugation, or vacuum dehydration.

Check Oxidation in Hydraulic Fluids
Changes in the colour, odour, or fluid acidity are signs of natural oxidation. The system's sludge, gum, or varnish indicates that oxidation has already occurred. Fluid analysis can detect the level of oxidation in the system. Any increase in the fluid's viscosity could indicate a higher level of oxidation or contamination in the system.

Check Presence of Excess Air
The ill effects of air contamination in hydraulic fluids can be reduced by properly designing the system reservoir, flooding the pump's suction side with the system fluid, and providing air bleeds and return-line diffusers.

Check Excess Heat/Temperature
It is necessary to regularly check the fluid's temperature during system operation. If the fluid is too hot, the associated cooling system may not be working correctly, or there may be pressure-related problems. Check the fluid cooler or reservoir in the system for any failure. Remove any dirt that inhibits airflow around them. The heating of the fluid in the system may cause it to break down.

Precautions While Handling Hydraulic Fluids

Hydraulic fluids must be handled properly to preserve their properties, prevent adverse chemical reactions, and protect them against contamination. Exposure of the hydraulic fluid to the eyes may cause severe pain. Wash the eye with water when exposed to hydraulic fluid. The exposure of fluids to the skin may irritate. The following general precautions must be observed while handling hydraulic fluids, apart from any other instructions specified by the fluid manufacturers:

- Do not keep the containers for hydraulic fluids open for more extended periods than necessary.
- Wear gloves that are impervious to hydraulic fluids when handling hydraulic fluids.
- Use eye protection when handling hydraulic fluids.
- Do not expose the hydraulic fluids, especially mineral-based ones, to high temperatures or open flames.
- Do not mix different categories of hydraulic fluids.
- If the skin is exposed to the fluid, wash the exposed part with soapy water.

Typical Fluid Analysis Procedure

Sending the fluid sample to a particle-counting laboratory can ascertain the health of the hydraulic fluid. Regular testing of the fluid not only indicates the condition of the fluid itself but also provides valuable information regarding the condition of the system components.

The laboratory determines the fluid's cleanliness level in accordance with the relevant standards. It also determines the fluid's chemical makeup to determine whether the additive package performs as initially planned. Suppose a higher level of contaminants is found in the fluid. In that case, the source of contamination should be identified and investigated to prevent future contamination of the fluid. If the fluid is contaminated beyond acceptable limits, it may be necessary to flush the system.

Fluid analysis should be carried out continuously, and sample results must be evaluated for trends that may indicate a likely change in the system's state. The frequency of fluid analysis has to be determined by the nature of the hydraulic application. In general, fluid sampling and testing are required every 500 hours of the power unit's operation or every three months, whichever comes first.

It is a maintenance technician's responsibility to collect fluid samples from critical locations in hydraulic systems, in accordance with the applicable regional standards. The following section discusses some general guidelines for fluid sampling in hydraulic systems.

General Guidelines for Fluid Sampling

Obtaining a fluid sample from a hydraulic system involves several steps to ensure a representative sample from the system reservoir. For complete fluid analysis, 200 to 500 ml of fluid may be required.

All parts of the tiny hand-operated vacuum-assisted bottle syringe used to extract the fluid sample must be washed and rinsed with a filtered solvent to remove contaminants. Inserting the bottle syringe to half the fluid height is required to take the sample fluid from the reservoir. Remember, dirty sampling devices and non-representative fluid samples lead to erroneous conclusions about the fluid and cost more in the long run.

The following bulleted lines give some general guidelines for taking the fluid sample from a hydraulic system:

- Operate the system for some time, say half an hour, before taking the sample to ensure uniform mixing of the fluid,
- Use an approved wide-mouth pre-cleaned sample bottle,
- Drain a sufficient quantity of the fluid from the reservoir to purge the line before collecting the sample,
- Fill the bottle with the fluid, leaving a small volume and
- Tag the sample bottle with pertinent data, including the machine number, date, fluid supplier, fluid type, and time elapsed since the last sample (if any).

Procedure for Replacing Hydraulic Fluids

Eventually, when the entire fluid in the hydraulic system is determined to be contaminated beyond serviceable limits, it must be replaced with a volume of new fluid. Remember that the correct replacement fluid, with the appropriate additive combination, must be used, usually as recommended by the manufacturer.

First, the contaminated fluid must be entirely drained from the reservoir. The reservoir should then be thoroughly flushed with a small quantity of clean hydraulic fluid many times to ensure the contaminated fluid has been completely removed. Then fill the reservoir with the new, filtered hydraulic fluid. In extreme cases, each component in the system may need to be cleaned, and filter elements may need to be changed to remove the contamination.

Maintenance of Hydraulic Filters/Strainers

Hydraulic filters/strainers are devices used to filter harmful particles from hydraulic systems and maintain the cleanliness of system fluids, as required by the application. At the same time, filters/strainers should be regularly maintained to keep them in good working condition.

The usual maintenance activities for filters/strainers in a hydraulic system include checking them for clogs and cleaning or replacing their media parts. The fluid must be drained entirely to access a strainer installed in the reservoir.

Cleaning the filter element begins by removing and shaking the dust by hand. If the filter is heavily clogged, rinse it thoroughly with a neutral solvent, brush off the debris and dirt, shake it to remove excess solvent, rinse it again with hydraulic fluid, and dry it in the shade.

A filter element must be replaced when its internal bypass valve opens, typically after the first 50 hours of use or after a severe system repair. Remember that impurities can be introduced into the filter during the system repair.

A suction strainer is cleanable and reusable. It can be cleaned by washing it with a non-caustic solvent, then blowing it dry with compressed air. A suction strainer prevents catastrophic system failure.

Maintenance and Servicing of Hydraulic Valves

As hydraulic valves are manufactured with tight-fitting, delicate parts, contaminants can pose significant problems. Small amounts of dirt, rust, and sludge can lodge between the valves' mating surfaces, causing their abrasion and internal fluid leakage. Next, vibration, moisture, and corrosion affect the coils of solenoid valves.

While servicing a hydraulic valve, it is essential to understand its parts and how to disassemble/assemble it correctly. Care should be taken to maintain the cleanliness of the valve and the surroundings while servicing. Once the valve is disassembled, check for worn or damaged parts, including the valve body, O-ring seals, plunger seats, and springs. Replace all the worn-out and damaged parts in the valve. Burrs and nicks are to be removed from the valve's internal parts to ensure smooth operation. All disassembled parts must be thoroughly cleaned before reassembly. Lubricate the parts properly with clean hydraulic fluid. Test the assembled valve to ensure it works correctly.

Maintenance of Hydraulic Cylinders

From a maintenance perspective, the most critical components of a hydraulic cylinder are its seals and piston rods. The surface of the piston-rod that enters the piston-rod gland must be kept smooth and clean to avoid failure of piston-rod seals and glands and to prevent wear on cylinder bearings. Impact forces can also damage the cylinder seals.

Other concerns in the cylinder include air presence, internal and external fluid leaks, and operation in harsh environmental conditions. Specialised components, such as metal rod scrapers or protective piston-rod boots, can be provided with the cylinder in harsh and dirty environments to protect the piston-rod's gland area.

Air contamination, fluid leakage, or the presence of an extremely viscous fluid medium in the cylinder may cause the cylinder to operate sluggishly or erratically.

Further, the cylinder must be aligned and mounted correctly with its associated load part to reduce side load on the cylinder, prevent consequent failure of the cylinder seals, minimise wear of the cylinder bushing, and prevent bending or breakage of the cylinder piston-rod. Ensure that appropriate equipment is available to handle the repair activities safely on the cylinder if it is heavy.

Essential Maintenance Activities for Hydraulic Cylinders

The following maintenance activities can be carried out on hydraulic cylinders:
- Check the piston-rod for straightness. Check for any dents or damage on the piston-rod
- Examine the piston-rod bearing for roundness
- Examine the barrel, the piston, and the piston-rod for nicks, scoring, and pitting
- Replace piston seals, piston-rod seals, and/or piston-rod bushings if leakages occur
- Check and control the internal and external fluid leakages in the cylinder
- Align the cylinder and its mating part inline to avoid side loads on the cylinder
- Check the cylinder mountings periodically for tightness or cracks
- Check for sluggish/erratic operation of the cylinder
- Check for the creeping of the cylinder
- Open the bleed ports provided in the cylinder to release the trapped air.

Maintenance of Hydraulic Motors
Most maintenance considerations for hydraulic motors are the same as those for hydraulic pumps. Dirt, corrosion, misalignment, loose bolting, and overload are the most common causes of hydraulic motor failure.

Maintenance of Hydraulic Accumulators
Following safe practices when working with accumulators and properly maintaining them are essential for ensuring the safety of the equipment and personnel.

Maintenance personnel should be familiar with the rules governing pressure vessels, including accumulators. It is also essential to take all safety precautions against hazardous stored energy in hydraulic accumulators, using built-in PRVs, shut-off valves, and solenoid- and/or manually operated bleed valves.

Therefore, an accumulator used in a hydraulic system should be designed to shut off, discharge trapped fluid, and protect the system in an emergency or during system shutdown. The solenoid-operated bleed valve in the accumulator automatically releases the energy trapped in the accumulator in an emergency or during the shutdown of the associated system.

General Guidelines for the Maintenance of Hydraulic Accumulators
The following lines list essential maintenance and safety guidelines for a hydraulic circuit with accumulators (especially bladder accumulators).

- Only qualified maintenance technicians must perform maintenance work in systems with accumulators.
- Attach a warning sign, such as 'ATTENTION: System with Accumulator', close to the accumulator.
- The accumulator should have the safety valve block with a PRV, a shut-off valve, and a bleed valve.
- Always depressurise and isolate the accumulator before servicing the system.
- Never start the system before charging the accumulator.
- Always maintain the accumulator's maximum working pressure, pre-charge pressure, and operating temperature within acceptable limits.
- If the accumulator is gas-charged, restrict the charging and discharging rates to reasonable values to avoid damage to the accumulator and other system components.
- Ensure that the accumulator is manufactured, tested, and certified as per the statutory standards and parameters.

Accumulator Installation
An accumulator in the hydraulic circuit should be installed very close to the source of shock or potential energy to minimise pressure loss along the line between them. It must be correctly installed in an easily accessible place using robust collars. The markings engraved on the accumulator must remain visible after its installation. Typically, an accumulator is installed vertically with the fluid connection port at the bottom. However, if the accumulator must be mounted horizontally due to space constraints, a loss of efficiency, reduced discharge volume, and a shorter service life are to be expected. Further, the horizontal arrangement could trap contaminants.

Accumulator Pre-charging Procedure

Before charging the accumulator, it is advisable to pour some fluid into it to allow the fluid to coat the inside of its shell. This fluid layer provides the initial lubrication between the bladder and the shell.

The usual pre-charge pressure for an accumulator is 80% to 90% of the minimum system pressure. For a hydraulic system utilising the accumulator as a shock absorber or pulsation dampener, a pre-charge pressure of 65% of the minimum system pressure can be used. Also, the pre-charge pressure should not fall below a specified value. If the pre-charge pressure in the accumulator exceeds its maximum or minimum limits, the elastomeric part is liable to be damaged.

If nitrogen gas is allowed to flow too rapidly into the accumulator, it can chill the polymeric material of the accumulator's diaphragm or bladder. This chilling effect may result in the polymeric material's immediate brittle failure.

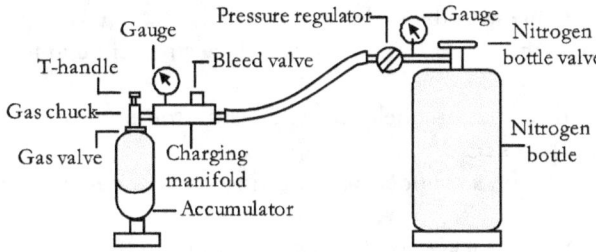

Figure 21.1 | A schematic diagram showing the setup for pre-charging a hydraulic accumulator

Initially, an accumulator should be pre-charged with clean, dry nitrogen gas of class 4.0 purity (N2 content 99.99% by volume) using a charging kit.

Figure 21.1 shows the schematic diagram indicating the arrangement for pre-charging the accumulator. This package consists mainly of a bottle of nitrogen gas, a charging manifold, a gas chuck, and connecting hoses. The manifold consists of a bleed valve and a pressure gauge. The nitrogen bottle has a gas valve and a pressure regulator. The accumulator is usually provided with a gas valve and a protective cap. The following lines list the procedure for pre-charging a hydraulic accumulator.

- Attach the charging kit hose to the gas chuck on one side and the regulator on the other side,
- Close the bleed valve on the charging manifold,
- Attach the chuck to the accumulator gas valve,
- Open the gas valve by turning the chuck's T-handle,
- Open the nitrogen bottle valve slowly and fill the accumulator to the desired pre-charge pressure,
- Close the nitrogen bottle valve,
- If the desired pre-charge is exceeded, open the bleed valve to relieve the excess pressure,
- Close the bleed valve,
- Close the gas valve, and
- Remove the gas chuck.

Maintenance of Hydraulic Seals

Seals are the most delicate parts of hydraulic components. They are exposed to contaminated fluids, chemical attacks, abrasive surfaces, side loads, high temperatures, high pressures, and rapid reciprocating and rotary motions. They may also be exposed to harmful oxygen, ozone, or sunlight. These factors affect seals, leading to premature wear. Seal materials undoubtedly change their physical properties with age. Seals play a significant role in the operation of high-performance hydraulic systems, so their proper installation and maintenance are paramount.

Proper installation and maintenance of seals are essential to ensuring their desired function and long working life. The following bulleted list outlines some of the most significant installation and maintenance activities for hydraulic seals.

- Lubricate the seals and their mating components with appropriate lubricant before their assembly to prevent excessive mechanical wear,
- Use cleaning agents compatible with the seals,
- Keep the clearance between the seal and the sealing surface small enough to prevent the extrusion of the seal,
- Install a metallic piston-rod scraper or a bellows-style piston-rod boot on a hydraulic cylinder, if the surrounding environment is polluted,
- Correctly install the wiper seal on a cylinder piston-rod to prevent contamination from entering the cylinder and
- Protect the seals from being cut, nicked, or rolled while installing them.

Installation, Routing & Maintenance of Fluid Conductors

The proper routing, installation, and maintenance of fluid conductors and their fittings are as necessary as any other component in a hydraulic system. The service life of the fluid conductors is reduced by their installation stress, abrasion, tight bends, and exposure to higher pressure, temperature, and corrosion. One major problem with the fittings is the risk of leakage due to loosening caused by system shock and vibration. The following sections look at the installation and maintenance of rigid hydraulic pipes, semi-rigid tubes, and flexible hoses, along with their fittings.

Installation of Hydraulic Conductors

It is essential to clean conductors and fittings before their installation. Remember to use a minimum number of fittings and connectors. The following bulleted lines give a few essential points for the proper installation of fluid conductors:

- Keep the length of conductors as small as possible to avoid tight bends,
- Route the fluid conductors optimally to minimise the pressure loss and leakage in the system and reduce their abrasion, rubbing, kinking, and excessive flexing,
- Restrain, support, protect, and guide fluid conductors using clamps at frequent intervals to prevent chafing against one another and minimise the vibration of the conductor system,
- Avoid crossing two hose lines. However, when the crossing is unavoidable, join the two lines at the junction point,
- Use proper tools for preparing a conductor for connecting to another conductor or a component, and
- Conductors must be flushed thoroughly with a suitable degreasing agent immediately after installation.

Hose Assembly Routing Tips

It is better to avoid sharp bends when routing hydraulic hoses. Using a bent or kinked hose causes severe backpressure in the associated system. Also, it may cause internal damage to the hose, leading to its premature failure. The following bulleted lines list some tips for routing a hose assembly, especially about its length, minimum bend radii, and multi-plane bending:

- The hose length must be slightly longer than the actual distance between two linear connections to accommodate the changes in the length with pressure changes,
- The bend radius of the hose must be as large as possible to avoid the hose collapse or flow restriction,
- As far as possible, bend a hose in one plane only. This precaution prevents twisting of its wire reinforcement and improves its pressure capacity. The multi-plane bending of a piece of the hose can often be avoided by rerouting the hose,
- If the multi-plane bending cannot be avoided, install a clamp between the bends and provide enough hose length on both sides of the clamp to relieve the strain on the hose's reinforcement wires,
- Use clamps to secure the hose's length in position and keep it from rubbing against adjacent surfaces. Ensure that the hose is slack on both sides of the clamp to compensate for contraction and expansion,
- The hose connected to a cylinder that undergoes a pivoting motion must be of proper length to avoid its kinking or bending beyond its minimum bend radius,
- Use appropriate swivel joints for the hose connection to reduce the bending transmitted to the hose assembly by the relative motion between the associated machine elements and
- The use of carriers keeps the hoses neatly nestled to prevent their rubbing against each other.

Maintenance of Hydraulic Conductors

The following points generalise the maintenance requirements for hydraulic conductors and fittings. For exhaustive maintenance activities of the conductors, the reader may refer to the manufacturer's catalogues.

- Inspect the conductors and their routing for damages, defects, and displacement,
- Examine the conductors and their joints for leakage, looseness, scratches, kinks, and burrs,
- Tighten any loose fittings or nut connections with the correct amount of torque to minimise leakage and reduce contamination and
- Repair or replace defective conductors or fittings. Ensure that any pipe replacement is the same length, size, and wall thickness as the original.

Troubleshooting Hydraulics

Preventive maintenance is performed on hydraulic systems to keep them in optimal working condition at all times. However, faults or breakdowns in hydraulic systems must be traced and corrected with minimal delay and expense. In general, hydraulic failures can be attributed to contaminants, filter clogging, and weak connections. Symptoms of failure include a lack of pressure, the development of high pressure and excessive heat, or the generation of unusual noise in these systems. In high-production industries, downtime can lead to significant losses. Unfortunately, the fault-finding is often performed randomly, leading to the replacement of components without proper justification.

General Troubleshooting Procedure

Figure 21.2 shows the flowchart of a typical troubleshooting procedure. The first step in troubleshooting a system that has developed trouble is understanding its operation and associated circuits. The circuit diagram of the system is virtually indispensable as a troubleshooting aid in all instances of system faults/breakdowns. In complex circuits, time constraints prevent the troubleshooter from studying the entire circuit. It is beneficial to consult the system operator and/or refer to the manufacturer's troubleshooting information to quickly learn how the system should operate. Use every available source of information to shorten the time required to find the source of the trouble. Once sufficient information is collected and evaluated, visualize all possible root causes of the fault.

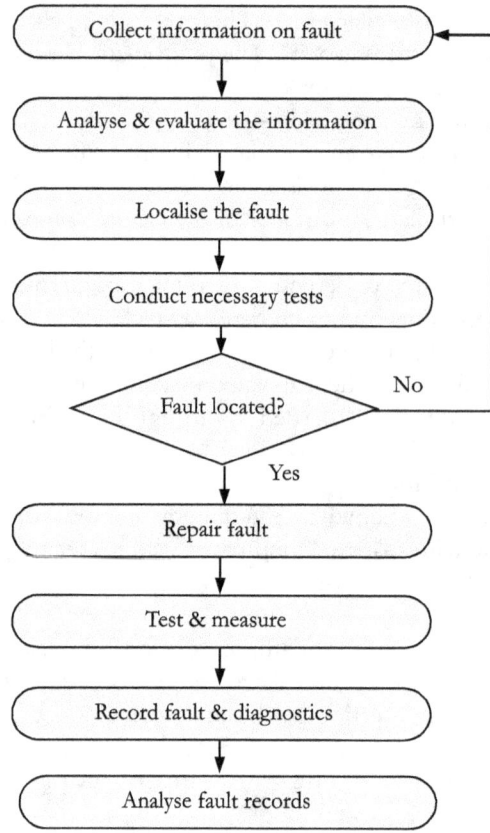

Figure 21.2 | Flow chart of a typical troubleshooting procedure

The most straightforward test is to determine which section contains the defective circuit component. A careful check of the components involved in this part may reveal the source of the trouble. If not, the cycle is repeated until the fault is traced and repaired. The most important rule in troubleshooting a system is to modify only one component at a time. Carry out the circuit analysis with appropriate test equipment, not merely by randomly checking components.

The final steps of troubleshooting involve fault recording and fault analysis to discover any recurring pattern of faults, design and application problems, or shortcomings in maintenance personnel's relevant knowledge. Proper documentation is critical to efficient troubleshooting.

Hydraulic Safety

When working directly with or near a hydraulic system, it is essential to understand the hazards involved. The key to success is assessing the system to determine hazards, applying administrative controls, and enforcing the use of proper personal protective equipment (PPE).

As we know, hydraulic systems operate at high pressure. A hydraulic system can trap pressurised oil in certain parts of its circuit, including accumulators, even when the system is shut down. When maintenance is to be carried out on a system with accumulators, the power source must be isolated from the system, the bleed valve opened, and the system pressure verified using a pressure gauge.

The pressurised oil in a hydraulic system may be discharged when a component fails or system maintenance is carried out without the correct isolation of the system power source. This discharge can cause the dangerous release of mechanical forces unexpectedly. Components or fittings can eject or move unexpectedly, hoses can whip around with great force, or high-pressure oil can be injected through pinholes at a very high speed.

Oil injection is a hazard when oil under pressure is injected into the skin or flesh. The severity of the injury also depends on how close one is to the source of the injected oil and on the size of the orifice through which the oil comes out. Oil injection can also cause severe infection, called gangrene. The injected oil eats away the flesh and fat in the body. If the infection spreads too far, removing the affected body part is necessary.

Hydraulic systems, including oil and components, are generally very hot. The temperature in a typical hydraulic system is around 70°C. Remember, certain oils can chemically burn the skin. Exposure to hot oil can cause blistering and even burns. Hydraulic oil is also flammable. When the oil is sprayed onto a hot surface, it can produce flashes. Atomised oil at higher temperatures can cause a fire hazard. Elevated temperatures may release toxic vapours from the oil, which are harmful when inhaled.

Some people can develop allergies when exposed to the oil or its additives. The oil can damage the skin with prolonged or repeated contact. Eyes are more sensitive to oil splash than other parts of the body. It can cause a severe reaction and permanent eye damage. Using Personal Protective Equipment (PPE) when working with hydraulic circuits is essential. Using chemical-resistant gloves and safety goggles while working with hydraulic systems is also important.

Objective-type Questions

1. Mark the correct statement:
    a) A hydraulic system may fail due to the wrong fluid, leakage, misalignment, or cavitation.
    b) A hydraulic system may fail due to insufficient fluid, faulty seals, and excessive pressure.
    c) A hydraulic system may fail due to fluid contamination, loose connections, and excessive heat.
    d) All the above statements are correct.

2. Mark the correct statement:
    a) The responsibility for the safety of a hydraulic system rests only with the operator.
    b) Troubleshooting a hydraulic system should consider first checking the costly part.
    c) A hydraulic accumulator should be maintained only after relieving trapped pressure.
    d) Preventive maintenance of the hydraulic system involves activities after the system fails.

Questions:
1) Define the term 'maintenance' regarding hydraulic systems and classify them.
2) List and explain various types of maintenance-related activities in hydraulic systems.
3) What are the essential requirements for the preventive maintenance of hydraulic systems?
4) What are the consequences of poor maintenance in hydraulic systems?
5) What are the essential maintenance activities regarding hydraulic power packs?
6) What are the useful design features of hydraulic reservoirs from the maintenance point of view?
7) Mention any five important activities for maintaining hydraulic reservoirs.
8) What are the main reasons for the premature failure of hydraulic pumps?
9) How can a hydraulic pump be maintained? Explain
10) Why is it essential to maintain hydraulic fluids properly?
11) List some important maintenance activities for preventing contamination from hydraulic systems.
12) List the precautions to be taken while handling fluids.
13) What is the procedure for collecting a sample of hydraulic fluid?
14) Explain the process of replacing the hydraulic fluid with new fluid.
15) Why is it essential to maintain the cleanliness level in hydraulic installations?
16) What actions should be taken to prevent the entry of contaminants into a hydraulic system during its maintenance or troubleshooting?
17) Briefly explain the three most significant maintenance activities for hydraulic filters.
18) Discuss the critical maintenance activities for hydraulic valves.
19) Why does fluid leak in cylinders?
20) How is air bled from a hydraulic cylinder?
21) What are the reasons for hydraulic cylinders' sluggish/erratic operation?
22) List essential maintenance aspects of hydraulic accumulators.
23) What are the essential points for the installation of hydraulic accumulators?
24) Explain the pre-charging procedure for a hydraulic accumulator.
25) Why is the maintenance of seals crucial in hydraulic systems? Explain.
26) What are the reasons for the premature failure of seals in a hydraulic system?
27) Write two essential maintenance activities for fluid conductors.
28) List six essential rules for the correct installation of fluid conductors
29) What are the causes of hydraulic hose failures?
30) List three critical points for correctly installing hydraulic hoses.
31) How is the hydraulic hose maintained?
32) Describe the general procedure to troubleshoot a hydraulic system.
33) List the requirements to troubleshoot a hydraulic system.
34) What are the major faults found in hydraulic systems?
35) Describe the possible consequences of neglecting the safety precautions in hydraulic systems.
36) Name three factors that the hydraulic safety circuits should ensure.
37) What are the common hazards found in hydraulic systems?
38) What are the probable causes and remedial actions for the following malfunctions in hydraulic systems: (1) Chattering noise, (2) Hot fluid, (3) excess noise, (4) Excess heat, (5) Absence of flow, (6) Insufficient pressure, and (7) Erratic or sluggish movement of actuators?
39) What are the probable causes and the remedial actions for the malfunctions of: (1) hydraulic cylinders, (2) accumulators, (3) hydraulic valves, (4) fluid conductors, and (5) seals?

Objective-type questions - answer key: 1-d, 2-c

Chapter 22 | Summary of Controls for Hydraulic Systems - Basic Level

It would be educational and insightful to summarise all control functions and methods of basic-level hydraulics in a single table. Table 22.1 highlights many control concepts in hydraulic systems.

Table 22.1 | Summary of controls for Hydraulic Systems - Basic level

Control Function	Control Valve / Method
To control a single-acting (s/a) cylinder or unidirectional hydraulic motor	3/2 Directional Control (DC) manually-operated valve Or 4/2 DC valve with B port blocked
To control the speed of a s/a cylinder	Throttle-check (one-way flow control) valve: For extension stroke: Meter-in method For retraction stroke: Meter-out method
To control the speed of a unidirectional motor	Throttle valve in series with or parallel to the motor
To control a double-acting (d/a) cylinder or bi-directional motor	4/2 DC valve, manually actuated
To lock the position of a d/a cylinder hydraulically for a long duration and unload the fixed-displacement pump manually for energy-saving purposes	Dedicate the pump to power the d/a cylinder exclusively and use a 4/3 DC valve with tandem centre position (Ports A and B blocked, and port P connected to port T) as the final control element
To control multiple d/a cylinders independently in a single pump system	4/3 DC valves, all-closed centre position (The specific requirement is that port P should be blocked)
To control a bi-directional hydraulic motor independently with a soft stop	4/3 DC valve, float centre position (Port P blocked, and ports A and B connected to port T)
To control an intermittently-operated hydraulic motor for soft stop and unload the associated fixed-displacement pump manually for energy-saving	Dedicate the pump to power the hydraulic motor exclusively and use a 4/3 DC valve, all-open centre position (Port P connected to port T, and ports A and B also connected to port T)
To hold a descending load in any stroke position and move the load	Pilot-operated check valve (Load holding circuit)
To hold and lock a descending load attached to a hydraulic cylinder in a leak-free manner and to reduce cavitation	Connect the dual pilot-operated check valve to the working lines of the cylinder
To control the speed of a d/a cylinder manually, where there is only a limited or no variation of the connected load (direction sensitive)	Throttle-check (one-way flow control) valve, pressure-dependent: - Meter-in method (for a horizontally mounted cylinder with a positive load) - Meter-out method (for a cylinder with a negative load). A PRV may be connected to the exhaust line to prevent pressure intensification in the line caused by excessive throttling. - Bleed-off method for less heat buildup and consequent energy saving. It cannot be used in circuits where inlet flow changes and overrunning loads are present.

To precisely control the speed of a d/a cylinder manually, where there is a wide variation in the inlet pressure or connected load/temperature	Throttle-check (one-way flow control) valve, pressure-compensated/temperature-compensated
To control the speed of a hydraulic motor	Throttle valve in series with or parallel to the motor
To get the same speeds for the forward and return strokes of a differential double-acting hydraulic cylinder	A regenerative circuit with a double-acting cylinder, with its piston area two times the piston-rod area
To get the synchronous movement of two hydraulic actuators under different load conditions	Use a flow divider
To get different pressures in a hydraulic circuit with a fixed-displacement pump (For example, a pressure of, say, 200 bar in one part and 100 bar in another part)	Pressure-reducing valve Note: A pressure-reducing valve with a relief feature (3-way type) can be used if an excessive load (shock pressure) is expected on the output side of the valve.
To realise the sequential operation of cylinders (For example: A+B+B-A-)	Pressure sequence valve
To dynamically unload a pump in a hydraulic system when a cylinder in the system reaches its end of stroke position or an accumulator in the system gets fully charged, to realise an energy storage function	Unloading valve
To control a high-volume cylinder with requirements of a high flow at low pressure initially and a high pressure at low flow subsequently, using two pumps	Use a hi-lo circuit with a high-volume, low-pressure pump controlled by an unloading valve and a low-volume, high-pressure pump
To prevent the uncontrollable operation of the load-coupled cylinder during the descending motion of the piston and the attached load	Counterbalance valve
To brake a hydraulic motor driving a loaded winch	Over-centre valve
To control a hydraulic cylinder with a heavy platen to get sufficient clamping pressure	Over-centre valve
To absorb shock pressures or dampen flow pulsations	Hydraulic accumulator
To store pressurised fluid during idle periods while clamping a large-volume hydraulic cylinder and feeding the pressurised fluid back to the system to supplement the associated pump flow during periods of high demand	In the pump-cylinder system, connect a properly-sized hydraulic accumulator with a check valve for containing the pressurised fluid in the accumulator and an unloading valve for unloading the pump when the accumulator is fully charged

Appendix 1

Graphic Symbols for Hydraulic Components as per ISO 1219

Symbol	Description	Symbol	Description
▲	Hydraulic power supply	——	Main line
⊥	Reservoir (Tank)	——	Pilot line
(symbol)	Fixed-displacement pump (Uni-directional)	——	Drain line
(symbol)	Fixed-displacement pump (Bi-directional)	(symbol)	Flexible hose
(symbol)	Variable-displacement pump (Uni-directional)	—+—	Connection point
(symbol)	Variable-displacement pump (Bi-directional)	—+—	Cross-over
(symbol)	Pressure relief valve (Direct-operated)	—×—	Closed
(symbol)	Pressure relief valve (Pilot-operated)	—⋇—	Branching (With connected piping)
(M)	Motor (Electric)	—>—<—	Fast coupling
[M]	Motor (Engine)	—◊◊—	Coupling with check valve

Symbol	Description	Symbol	Description
	Hydraulic accumulator		2/2-way DC valve
	Filter		3/2-way DC valve
	Heater		4/2-way DC valve
	Cooler		4/3-way DC valve
	Liquid-operated cooler		Push button actuation
	Pressure gauge		Lever actuation
	Pressure switch		Cam operated
	Rotating shaft (one direction)		Spring actuated
	Rotating shaft (Two direction)		Detent
	Cylinder with fixed stroke end cushioning Cushioning on one side		Hydraulically actuated

⌐┐	Electric (Solenoid)	⌀	Compensated flow regulator, two way
▭▶	Electro-hydraulic	⌀	Compensated flow regulator, three way
⌐	Electric (Proportional)		Sequence valve, direct-operated
▭▶	Electro-hydraulic (Proportional)		Pressure reducing valve, direct-operated
◇	Check valve (Without spring)		Pressure reducing valve, pilot-operated
≢	Check valve (With spring)		Counter balance valve
	Pilot-operated check valve		Single-acting cylinder, return stroke by external force
	Pilot-operated check valve (with drain)		Single-acting cylinder, spring return
✶	Throttle valve, variable, two-way		Double-acting cylinder, single rod
	Throttle valve, variable, one-way		Semi-rotary actuator

	Double-acting cylinder, double rod		Telescopic cylinder, double-acting
	Cylinder with fixed stroke, end-cushioning (one side)		Hydraulic motor. Fixed displacement, uni-directional
	Cylinder with fixed stroke, end-cushioning (Two sides)		Hydraulic motor. Fixed displacement, bi-directional
	Cylinder with adjustable end-cushioning (Two sides)		Hydraulic motor. Variable displacement, uni-directional
	Telescopic cylinder, single-acting		Hydraulic motor. Variable displacement, bi-directional

Appendix 2

Unit Conversions

2.A | Metric to English

Length and Area

Table A2.1 | Metric unit to English unit conversions (Length and Area)

Length		Area	
Metric unit	English unit	Metric unit	English unit
1 mm	0.03937 inches	1 mm²	0.001550 sq. inches
1 cm	0.3937 inches	1 cm²	0.1550 sq. inches
1 m	3.281 feet 1.094 yards	1 m²	10.76 sq. feet 1.196 sq. yards
1 km	0.6214 miles	1 km²	247.1 acres 0.3861 sq. miles

Volume and Weight

Table A2.2 | Metric unit to English unit conversions (Volume and Weight)

Volume		Weight	
Metric unit	English unit	Metric unit	English unit
1 ml	0.03381 fl. ounces	1 g	0.03527 ounces
1 l	2.113 pints 1.056 quarts 0.2641 gallons[1]	1 kg	35.27 ounces 2.205 pounds
		1 t	2205 pounds 1.102 tons

2. B | English to Metric

Length and Area

Table A2.3 | English unit to Metric unit conversions (Length and Area)

Length		Area	
English unit	Metric unit	English unit	Metric unit
1 inch	25.4 mm	1 sq. inch	645.16 mm² 6.4516 cm²
1 foot	30.48 cm	1 sq. foot	929.0 cm² 9.290 dm²
1 yard	91.44 cm 0.9144 m	1 sq. yard	0.8361 m²
1 mile	1.609 km	1 acre	4'046 m²
1 nautical mile	1.852 km	1 sq. mile	2.590 km²

Volume and Weight

Table A2.4 | English unit to Metric unit conversions (Volume and Weight)

Volume		Weight	
English unit	Metric unit	English unit	Metric unit
1 teaspoon	4.929 ml	1 ounce	28.35 g
1 tablespoon	14.79 ml	1 pound	453.6 g 0.4536 kg
1 fl. ounce	29.57 ml	1 ton	907.2 kg 0.9072 t
1 cup	0.2365 l		
1 pint	0.4732 l		
1 quart	0.9464 l		
1 gallon	3.785 l		

2.C | psi to bar and bar to psi Conversions

Table A2.5 | psi to bar and bar to psi conversions

psi to bar		bar to psi	
psi	bar	bar	psi
14.5	1	1	14.50
100	6.89	10	145.04
500	34.47	50	725.19
1000	68.95	100	1450.38
1500	103.42	150	2175.57
2000	137.9	200	2900.75
2500	172.37	250	3625.94
3000	206.84	300	4351.13
3500	241.32	350	5076.32
4000	275.79	400	5801.51
4500	310.26	450	6526.70
5000	344.74	500	7251.89
6000	413.69	600	8702.26
7000	482.63	700	10152.64
8000	551.58	800	11603.02
9000	620.53	900	13053.40
10000	689.48	1000	14503.77

2.D | Conversion Factors for Units of Pressure

Table A2.6 | Conversion factors for units of pressure

To convert	To	Multiply with
bar	kg/m^2	1.0197
bar	N/m^2	100,000
bar	Pa	100,000
bar	psi	14.504
kg/m^2	bar	0.00009807
kg/m^2	N/m^2	9.8067
kg/m^2	Pa	9.8067
kg/m^2	psi	0.001422
MPa	psi	145.038
N/m^2	Pa	1.0
N/m^2	psi	0.0001450
Pa	bar	0.000,01
Pa	N/m^2	1.0
Pa	psi	0.0001450
psi	bar	0.06895
psi	N/m^2	6894.7574
psi	Pa	6894.7574

2.E | Viscosity Unit Conversions

Table A2.7 | Viscosity unit conversions - Absolute (Dynamic) viscosity

To convert	To	Multiply with
lbf-sec/ft²	Centipoises	47880.26
lbf-sec/ft²	Pascal-sec	47.8803
centipoises	kg-sec/m²	0.000102
centipoises	lbf-sec/ft²*	0.00000208854
centipoises	Pascal-sec	0.001
Pascal-sec	lbf-sec/ft²	0.0208854
Pascal-sec	centipoises	1000

Table A2.8 | Viscosity unit conversions - Kinematic viscosity

To convert	To	Multiply with
ft²/sec	centistokes	92903.04
ft²/sec	m²/sec	0.092903
m²/sec	ft²/sec	10.7639
m²/sec	centistokes	1000000.0
centistokes	m²/sec	0.000001
centistokes	ft²/sec	0.0000107639

Table A2.9 | Viscosity unit conversions - Absolute to Kinematic Viscosity

To convert	To	Multiply with
centipoises	centistokes	1/density (g/cm³)
centipoises	ft²/sec	0.00067197/density (lb/ft³)
lbf-sec/ft²	ft²/sec	32.174/density (lb/ft³)
kg-sec/m²	m²/sec	9.80665/density (kg/m³)
Pascal-sec	centistokes	1000/density (g/cm³)

Table A2.10 | Viscosity unit conversions - Kinematic to Absolute Viscosity

To convert	To	Multiply with
centistokes	centipoises	density (g/cm³)
sq meters/sec	kg-sec/m²	0.10197 x density (kg/m³)
ft²/sec	lbf-sec/ft²	0.03108 x density (lb/ft³)
ft²/sec	centipoises	1488.16 x density (lb/ft³)
centistokes	Pascal-sec	0.001 x density (g/cm³)

The relationships between lpm and gpm are as follows:
  1 gpm = 3.785 lpm and
  1 lpm = 0.264 gpm

Appendix 3

3.A | Viscosity Grades and Viscosity Ranges as per ISO 3348

Table A3.1 | ISO 3348 Viscosity Grades and viscosity ranges

ISO VG	Mid-Point	Minimum	Maximum
2	2.2	1.98	2.42
3	3.2	2.88	3.52
5	4.6	4.14	5.06
7	6.8	6.12	7.48
10	10	9.0	11.0
15	15	13.5	16.5
22	22	19.8	24.2
32	32	28.8	35.2
46	46	41.4	50.6
68	68	61.2	74.8
100	100	90	110
150	150	135	165
220	220	198	242
320	320	288	352
460	460	414	506
680	680	612	748
1000	1000	900	1100
1500	1500	1350	1650

3.B | Viscosity Comparison

Table A3.2 | Viscosity Comparison Table

ISO-VG	SSU	CentiPoise	CentiStoke	Typical Brands/ Liquids at 100°F
2	31	0.876	1.0	Water
3	35	2.19	2.5	–
5	40	3.68	4.2	-
5/7	45	5.17	5.9	-
7	50	6.57	7.5	Kerosene
7/10	55	7.71	8.8	Atlantic Richfield/ Duro 55 Hyd. Oil
10	60	9.20	10.5	Monsanto/Skydrol - 500 A
10/15	70	11.56	13.2	Mobil/Aero HFA Hydraulic Oil
15	80	13.75	15.7	No 4 Fuel Oil
22	90	15.94	18.2	Stauffer Chemical/ Fyrquel 90
22	100	18.05	20.6	Conoco/Syncon Synthetic AW Hyd. Oil
32	150	28.03	32.0	Mobile/DTE 24 Hydraulic Oil
46	200	37.84	43.2	Citco/Glycol FR-40XD (Oil in Water)
68	300	56.94	65.0	SAE 20 Crankcase Oil
68/100	400	75.34	86.0	Sunoco/Sunvis 41 Hydraulic Oil
100	500	94.61	108	SAE 30 Crankcase Oil
150	750	141.91	162	SAE 40 Crankcase Oil
220	1000	189.22	216	Mobil/Paper Machine Oil - Type K
320	1500	282.95	323	SAE 50 Crankcase Oil
460	2000	377.56	431	Amoco/American Industrial Oil - No. 460
680	3000	567.65	648	SAE 140 Gear Oil
1000	4000	755.11	862	SAE 250 Gear Oil

Appendix 4

Important Standards for Hydraulic Systems

Table A4.1 | Important standards for hydraulic systems

Standard	Description
ISO 3448:1992	Establishes a system of viscosity classification for hydraulic fluids
ISO 2909:2002 (ASTM D2270–04)	'This standard describes two procedures for calculating the viscosity index (VI) of petroleum'[10] oils from their kinematic viscosities at 40°C and 100°C
ISO 11171:2010	This standard specifies procedures for 'primary particle-sizing calibration, sensor resolution, and counting performance of automatic particle counters (APCs) for liquids capable of analyzing bottle samples'[11]
ISO 4406:1999	Hydraulic Fluids - Method for coding the level of contamination by solid particles
NAS 1638, Rev 4	This standard describes the cleanliness requirements of parts used, especially in aircraft hydraulic systems (inactive for new designs)
SAE AS 4059 E	This aerospace standard defines cleanliness classes for particulate contamination of hydraulic fluids
NFPA/T3.6.11 R1-1998 (r2024)	Fluid power systems and components - cylinder bore and rod size combinations - rod end configurations, dimensional identification code - Mounting dimensions for bore sizes less than 1.5 inch bore catalogued square head tie rod type
NFPA/T3.4.7 R2-2000 (R2024)	Accumulator - Pressure rating supplement to NFPA/T2.6.1 R2-2000, - method for verifying the fatigue and burst pressure ratings of the pressure containing envelope of a metal fluid power accumulator
ISO 1219-1:2006	'Fluid power systems and components - Graphic symbols and circuit diagrams - Part 1: Graphic symbols for conventional use and data-processing applications'[12]
ISO 1219-2:1995	'Fluid power systems and components - Graphic symbols and circuit diagrams - Part '[13]2: Circuit diagram.
ISO 5598:2008	Fluid power systems and components -- Vocabulary
ISO 4401:1994	Hydraulic fluid power - Four-port directional control valves - Mounting surfaces
ISO 6020–1:2007	Hydraulic fluid power - Mounting dimensions for single rod cylinders, 160 bar series - Part 1: Medium series
ISO 6020-2:2006	Hydraulic fluid power - Mounting dimensions for single rod cylinders, 160 bar series - Part 2: Compact series
ISO 6020-3:1994	Hydraulic fluid power - Mounting dimensions for single rod cylinders, 160 bar series - Part 3: Compact series with bores from 250 mm to 500 mm
ISO 4395:2009	Fluid power systems and components – Types and dimensions of piston rod ends in cylinders
ISO 5596-1999	Hydraulic fluid power - Gas-loaded accumulators with separator - Ranges of pressures, volumes, and characteristic quantities.
ISO 4413:2010	Hydraulic fluid power - General rules and safety requirements for systems and their components

Appendix 5

5.A | Hydraulic Fluid Additives and Elements

Table A5.1 | Hydraulic fluid additives

Additives	Elements
VI improvers	Polyalphaolefins, Polymethacrylates, and polyalkylstyrenes
EP additives	Organic sulfur-, phosphorus-, and chlorine-containing compounds
Anti-wear additives	Zinc dithiophosphate (ZDP), ashless additives
Oxidation inhibitors	Phenols, amines, and sulfides
Corrosion inhibitors	'Fatty acids, sulfonates and salts of fatty acids'[1]
Antifoam agents	Silicone oils
Demulsifiers	Ionogenic and nonionogenic polar compounds
Pour point depressants	Polymethacrylates and condensation products

Reference
1. PRODUCTION, IMPORT/EXPORT, USE, AND DISPOSAL 4.1 PRODUCTION. (n.d.). Retrieved from http://www.atsdr.cdc.gov/toxprofiles/tp99-c4.pdf_br

5.B | Properties of Some Hydraulic Fluids

Table A5.2(a) | Mineral-based hydraulic fluid – ISO VG 32

Property	Value in metric unit		Value in English unit	
Density at 15.6°C (60°F)	0.868×10^3	kg/m³	54.2	lb/ft³
Kinematic viscosity at 40°C (104°F)	32.2	cSt	32.2	cSt
Kinematic viscosity at 100°C (212°F)	5.52	cSt	5.52	cSt
Viscosity index	108		108	
Flash point	212	°C	414	°F
Pour Point	-33	°C	-27	°F

Table A5.2(b) | Bio-degradable synthetic ester-based hydraulic fluid - ISO VG 46 (SAE 10W)

Property	Value in metric unit		Value in English unit	
Density at 60°F (15.6°C)	0.921×10^3	kg/m³	57.5	lb/ft³
Kinematic viscosity at 40°C (104°F)	48.7	cSt	48.7	cSt
Kinematic viscosity at 100°C (212°F)	8.7	cSt	8.7	cSt
Viscosity index	160		160	
Flash point	220	°C	428	°F
Pour Point	-58	°C	-72	°F
Zinc	max.5	ppm	max.5	ppm

Appendix 6

6.A | Contamination Code Rating

Table A6.1 | Contamination code rating system as per ISO 4406: 1999

Range Code	Number of particles	
	>	<=
1	0	0.02
2	0.02	0.04
3	0.04	0.08
4	0.08	0.15
5	0.15	0.3
6	0.3	0.6
7	0.6	1.3
8	1.3	2.5
9	2.5	5
10	5	10
11	10	20
12	20	40
13	40	80
14	80	160
15	160	320
16	320	640
17	640	1,300
18	1,300	2,500
19	2,500	5,000
20	5,000	10,000
21	10,000	20,000
22	20,000	40,000
23	40,000	80,000
24	80,000	160,000
25	160,000	320,000
26	320,000	640,000
27	640,000	1,300,000
28	1,300,000	2,500,000
29	2,500,000	5,000,000
30	5,000,000	10,000,000

6.B | Recommended Fluid Cleanliness Levels

Table A6.2 | Typical cleanliness levels, using petroleum oil, for hydraulic components.
Courtesy: Eaton Hydraulics

Components	System Pressure Level		
	<140 bar	140–207 bar	>207 bar
Vane pumps, fixed	20/18/15	19/17/14	18/16/13
Vane pumps, variable	18/16/14	17/15/13	--
Piston pumps, fixed	19/17/15	18/16/14	17/15/13
Piston pumps, variable	18/16/14	17/15/13	16/14/12
Directional valves	20/18/15	20/18/15	19/17/14
Proportional valves	17/15/12	17/15/12	15/13/11
Servo valves	16/14/11	16/14/11	15/13/10
Pressure/Flow controls	19/17/14	19/17/14	19/17/14
Cylinders	20/18/15	20/18/15	20/18/15
Vane motors	20/18/15	19/17/14	18/16/13
Axial piston motors	19/17/14	18/16/13	17/15/12
Radial piston motors	20/18/14	19/17/13	18/16/13

Appendix 7

The SAE Aerospace Standard AS4059

The NAS 1638 standard is no longer applicable to new components or systems after May 30, 2001, due to changes in ISO standards for calibrating automatic particle counters (APCs). The SAE aerospace standard AS4059, for specifying particulate contamination in hydraulic fluids in different classes, was developed in 1988 as a replacement for the NAS 1638 standard. Since then, this standard has undergone many revisions.

This standard AS4059 offers two classifications:
- One classification, based on microscopic counting, applies to those currently using NAS 1638 classes and desiring to maintain the same NAS format.

- The second one, based on automatic particle counting, applies to those using the methods of previous revisions of AS4059 and/or cumulative particle counts.

Method of Particle Counting

The introduction of automatic particle counters (APCs) during the 1960s revolutionized the measurement of the size distribution of dirt particles.

As per ISO 4402, the method for calibrating APCs was based on the size distribution of the silica-based A.C. Fine Test Dust (ACFTD), derived from optical microscope measurements.

However, as the supply of ACFTD ceased in 1992, this method was replaced by another method in accordance with ISO 11171. The National Institute of Standards and Technology (NIST) replaced ISO Medium Test Dust (MTD). This method uses a scanning electron microscope (SEM) with image analysis software to precisely identify particle size and number down to 1 μm.

Particle Size Classification

The particle size classification based on differential size ranges and cumulative sizes is given below:

SAE AS 4059 specifies the following differential size ranges of particles for the optical counting method, similar to that used in the NAS standard: (1) 6 -14 μm(c), (2) 14 -21 μm(c), (3) 21 -38 μm(c), (4) 38 -70 μm(c), and (5) >70 μm(c).

SAE AS 4059 specifies the following cumulative sizes of particles for the automatic particle counting method using electron microscopes meant for new systems: (1) > 4 μm(c) (Code A), (2) > 6 μm(c) (Code B), (3) > 14 μm(c) (Code C), (4) > 21 μm(c) (Code D), (5) > 38 μm(c) (Code E), and (6) > 70 μm(c) (Code F).

Contamination Concentration Levels

SAE AS 4059 specifies the cleanliness level of a given fluid sample by a single figure representing the maximum allowed differential or cumulative particle counts (i.e., worst case) present in 100 ml of the fluid for the designated particle sizes according to the particle counting method.

Cleanliness Classes for Differential Particle Counts

The cleanliness classes for the differential particle counts are given in Table A7.1.

Table A7.1 | Cleanliness Classes for Differential Particle Counts

Size		Maximum contamination limits, particles/100 ml				
		6 – 14 μm(c)	14 – 21 μm(c)	21 – 38 μm(c)	38 – 70 μm(c)	>70 μm(c)
Class	00	125	22	4	1	0
	0	250	44	8	2	0
	1	500	89	16	3	1
	2	1000	178	32	6	1
	3	2000	356	63	11	2
	4	4000	712	126	22	4
	5	8000	1425	253	45	8
	6	16000	2850	506	90	16
	7	32000	5700	1012	180	32
	8	64000	11400	2025	360	64
	9	128000	22800	4050	720	128
	10	256000	45600	8100	1440	256
	11	512000	91200	16200	2880	512
	12	1024000	182400	32400	5760	1024

Cleanliness Classes for Cumulative Particle Counts

The cleanliness classes for the cumulative particle counts are given in Table A7.2.

Table A7.2 | Cleanliness Classes for Cumulative Particle Counts

Size		Maximum contamination limits, particles/100 ml					
		>4 μm(c)	>6 μm(c)	>14 μm(c)	>21 μm(c)	>38 μm(c)	>70 μm(c)
		A	B	C	D	E	F
Class	000	195	76	14	3	1	0
	00	390	152	27	5	1	0
	0	780	304	54	10	2	0
	1	1560	609	109	20	4	1
	2	3120	1217	217	39	7	1
	3	6250	2432	432	76	13	2
	4	12500	4864	864	152	26	4
	5	25000	9731	1731	306	53	8
	6	50000	19462	3462	612	106	16
	7	100000	38924	6924	1224	212	32
	8	200000	77849	13849	2449	424	64
	9	400000	155698	27698	4898	848	128
	10	800000	311396	55396	9796	1696	256
	11	1600000	622792	110792	19592	3392	512
	12	3200000	1245584	221584	39184	6784	1024

Note: The information reproduced in Tables A7.1 and A7.2 is a brief extract from SAE AS4059. For further details and explanations, refer to the full standard.

Appendix 8

Mesh to Micron Conversion

Table A8.1 | Mesh to Micron conversion chart

U. S. Mesh	Microns	Millimeter	Inches
3	6730	6.730	0.2650
4	4760	4.760	0.1870
5	4000	4.000	0.1570
6	3360	3.360	0.1320
7	2830	2.830	0.1110
8	2380	2.380	0.0937
10	2000	2.000	0.0787
12	1680	1.680	0.0661
14	1410	1.410	0.0555
16	1190	1.190	0.0469
18	1000	1.000	0.0394
20	841	0.841	0.0331
25	707	0.707	0.0280
30	595	0.595	0.0232
35	500	0.500	0.0197
40	400	0.400	0.0165
45	354	0.354	0.0138
50	297	0.297	0.0117
60	250	0.250	0.0098
70	210	0.210	0.0083
80	177	0.177	0.0070
100	149	0.149	0.0059
120	125	0.125	0.0049
140	105	0.105	0.0041
170	88	0.088	0.0035
200	74	0.074	0.0029
230	63	0.063	0.0024
270	53	0.053	0.0021
325	44	0.044	0.0017
400	37	0.037	0.0015

Appendix 9

9. A | Theoretical Cylinder Forces

In the English Units: Force (lb) = Pressure (psi) x Piston Area (in²)

Table A9.1 | Theoretical Cylinder Forces in the English Units

Bore size (in)	Rod dia. (in)	Force (lb)	System pressure (psi)					Volume (Gallons per inch of stroke)
			1000	2000	3000	4000	5000	
1½	¾	Thrust	1,770	3,530	5,300	7,070	8,830	0.0076
		Pull	1,320	2,650	3,970	5,300	6,630	0.0057
2	1	Thrust	3,140	6,280	9,420	12,560	15,700	0.0136
		Pull	2,360	4,710	7,070	9,420	11,780	0.0102
3	2	Thrust	7,070	14,130	21,200	28,270	35,340	0.0306
		Pull	3,930	7,850	11,780	15,700	19,630	0.0170
4	3	Thrust	12,550	25,130	37,700	50,260	62,830	0.0544
		Pull	5,500	11,000	16,490	21,990	27,490	0.0238
6	4	Thrust	28,270	56,550	84,820	113,100	141,370	0.1224
		Pull	15,700	31,410	47,120	62,830	78,540	0.0680
8	5	Thrust	50,260	100,530	150,800	201,060	251,330	0.2176
		Pull	30,630	61,260	91,890	122,520	153,150	0.1326
10	6	Thrust	78,540	158,080	235,620	314,160	392,700	0.3400
		Pull	50,260	100,530	150,800	201,060	251,330	0.2176

9.B | NFPA Standards for Hydraulic Cylinders

Table A9.2 | NFPA Standards for Hydraulic cylinders

NFPA/T3.6.11 R1-1998 (r2024)	Fluid power systems and components - cylinder bore and rod size combinations - rod end configurations, dimensional identification code - Mounting dimensions for bore sizes less than 1.5 inch bore catalogued square head tie rod type
NFPA/T3.6.29 R2-2000(R2024)	Tie rod or bolted cylinder - Pressure rating supplement to NFPA/T2.6.1 R2-2000, Fluid power components - method for verifying the fatigue and burst pressure ratings of the pressure containing envelope of a tie rod or bolted cylinder
NFPA/T3.6.1-1984 (R2024)	Fluid power systems and products - cylinder bores and piston rod diameters - inch series
NFPA/T3.6.37 R1-2010 (R2024)	Hydraulic fluid power - cylinders - method for determining the buckling load
NFPA/T3.6.31 R2-2000 (R2024)	Telescopic cylinders and cylinders of non-bolted end construction - Pressure rating supplement to NFPA/T2.6.1 R2-2000, Fluid Power components - method for verifying the fatigue and burst pressure ratings of the pressure containing envelope of a tie rod or bolted cylinder
NFPA/T3.6.59-1993 (R2024)	Hydraulic fluid power - Cylinders - Cushion performance
NFPA/T3.6.54 R1-1997 (R2024)	Hydraulic fluid power - Cylinder ports - SAE straight thread O-ring and four-bolt flange ports - Heavy-duty and light-duty square-head tie rod cylinders
NFPA/T3.6.8 R3-2010 (R2024)	Fluid power systems - Cylinders - Dimensions for accessories for cataloged square head industrial types

Appendix 10

Seal Materials and their Temperature Ranges

Table A10.1 | Seal materials and their temperature ranges

Material	Temperature Range	
	°C	°F
Nitrile	-30°C to 100°C	-22°F to 212°F
H-Nitrile (Hydrogenated Nitrile)	-35°C to 150°C	-30°F to 302°F
Viton	-25°C to 262°C	-4°F to 400 °F
Silicone	-60°C to 232°C	-75°F to 450°F
EPDM	-45°C to 150°C	-65°F to 350°F
Polyurethane	-30°C to 110°C	-22°F to 230°F
Nylon	-40°C to 120°C	-40°F to 248°F
Teflon, virgin	-200°C to 260°C	-328°F to 500°F
Teflon, filled	-200°C to 260°C	-328°F to 500°F

Appendix 11

11.A | Pipe Specifications

(i) General Specifications for Steel Pipes

Table A11.1 | General pipe specifications

Nominal Pipe Size		Outside Diameter		Wall thickness					
				Schedule 40		Schedule 80		Schedule 160	
inch	mm	inch	mm	inch	mm	inch	mm	inch	mm
⅛	3.175	0.125	10.29	0.405	1.72	0.068	2.41	0.095	--	--
¼	6.350	0.250	13.72	0.540	2.24	0.088	3.02	0.119	--	--
⅜	9.525	0.375	17.15	0.675	2.31	0.091	3.20	0.126	--	--
½	12.70	0.500	21.34	0.840	2.77	0.109	3.73	0.147	4.78	0.188
¾	19.05	0.750	26.67	1.050	2.87	0.113	3.91	0.154	5.56	0.219
1	25.40	1.000	33.35	1.315	3.38	0.133	4.55	0.179	6.35	0.250
1¼	31.75	1.250	42.16	1.660	3.56	0.140	4.85	0.191	6.35	0.250
1½	38.10	1.500	48.26	1.900	3.68	0.145	5.08	0.200	7.14	0.281
2	50.80	2.000	60.33	2.375	3.91	0.154	5.53	0.218	8.74	0.344
2½	63.50	2.500	73.03	2.875	5.16	0.203	7.01	0.276	9.53	0.375
3	76.20	3.000	88.90	3.500	5.49	0.216	7.62	0.300	11.1	0.438
3½	88.90	3.500	101.6	4.000	5.74	0.226	8.07	0.318	--	--
4	101.6	4.000	114.3	4.500	6.02	0.237	8.56	0.337	13.5	0.531
5	127.0	5.000	141.3	5.563	6.55	0.258	9.53	0.375	15.9	0.625
6	152.4	6.000	168.3	6.625	7.11	0.280	10.8	0.432	18.3	0.719
8	203.2	8.000	219.1	8.625	8.18	0.322	12.7	0.500	23.0	0.906
10	254.0	10.00	273.1	10.75	9.27	0.365	15.1	0.594	28.6	1.125
12	304.8	12.00	323.9	12.75	10.3	0.406	17.5	0.688	33.3	1.312
14	355.6	14.00	355.6	14.00	11.1	0.438	19.1	0.750	35.7	1.406
16	406.4	16.00	406.4	16.00	12.7	0.500	21.4	0.844	40.5	1.594
18	457.2	18.00	457.2	18.00	14.2	0.562	23.8	0.938	45.2	1.781

11.B | Tube Specifications

(i) Size and Pressure Chart for Carbon Steel Tubing

Table A11.2 | Size and Pressure Chart for Carbon Steel hydraulic tubing (Inch Sizes)
(Courtesy: HSC Hydraulic Supply Company)

Tube OD (inch)	Wall thickness (inch)	Max. Working Pressure @ 6:1 SF (psi)	Burst Pressure (psi)
3/16	.035	3422	20533
1/4	.035	2567	15400
1/4	.049	3593	21560
5/16	.049	2875	17248
5/16	.065	3813	22880
3/8	.049	2396	14373
3/8	.065	3178	19067
1/2	.049	1797	10780
1/2	.065	2383	14300
1/2	.083	3043	18260
5/8	.065	1907	11440
5/8	.095	2787	16720
3/4	.049	1198	7187
3/4	.065	1589	9533
3/4	.095	2322	13933
3/4	.109	2664	15987
1	.065	1192	7150
1	.095	1742	10450
1	.120	2200	13200
1-1/4	.095	1393	8360
1-1/4	.120	1760	10560

11. C | Hose Specifications

(i) Dash Numbers and Corresponding Hose IDs

Table A11.3 | Dash numbers and corresponding Hose IDs

Dash No	ID	
	Other than SAE100 R5	For SAE 100 R5
-2	1/8	--
-3	3/16	--
-4	1/4	3/16
-5	5/16	1/4
-6	3/8	5/16
-8	1/2	13/32
-10	5/8	1/2
-12	3/4	5/8
-14	7/8	--
-16	1	7/8
-20	1-1/4	1-1/8
-24	1-1/2	1-1/4
-32	2	1-13/16
-36	2-1/4	--
-40	2-1/2	2-3/8
-48	3	--
-56	3-1/2	--
-64	4	--

Note that the dash number applies to the hose ID for all types of hoses except SAE 100R5. In the case of a single wire braid (SAE 100R5), the dash number is equal to a relevant tube outside diameter (OD). The dash sizes of rigid tubes determined with the tube OD form the basis for determining the dash sizes for single-wire braid (SAE 100R5) hose.

(ii) Hose Specifications in Metric and Inch Sizes

Table A11.4 | Hose specifications in metric and inch sizes
Courtesy: PARKER / PAGE International Hose, Texas

Dash size	ID		OD		Working pressure @ 22°C (72°F)		Min. burst pressure		Min. bend radius	
	mm	inch	mm	inch	bar	psi	bar	psi	mm	inch
-4	5.59	0.22	9.53	0.38	340	5000	1088	16000	38.1	1.5
-6	7.95	0.31	12.32	0.49	340	5000	1088	16000	63.5	2.5
-8	10.31	0.41	15.62	0.62	340	5000	1088	16000	73.7	2.9
-10	12.70	0.50	18.80	0.74	340	5000	1088	16000	83.8	3.3
-12	15.88	0.62	24.64	0.97	340	5000	1088	16000	101.6	4.0
-16	22.23	0.88	32.89	1.30	340	5000	1088	16000	127.0	5.0
-20	28.58	1.13	40.64	1.60	340	5000	1088	16000	304.8	12.0
-24	34.93	1.38	46.99	1.85	340	5000	816	12000	355.6	14.0

11.D Standards Relevant to Hydraulic Fluid Conductors

Table A11.5 | Standards relevant to hydraulic fluid conductors

Standard	Description
ISO 1179	Tube connections, threaded to ISO-228/1, for plain-end steel and other metal tubes in industrial applications
ISO-3862-1	Rubber hose and assemblies
ISO 1436	Rubber hose and assemblies, wire-braid reinforced hydraulic types specification, part 1: oil-based applications
ISO 4079	Rubber, textile-reinforced hydraulic type hose (R3 and R6)
ISO 4939	Thermo-plastic, textile-reinforced hydraulic-type hoses
ISO 3862	Rubber, rubber-covered, spiral-wire-reinforced hose
ISO 11237	Rubber, wire-braid-reinforced compact hoses
ISO 9329-4	Seamless steel tubes for pressure purposes- Technical delivery conditions, Part 4: Austenitic stainless steel
ISO 3304	Hydraulic fluid power -Plain end seamless precision steel tube-Technical conditions for delivery
ISO 4042	Fluid power for surface protection coating on tube Trivalent Chrome (Chrome Hex Free)

22 | References

1) ANDERSON GREENWOOD 9290 POSRV INSTALLATION AND MAINTENANCE ... (n.d.). Retrieved from http://valves.pentair.com/valves/Images/ANGMC-6012-US.pdf_br
2) Andrew Parr, Hydraulics & Pneumatics, A technician's and Engineer's Guide, 2nd Edition, Butterworth, Heinemann, 1998
3) Anthony Esposito, Fluid power with applications, 6th Ed., Prentice-Hall of India, New Delhi, 2006
4) Article on: (1) 'About Hydraulic Pumps' and (2) 'About Hydraulic Reservoirs', GlobalSpec Inc., Jordan Rd, Troy, NY, USA.
5) Article on 'About Hydraulic Motors', GlobalSpec Inc. 350 Jordan Rd, Troy, NY, USA.
6) Article on 'anti-cavitation valve', Sunfab Hydraulics AB, Box 1094, SE-824 12 Hudiksvall, Sweden, www.sunfab.com
7) Article on 'Bigger Isn't Always Better When It Comes to Sizing Your Hydraulic Motors for Efficiency', By Phillip Groves, White Hydraulics Inc., Hopkinsville, Ky., in Compact Equipment, October 2003.
8) Article on 'Clamping elements', HYDROKOMP®, Mücke, Germany, www.hydrokomp.de
9) Article on 'Cylinders-Part 1', Hydraulics & Pneumatics Magazine, Penton Media, Inc.
10) Article on 'Detecting Pump Cavitation with a Vibration Sensor', Renard Klubnik, Wilcoxon Research, PUMPS & SYSTEMS, JUNE 2007.
11) Article on 'How Does a Hydraulic Motor Work', by Shashank Nakate in Buzzle.com
12) Article on 'HYDRAULIC CLAMPING', VEKTEK LLC, U.S.A., www.vektek.com
13) Article on 'Hydraulic cylinder anatomy', by Frank J. Bartos, Control Engineering Resource Center
14) Article on 'Hydraulic cylinder', HYDRAULIC EQUIPMENT & TOOLS MARKETPLACE, Hydraulic Equipment Manufacturers.
15) Article on 'Hydraulic motors – Part 1' and 'Hydraulic motors – Part 2', Penton Media, Inc. & Hydraulics & Pneumatics magazine.
16) Article on 'Hydraulic Pumps', by Myounggyu Noh, Dept. Mechatronics Engineering, Chungnam National University.
17) Article on 'Hydraulic swing clamp', KOSMEK, Japan
18) Article on 'Hydraulic Swing Cylinders – Top Mounted, Double-acting', Precision Engineering Accessories, Bangalore, India, www.preacindia.com
19) Article on 'Hydraulics: An inside look at pumps', by Dr Robert M. Gresham, in Tribology & Lubrication Technology magazine, August 2004 issue
20) Article on 'Mini Swing Clamps with Sturdy Swing Mechanism, threaded-body type, double-acting, max. operating pressure 150 bar', Issue B1.848/12-10E, ROEMHELD GmbH.
21) Article on 'THEORY OF PISTON PUMPS -AXIAL-PISTON PUMPS', DELTAQ, 10827 Tower Oaks Blvd., Houston, TX, USA, www.deltaq.com
22) Article on Hydraulic contamination - part 1&2', Penton Media, Inc. & Hydraulics & Pneumatics magazine.
23) Article on: 'A Quick and Easy Guide to Hydraulic Pump Technology and Selection', by Rodney B. Erickson, Senior Training Specialist, Eaton Corp.
24) Article on: 'Clean up hydraulic circuits' by Phillip Johnson, Plant Services, Itasca, IL.
25) Article on: 'Clean up hydraulic circuits', by Phillip Johnson, PlantServices.com
26) Article on: 'Comparing 4 Types of PD Pumps', Viking Pump Inc., Cedar Falls, Iowa, U.S.A.
27) Article on: 'Condition Monitoring for Hydraulic and Lubricating Fluids' HYDAC International GmbH, Hauptstrasse, Saarbrücken, www.hydac.com
28) Article on: 'Contamination Control - A Hydraulic OEM Perspective', by R.W. Park, BE (Hons), MIE Aust., CP Eng., Managing Director, Moog Australia Pty Ltd.
29) Article on: 'Directional Control Valves', Penton Media Inc., Hydraulics & Pneumatics Magazine
30) Article on: 'Element Technical Data Fundamentals', SCHROEDER INDUSTRIES LLC.
31) Article on: 'Facts worth knowing about hydraulics', Danfoss Hydraulics
32) Article on: 'Filtration' Stauff Corporation, Inc., 7 Wm. Demarest Place, Waldwick, New Jersey, USA.
33) Article on: 'Flow control valves', Hydraulics & Pneumatics webmaster, Penton Media Inc., USA
34) Article on: 'Fluid power pumps', Hydraulics and Pneumatics Magazine, Penton Media Inc
35) Article on: 'Highlights in the History of Hydraulics' by Hunter Rouse.

36) Article on: 'How to maintain a hydraulic valve', Valve Products, http://valveproducts.net/
37) Article on: 'Hydraulic & Lubrication Filters, Part I: Filter Types and Locations and Part II: Proper Filter Sizing, HY-PRO Filtration', www.filterelement.com
38) Article on: 'Hydraulic accumulators, Book 2, Chapter 1, Part 1, Part 2, Part 3', Hydraulics & Pneumatic Magazine
39) Article on: 'Hydraulic filtration - part 1, 2, 3, 4', Penton Media, Inc. & Hydraulics & Pneumatics magazine.
40) Article on: 'Hydraulic Filtration and Contamination', Filter Manufacturers Council, NC, USA, www.filtercouncil.org
41) Article on: 'Hydraulic Fluid Care Guide', MTS Systems Corporation, Minnesota, USA.
42) Article on: 'Hydraulic System Filters New ISO Fluid and Cleanliness Rating Standards', Moog Inc., NY USA, www.moog.com
43) Article on: 'Hydraulic systems and fluid selection', Machinery Lubrication, NORIA
44) Article on: 'Hydraulic systems safety', Paul D. Ayers, Service in Action 5.017, Cooperative Extension, Colorado State University.
45) Article on: 'Hydraulics provides a versatile solution', by Jason Christopher, Field Editor Control Design, Itasca, IL.
46) Article on: 'Important facts about seal materials ELASTOMERS', MSO seal and gasket, Huston, Texas.
47) Article on: 'Maintenance of Hydraulic Systems', Ricky Smith, GP Allied, www.gpallied.com
48) Article on: 'Multi-Element Sealing System for Hydraulic Applications to Improve Performance and Service Time under Highest Loads', Tech Directory 2008, Fluid Power Journal.
49) Article on: 'Oil Analysis 101, Part 1 of 2', by Daniel P. Walsh, Business Development Manager, National Tribology Services Inc.
50) Article on: 'Operation principle of check valve, Valve Products', http://valveproducts.net/
51) Article on: 'Pressure Compensated Variable displacement Internal Gear Pump' by Mark Swaney, University of Arkansas, Technology Licensing Office.
52) Article on: 'Pressure valves, Pressure control circuits', Hydraulics & Pneumatics Magazine, The Penton Media Building, Cleveland, OH, USA.
53) Article on: 'Prevent Pump Damage through Automatic Detection of Cavitation', by Todd Reeves, Emerson Process, Knoxville, TN, USA, www.emersonprocess.com
54) Article on: 'Preventive maintenance for hydraulic presses', Carl Jean, FMA Communications, Inc., Featherstone Road, Rockford, IL, USA.
55) Article on: 'Principles of Hydraulics v 2.0', Content provider U.S. Army, Web Developer: David L. Heiserman, Publisher: SweetHaven Publishing Services
56) Article on: 'Pump Maintenance & Reliability-Digging to the Root of the Problem', Jason W. Bitting and Hardin T. Wells, Albemarle Corporation, PUMPS & SYSTEMS, SEPTEMBER 2007.
57) Article on: 'Pump selection 101', by Tom Nash, Hydraulic Product Manager for Applied Industrial Technologies, retrieved from MRO Today.
58) Article on: 'Reclaiming hydraulic oil eliminates disposal problems', by W. Stofey and M. Horgan, assistant editor.
59) Article on: 'Sure Signs Your Pump Needs a Tune-Up!', Mike Whitaker, Service Supervisor, InnoCal® Calibration Services, Cole-Parmer Technical Library.
60) Article on: 'The Evolution of Hydraulics', HYDRAULIC TALK, Issue 45, November 2006.
61) Article on: 'Understanding Truck Mounted Hydraulic Systems Sixth Edition, Written and published by Muncie Power Products.
62) Article on: 'What is an accumulator?' & TOBUL_Int_Catalog_101912v2, Tobul Accumulator, Inc., USA.
63) Article on: 'What is fluid power?', National Fluid Power Association, Milwaukee, WI.
64) Article on: 'WHY OIL NEEDS ON-LINE MONITORING', by Zhang Qisheng, Zhao Jingyi and Li Shuli, Fluid Transmission and Control Institute, Yanshan University, Qinhuangdao, P.R.China.
65) Article on: 'Why the Tank May Well Be a Hydraulic Fluid's Best Friend', Brendan Casey, in the publication 'Machinery Lubrication' of NORIA
66) Article on: "Detecting and Managing Hydraulic System Leakage" by Kevan Slater, Schematic Approach, Machinery Lubrication Magazine, July 2001.

67) Articles on: (1) 'The Micron Rating for Media in Fluid Filters', Revised October 2005, and (2) 'Hydraulic Filter Performance Criteria', Filter Manufacturers Council, Research Triangle Park, NC, USA.
68) Articles on: (1) Rotary pump handbook and (2) High-Performance Rotary Screw Pumps
69) Articles on 'Hydraulic Cylinder Seal Selection', 'Hydraulic Cylinder Repair Tutorial', 'The Advantages of Hydraulic Cylinders', 'Hydraulic Cylinders- Design Considerations for Hydraulic Cylinders', 'Hydraulic Cylinders-Engineering and Design Tips', 'Hydraulic Oil Tutorial, Hydraulic Cylinder Safety Tutorial', and 'Advantages of Welded Body Hydraulic Cylinders', HYCO ULTRAMETAL, Kitchener, Strasburg Road, Unit C Kitchener, Ontario.
70) Articles on 'Hydraulic reservoirs' and 'Heat exchangers' Penton Media, Inc. & Hydraulics & Pneumatics magazine
71) Brochure on 'Cylinders Capabilities', Document No. M-CYOV-MR001-E, January 2007, Eaton, USA.
72) Bulletin on: 'Directional Valve Features, Selection and Operating Recommendations' DYNEX High-Pressure Hydraulics, Dynex/Rivett Inc., Pewaukee, WI, U.S.A.
73) Bulletin on: 'HANDBOOK FOR CONTROL VALVE SIZING', PARCOL S.p.A. ITALY.
74) Cartridge Filtration, Schroeder Industries, Leetsdale, PA, USA.
75) Catalog on 'Flow Control Valves' No. HY 15-3502/US, Parker Hannifin Corporation, Hydraulic Cartridge Systems, USA.
76) Catalog on 'High-pressure hydraulics', EURO PRESS PACK, Via M. Disma, Carasco (GE), ITALY
77) Catalog on 'Hydraulic Components – Directional Control Valves', POCLAIN HYDRAULICS s. a., www.poclain-hydraulics.com
78) Catalog on 'Hydraulic cylinder, Type CDL1, Series 1X, Nominal pressure 160 bar (16 MPa), RE 17325/09.05', Bosch Rexroth AG Hydraulics
79) Catalog on 'Medium Duty Hydraulic Cylinders, Series 3L', 4/10 / No. HY08-1130-1/NA, Parker Hannifin Corporation, Cylinder Division, USA. www.parker.com/cylinder
80) Catalog on 'Metric Hydraulic Cylinders, Series HMI, Conforms to ISO 6020/2 (1991), for working pressures up to 210 bar', Parker Hannifin Corporation, Cylinder Division, USA.
81) Catalog on Series H, Milwaukee Cylinder, Milwaukee, USA, www.milwaukeecylinder.com
82) Catalog on: 'Accumulators V-FIFI-MC003-E' July 2005, EATON, Eden Prairie, MN, USA, www.hydraulics.eaton.com
83) Catalog on: 'Flow Dividers', Eaton Hydraulics Operations USA, Eden Prairie, MN, USA
84) Catalog on: 'Pressure controls', ROBERT BOSCH FLUID POWER CORPORATION, U.S.A
85) Catalog on: 'Sizing and Selection – Piston Accumulators, Bladder Accumulators, Kleen Vent', Catalog No. HY10-1630/US, Parker Hannifin Corporation, Hydraulic Accumulator Division, Illinois USA.
86) Catalog: MAHLE Industrial Filters, MAHLE Filtersysteme GmbH, Industriefilter Schleifbachweg 45 D-74613 Öhringen.
87) Catalogues on 'Precautions for use and maintenance recommendations', 'energy, silence, comfort, service life…', 'What are accumulators used for?', HYDRO LEDUC, FRANCE, www.hydroleduc.com
88) Catalogues on 'Reservoirs', 'In tank-coolers (water-cooled)', 'Suction strainers', 'Tank magnets', VESCOR, LDI Industries, Manitowoc, WI.
89) Catalogues on: '3 Easy Steps to Select A Seal', 'Hydraulic & Fluid Seals – Backup rings, Buffer seals, Piston seals, Rod seals, Static seals, Symmetrical seals, Wear rings, Wipers', American High Performance Seals, Inc., Oakdale, PA, USA.
90) Catalogues on: 'Desiccant Breathers', 'Filler Breather Caps', 'Filler Breather Mini Cap', 'Sight level Gauge', 'Magnetic tank cleaners', 'Slim Line Magnet Tank Cleaners' and 'Fluid Level Sight Windows' and other product
91) Catalogues, Lenz inc., Ohio, US A., www.lenzinc.com
92) Catalogues on: 'Designing Rubber Components', 'Elastomers/Materials', 'Rubber/Standard Products', 'Selecting an Elastomeric Material', 'Plastic & Thermoplastic Elastomer Materials', 'Designing Plastic Components', Minnesota Rubber and QMR Plastics, Minneapolis, MN, USA.
93) Catalogues on: 'HYDRAULIC SEALS PISTON SEALS', 'HYDRAULIC SEALS SYMMETRIC SEALS', Busal+Shamban.
94) Catalogues on: 'POLYPACK SEALS, PISTON SEALS, ROD SEALS, T SEALS, U-CUPS, V-PACKING, WEAR RING & BEARINGS, WIPER & SCRAPER', BEETLE TECHNOLOGY

94) CHAPTER 8: Air and Hydraulic Pumps (part 1), December 2006 Hydraulics & Pneumatics.
95) Charles S Hedges and Robert C Womack, Fluid power – in plant and field, second edition, Womack Educational Publications, July 2000
96) Document on 'MAHLE Industrial Filters, Filtration processes for the metalworking industry', MAHLE Filtersysteme GmbH, Industriefilter, Schleifbachweg, Öhringen, www.mahle.com
97) Document on Anderson Greenwood Type 81P SOPRV Installation and Maintenance Instructions of Spring Operated Safety Relief Valves (SOPRV), www.tycovalves.com
98) Document on Cat® Hydraulic Systems, Management Guide, Caterpillar, www.cat.com
99) Document on Crosby® Pressure Relief Valve Engineering Handbook, Technical Document No. TP-V300, Anderson Greenwood, Crosby Valve Inc.
100) Document on: 'A guide to contamination control for hydraulic and lubrication systems Brochure: FDHB138GB1', www.parker.com
101) Document on: 'Accumulator sizing' EPE ITALIANA Srl, COLOGNO MONZESE (MI).
102) Document on: 'Accumulator station, RE 50135/07.11', Bosch Rexroth AG, Hydraulics, Germany
103) Document on: 'Catalogues on Hydraulic Seals, and 'O' Rings, Rotary Seals', James Walker & Co Ltd., Cheshire, UK.
104) Document on: 'Eaton® Hydraulic Fluid Recommendations' Eaton Corporation, U. S. A.
105) Document on: 'ENGINEERING INFORMATION - flow data flow factor and orifice size' ASCO Numatics, France.
106) Document on: 'ENGINEERING INFORMATION flow data, flow factor, and orifice size', ASCO, JOUCOMATIC www.ascojoucomatic.com
107) Document on: 'Filter Element Beta Ratio Information', Swift Filters, Inc., Ohio, USA.
108) Document on: 'Filters & contamination controls', Fluidea
109) Document on: 'FUNDAMENTAL TECHNICAL HYDRAULIC CLAMPING INFORMATION', VEKTEK, INC., Emporia, KS.
110) Document on: 'General instructions for the installation, operation, and maintenance of Albany Rotary Gear Pumps', ALBANY STANDARD PUMPS, Bradford
111) Document on: 'GR3050 & GR3100 Preventative Maintenance', The Toro Company.
112) Document on: 'Hardness', R. L. Hudson & Company.
113) Document on: 'Hydraulic & Lubrication Filters, Part I: Filter Types and Locations', HY-PRO FILTRATION, U. S. A., www.filterelement.com
114) Document on: 'Hydraulic Hints &Trouble Shooting Guide', Vickers - General Product Support
115) Document on: 'HYDRAULIC PRESSURE INTENSIFIERS, AIR-OIL AND OIL-OIL, SCREW PUMPS, SCREW PISTONS, PRESSURE ACCUMULATORS', KOSTYRKA GmbH, Germany, www.kostyrka.com
116) Document on: 'Hydraulic Pressure Intensifiers', ScanWill Fluid Power ApS, Roholmsvej 10L, Albertslund, DK-2620 Denmark, www.scanwill.com
117) Document on: 'Hydraulic Seals - Types and Definitions', American Seal & Packing.
118) Document on: 'HYDRAULIC SYSTEM DESIGN CONSIDERATIONS', MRL HYDRAULICS LLC.
119) Document on: 'Hydraulic Troubleshooting Chart & Common Hydraulic Symptoms Problem Guide | Learn Hydraulics', MACHINETOOLHELP.COM
120) Document on: 'Hydraulics (FM 5 – 499)', Headquarters, Department of the Army, USA.
121) Document on: 'Hydraulics', Headquarters, Department of the Army, Washington, DC, USA.
122) Document on: 'Industrial Hydraulics', Donaldson Europe B.V.B.A., www.donaldson.com
123) Document on: 'INFORMATION AND TROUBLESHOOTING GUIDE FOR MONARCH 'M' Series D.C. HYDRAULIC POWER UNITS', MONARCH HYDRAULICS, INC., MI, U.S.A., http://www.monarchhyd.com
124) Document on: 'Installation, operation and maintenance manual TWO POST LIFT HCT TECHNOLOGY', GARAGE EQUIPMENT PROFESSIONALS.
125) Document on: 'ISO Cleanliness Levels, Fluid Service Catalog', HYDAC
126) Document on: 'MAINTENANCE MANUAL FOR BETTS HYDRAULIC PUMP HP46982SL', Betts Industries, Inc. Warren, PA, USA.

127) Document on: 'Oil Cleanliness, Oil Sampling and Oil Analysis', JLM Systems Limited, Richmond, BC Canada.
128) Document on: 'Operating Manual, 82000 HPU, Hydraulic Power Unit, 372-82000-01', ROMAC INDUSTRIES, INC.
129) Document on: 'Overview of Reservoir Accessories' HYDAC, Germany
130) Document on: 'POWER PACK-OPERATION & MAINTENANCE', General Machine Products Co., Inc., Old Lincoln Hwy, Trevose, PA, USA.
131) Document on: 'Pressure Boosters', Milwaukee, Cudahy, Wisconsin, SA, www.milwaukeecylinder.com
132) Document on: 'The Handbook of Hydraulic Filtration', Parker Hannifin Corporation, Metamora, OH, USA.
133) Document on: 'Unloading valves', DENISON HYDRAULICS, Marysville OH, USA
134) Document on: 'Water Based and Synthetic Fire-Resistant Fluids', RA 09 296/06.98, Rexroth Bosch Group.
135) Document on: MODEL A, HYDRAULIC INTENSIFIER, Max. Capacity: 5,000 PSI', SPX Corporation, Rockford, IL, USA, http://www.hytec.com
136) Documents on: 'Hydraulic Filtration Technical Reference' (Doc. No. F115354 rev.1), and 'Hydraulic Filtration Product Guide', Donaldson Company, Inc., Minneapolis, MN, USA
137) Donaldson Technical Reference Guide "The Blue Pages," Donaldson Company, Inc., U. S. A., www.donaldson.com
138) Filtration Catalog Technical Catalog, Eaton, Eden Prairie, MN, USA.
139) Fluid Condition Handbook, MP FILTRI S.p.A.
140) Fluid Power, by MMC Albert Beasley, Jr.1990 Edition published by Naval Education and Training Development and technology center
141) Hydraulic Cylinder Troubleshooting - Parker Hannifin. (n.d.). Retrieved from http://parker.com/literature/Literature%20Files/euro_cylinder/v4/Trouble_1242-1-gb.pdf_br
142) Hydraulic Intensifiers | Hytec | Product Detail. (n.d.). Retrieved from http://www.spx.com/en/hytec/pd-hydraulic-intensifers/_br
143) Hydraulic Talk, Issue 45, Nov 2006, presented by Consolidated Fluid Power, www.cfp.ns.ca
144) Intelligent Hydraulics in New Dimensions. Industrial Hydraulics from Rexroth, Bosch Rexroth Corporation, www.boschrexroth-us.com
145) Joji P., Pneumatic controls, Wiley India Pvt Ltd, New Delhi, 2008
146) N. M. Morris, Control Engineering, McGraw-Hill, LONDON.
147) Owner's Manual - Hydraulic Reservoirs, Parker Hannifin Corporation, Mississippi, USA.
148) Paper on: 'Hydraulic seals - linear', TRELLEBORG, www.tss.trelleborg.com
149) Powerpoint presentation for 'Fluid Power' by James R. Daines, The Goodheart-Willcox Co. Inc, Illinois, USA
150) Relief valve application notes retrieved from Sterling Hydraulics
151) S R Majumdar, Oil hydraulic systems, principles and maintenance, Tata McGraw-Hill Publishing Company Limited, New Delhi, 2001
152) Technical Principles of Valves, OMEGA ENGINEERING, Inc., USA
153) Vickers literature #510, 'Noise Control in Hydraulic Systems, for design guidelines'
154) Vocational Training Course, HYDRAULICS – 17 Exercises with Instructions, published by Bundesinstitut fur Berufsbildungsforschung, Berlin, 1973
155) William D. Wolansky et al., Fundamentals of Fluid Power, Houghton Mifflin Company, Boston, 1977

Fluid Power Educational Series Books

1. Pneumatic Systems and Circuits -Basic Level (In the SI Units)
2. Industrial Pneumatics -Basic Level (In the English Units)
3. Pneumatic Systems and Circuits -Advanced Level
4. Electro-Pneumatics and Automation
5. Design of Pneumatic Systems (In the SI Units)
6. Design Concepts in Pneumatic Systems (In the English Units)
7. Maintenance, Troubleshooting, and Safety in Pneumatic Systems
8. Industrial Hydraulic Systems and Circuits -Basic Level (In the SI Units)
9. Industrial Hydraulics -Basic Level (In the English Units)
10. Hydraulic Fluids
11. Hydraulic Filters: Construction, Installation Locations, and Specifications
12. Hydraulic Power Packs (In the SI Units)
13. Power Packs in Hydraulic Systems (In the English Units)
14. Hydraulic Cylinders (In the SI Units)
15. Hydraulic Linear Actuators (In the English Units)
16. Hydraulic Motors (In the SI Units)
17. Hydraulic Rotary Actuators (In the English Units)
18. Hydraulic Accumulators and Circuits (In the SI Units)
19. Accumulators in Hydraulic Systems (In the English Units)
20. Hydraulic Pipes, Tubes, and Hoses (In the SI Units)
21. Pipes, Tubes, and Hoses in Hydraulic Systems (In the English Units)
22. Design of Industrial Hydraulic Systems (In the SI Units)
23. Design Concepts in Industrial Hydraulic Systems (In the English Units)
24. Maintenance, Troubleshooting, and Safety in Hydraulic Systems
25. Hydrostatic Transmissions (HSTs) (In the SI Units)
26. Concepts of Hydrostatic Transmissions (In the English Units)
27. Load Sensing Hydraulic Systems (In the SI Units)
28. Concepts of Load Sensing Hydraulic Systems (In the English Units)
29. Electro-hydraulic Proportional Valves
30. Electro-hydraulic Servo Valves
31. Cartridge Valves
32. Electro-hydraulic Systems and Relay Circuits
33. Practical Book: Pneumatics - Basic Level
34. Practical Book: Electro-pneumatics - Basic Level
35. Practical Book: Industrial Hydraulics – Basic Level
36. Programmable Logic Controllers and Programming Concepts
37. Compressed Air Dryers
38. Hydraulic Circuits – Identification of Components and Analysis

For more details, please visit **https://jojibooks.com.**

www.ingramcontent.com/pod-product-compliance
Lightning Source LLC
Chambersburg PA
CBHW080451220526
45465CB00006B/2234